Seismic Design of Building Structures

A Professional's Introduction
to Earthquake Forces
and Design Details

Eighth Edition

Michael R. Lindeburg, PE
with **Majid Baradar, PE**

Professional Publications, Inc.
Belmont, CA

How to Get Online Updates for This Book

I wish I could claim that this book is 100% perfect, but 25 years of publishing have taught me that engineering textbooks seldom are. Even if you only took one engineering course in college, you are familiar with the issue of mistakes in textbooks.

I am inviting you to log on to Professional Publications' web site at **www.ppi2pass.com** to obtain a current listing of known errata in this book. From the web site home page, click on "Errata." Every significant known update to this book will be listed as fast as we can say "HTML." Suggestions from readers (such as yourself) will be added as they are received, so check in regularly.

PPI and I have gone to great lengths to ensure that we have brought you a high-quality book. Now, we want to provide you with high-quality after-publication support. Please visit us at **www.ppi2pass.com**.

> Michael R. Lindeburg, PE
> Publisher, Professional Publications, Inc.

The ICBO assumes no responsibility for the accuracy or completion of items in this book that were obtained with permission from the *Uniform Building Code*.

SEISMIC DESIGN OF BUILDING STRUCTURES
Eighth Edition

Printed in the United States of America

Professional Publications, Inc.
1250 Fifth Avenue, Belmont, CA 94002
(650) 593-9119
www.ppi2pass.com

Current printing of this edition: 2

Library of Congress Cataloging-in-Publication Data
Lindeburg, Michael R.
    Seismic design of building structures: a professional's introduction to earthquake
forces and design details / Michael R. Lindeburg with Majid Baradar. -- 8th ed.
        p.   cm.
    Includes index.
    ISBN 1-888577-52-5 (pbk.)
    1. Earthquake engineering--California--Problems, exercises, etc.
2. Civil engineering--California. I. Baradar, Majid, 1953- II. Title.
TA654.6.L56   2000
624.1' 762' 076--dc21
                                                    00-045755

TABLE OF CONTENTS

PREFACE TO THE EIGHTH EDITION

This is the eighth edition of *Seismic Design of Building Structures*. It has been extensively rewritten to conform to the 1997 Uniform Building Code as corrected by ICBO errata through March 2001, referred to in this book as "UBC-97". The seismic design provisions of UBC-97 are considerably different from those in previous codes, most notably in the use of strength theory rather than working stress theory.

This book is a concise introduction to basic seismic concepts and principles, and is particularly intended for individuals taking the engineering (PE) and architectural (ARE) licensing exams. I want to emphasize the descriptive word "introduction," as this book consistently summarizes subjects that other entire books have been written about.

Unlike in previous editions of this book, *Lateral Force Requirements* (the "Blue Book") published by the Structural Engineers of California (SEAOC) was not used as the primary source document for this edition. Therefore, all Blue Book references are to earlier editions, and as such, are primarily limited to historical notes.

In addition to bringing the code-related sections into conformance with the 1997 UBC, the contents of this book have been extensively rewritten. Many of the changes were suggested by readers and review course instructors who questioned what I was trying to say or who simply took the time to tell me how to make the book more useful. However, many things just needed to be changed because the UBC-97 is so completely different from previous editions in its seismic provisions.

As in the previous edition of *Seismic Design of Building Structures*, considerable effort was made to present the material in both traditional English and SI units. All of the examples and problems continue to give you a choice of units: You can choose to work solely in SI, solely in traditional English, or in both (for twice the practice). This edition reverses the sequence of

many dual-dimensioned items, however, placing the traditional English values first to correspond to the unit system typically used in actual seismic design.

Supporting pages from the Simpson Strong-Tie Company product catalog are taken from the most recent catalog available.

I have omitted the index of problem types that I added to the seventh edition. It is not my desire to produce a book that is useful to engineers who wish to "design by example." However, the detailed subject index (which is characteristic of my books), the index of figures and tables, and index of referenced UBC sections have been retained.

Other omitted appendices include the cross reference between current and older UBC equation numbers (since users of this book are unlikely to be working with two simultaneous editions) and the bibliography (because the books listed largely refer to previous UBCs).

Since the current seismic examination was originally implemented many years ago, the California Board for Professional Engineers and Land Surveyors has received numerous comments from nonstructural examinees. These examinees often commented that the depth of structural analysis experience and code-related knowledge expected of them in the exam far exceeds anything they will ever be faced with in practice.

I don't know if these comments are the reason, but there has been a shift in the exam emphasis away from highly technical, code-detailed, calculation-intensive questions, and toward conceptual and general-familiarity questions. Inasmuch as studying for the right kind of seismic test could be described as hitting a fly in a moving train with a rifle bullet from two miles away, this book is intended to prepare you for both types of questions.

When the earliest editions of this book were written, the concept of base isolation was new and essentially unproven. However, base isolation has become a common element in new and retrofit designs, and the UBC has contained design provisions for several editions. Therefore, this subject has been slightly "toned down" in this edition of *Seismic Design*.

In the 1994 Northridge earthquake, welded connections in steel-framed buildings that were designed and constructed to the provisions of earlier codes failed with an alarmingly consistent rate. Accordingly, the provisions regarding rigid beam-to-column connections have either been changed or omitted. Various practical and economical fixes to the design methodology have yet to be tested in a truly severe earthquake. But, such is "the way" in seismic engineering. We have to wait and see how buildings designed according to the latest seismic codes and constructed using current technology fare.

PPI produced this book knowing that it was an "interim" edition. (Actually, all editions are "interim," since they are between the previous and subsequent editions. Publishers just hope that an edition can last for a couple of years before it becomes obsolete due to code changes.) However, the 1997 UBC is essentially the last UBC that will be prepared. In the future, we can expect seismic design to follow the provisions of the new International Building Code (which is discussed in this book). I've already accumulated numerous ideas for the next edition. So, someday there will be a "SEIS9".

Approximately 20 years'-worth of examinees have shaped this book considerably. I will be grateful if you, too, take the time to tell me how you think this book can be improved. A postage-paid comment card appears in the back for that purpose. You may also contact me via email or PPI via the web at **www.ppi2pass.com**. I will humbly accept any suggestions you think might help future engineers.

Michael R. Lindeburg
mlindeburg@ppi2pass.com

ACKNOWLEDGMENTS

Production of this eighth edition of *Seismic Design of Building Structures* was a huge team effort. Most of the editorial changes were completed by Majid Baradar, PE (California Department of Transportation). Mr. Baradar, who is also the author the *345 Solved Seismic Design Problems* and *Seismic Principles Practice Exams for the California Special Civil Engineer Examination* (both published by Professional Publications), updated the UBC code section references, added additional clarifying comments, extensively rewrote Chap. 6 and the other code-intensive sections, and provided new numerical examples. Mr. Baradar is not only a registered civil engineer but also a licensed general contractor. He brings knowledge of the practical and actual to what otherwise could be very theoretical and ideal. Without him, this new edition wouldn't have been nearly as complete and would have been a long time coming. What is even more amazing is that Mr. Baradar did all of his work on his back, recovering from a back injury.

The revisions to this edition were extensively reviewed by Shari Day, PE (previously with Exponent Failure Analysis). Ms. Day, who has a BSCE from Virginia Tech and an MSCE from University of Colorado, Boulder, drew upon her $15^1/_2$ years of engineering investigation work, which included seismic design reviews of structures of various construction including wood frame, concrete tilt-up, steel frame, and masonry. Ms. Day's involvement with seismic code provisions is much greater than mine. Reading her comments was sometimes a humbling experience, but considering the improvements she suggested, I wouldn't have wanted it any other way. I have also "saved" a significant number of Ms. Day's suggestions for the next edition, which is already taking shape in my mind.

A new edition with changes this extensive is essentially a new publication. Producing a top-quality book on a short schedule is comparable in difficulty to producing a monthly magazine. The task of bringing out this publication fell to the team of Jessica Holden (editing), Cathy Schrott (typesetting), Yvonne Sartain (illustrating), Tracy Berdichevskiy (proofreading), and Aline Magee (acquisitions). Tamer Hadi, Professional Publications' staff engineer, checked all of the calculations and units in this book.

I am grateful also to the International Conference of Building Officials (ICBO) for its permission to reproduce so many of the tables found the the UBC. These tables are critical to the understanding and use of the code-related provisions of this book. The value of this book has been greatly enhanced by ICBO's generosity.

As in previous editions, Simpson Strong-Tie kindly gave its permission to reprint various pages of its "Connectors for Wood Construction," 2000 catalog. These pages illustrate the complexity, design, and installation of typical seismic appliances, as well as support some of the practice problems in this book.

Other publishers also cooperated in granting permission to reproduce important tables and figures. They are credited where their material appears.

Grateful acknowledgement is given to the contributions of the following individuals: Robert H. Sydnor, RG, senior engineering geologist, California Division of Mines and Geology, Sacramento, CA (background and early documents describing the 1997 seismic provisions before they were actually published); David L. Houghton, SE, president, SidePlate™ Systems, Inc., Long Beach, CA (contributions on developments in the detailing of beam-to-column connections, most of which I have retained for use in the next edition of this book); and Shannon King, P.Eng., PE (generous loan of numerous technical, historical, and supporting documents during the preparation of this manuscript).

Previous editions drew upon the expertise of many individuals. Even though their work has spanned a decade or more, their efforts continue to contribute to the scope and organization of this edition. I am lucky to have had earlier editions reviewed by several outstanding specialists in the earthquake field, including Vitelmo V. Bertero, director of the Earthquake Engineering Research Center at the University of California at Berkeley; John G. Shipp, PE, senior technical manager of EQE, Inc., Costa Mesa; Albert Tung, PhD, PE of Stanford's Civil Engineering Department; and James E. Onderka, PE, Orinda, CA.

Of course, the material in this publication is ultimately my responsibility. And, while I hope they are few, the inevitable errors are attributable to me and me alone. You can log onto Professional Publications' web site (**www.ppi2pass.com**) to see a cumulative listing of errata for this book.

Although the problems in *Seismic Design* have been changed significantly to reflect current codes, acknowledgement is made that several of the practice and example problems in this book are derived from problems that appeared in previous years' California engineering licensing examinations. The original problems are used with permission of the California Board for Professional Engineers and Land Surveyors, which copyrighted the original material.

More than 10 years ago, I wrote my acknowledgments for the 5th edition of this book. At that time, my children were only 7 and 10 years old. Now they are at an age when they no longer "need" their parents and certainly don't complain about being absent from us, but I include the following paragraph for its historical value.

Finally, I acknowledge the unwavering support of my family which, as usual, has had to put up with my habitual writing. My wife, Elizabeth, and my two daughters, Jenny and Katie, lost the family time that went into this book. They may be beginning to understand how I think and what is important to me. If not, they never seem to complain.

Michael R. Lindeburg
mlindeburg@ppi2pass.com

HOW TO USE THIS BOOK

If you are the type who never reads instructions, here are my "Quickstart" theorems on how to get the most from *Seismic Design of Building Structures* during your exam preparation.

1. Get copies of UBC-97 and its errata.

2. Start reading this book from the first chapter. Don't skip around.

3. Read slowly; a page or two a day is plenty. Look up every code section referred to.

4. Work all of the example and practice problems.

Here are my "Quickstart" corrolaries on how to get the most from this book during the exam.

1. Don't forget to take it with you.

2. Put lots of tabs on the UBC tables.

3. Use the indexes extensively.

Seismic Design was originally written specifically for exam review. But as the exam vacillated each year between areas of emphasis, the scope and depth of this book has increased. By my own pronouncement, it is now one of the best quick-primers on seismic design in print. However, even though its scope and depth have increased by several orders of magnitude, I suspect this book will remain typecast in its leading role—that of an exam review book. Therefore, I am writing this section assuming that you are using *Seismic Design* book for exam review.

The actual exam is based on UBC-97. You are much more likely to need to pull a number out of the UBC than to read a code section. Not surprisingly, the only place where a complete compilation of all of the constants and other numerical values can be found is in UBC-97. The numerous tables and figures in this book give a false impression of completeness. True, every detail needed to solve the example and practice problems is contained somewhere in this book. However, the actual exam will not be so kind. It's always a shame to lose points simply because you couldn't perform a simple table look-up in "the code." So, buy UBC-97.

Although it develops subjects gradually, gently, and linearly, this book crams innumerable concepts onto every page. If you don't have a seismic background, you'll pretty much have to start at page 1 and work your way through the book, page by page. I don't assume that you know anything. And I've tried to write and edit the material in such a way as to provide instruction that is intuitive. However, you'll still need to go slowly.

Some people will buy this book just to take it into the exam. They'll use the index a lot, hoping to hit the jackpot. You can't pass the California special seismic exam that way. Since the material is pretty much different from most everything else civil engineers do, even the language is different. What's the difference between a "drag strut" and a "collector"? (Answer: There is no difference.) What's the difference between a "space frame" and a "ductile moment-resisting space frame"? (Answer: There can be a lot of difference.) You can read the words, but without having studied this material, they won't mean anything to you.

The example and practice problems in this book have been selected to make or emphasize certain points. These problems do not try to mimic the exam complexity. (Professional Publications has produced two other problem-oriented books authored by Majid Baradar that offer hundreds of additional practice problems in exam format. You can get additional practice by working problems in these two well-organized books.)

A LITTLE ABOUT THE EXAM FORMAT

In California, examinees take the NCEES civil engineering PE exam on a Friday. Then they come back the next day, on Saturday, for the California "special" seismic exam. (They may also take the California "special" surveying exam on the same day.)

The format of this exam has evolved considerably. The current seismic exam is $2^1/_2$ hours in length, which you take in one single sitting. The exam consists of 50 questions, all of which are multiple choice, each with four answer selections. Some of the questions are conceptual, some are theoretical, some are practical, some are straight look-up, and some require simple calculations. By now, you have probably figured out that the average time allowed per question is about 2 minutes. So you can see the level of difficulty cannot be insurmountable. You just have to go, go, go. For most people, it's pretty easy to know when you have to guess at a question and then move on.

Each question is worth from 2 to 8 points, with an average value of 6 points, making a total score of 300 points possible. The minimum passing score (i.e., the "cut score") varies somewhat, but seems to hover around 60%. The passing rate varies considerably: 40 to 50% passing is typical. There is only moderate correlation between the people who pass the NCEES 8-hour exam, the people who pass the special seismic exam, and those who pass the special surveying exam. The percentage of people who pass all three exams at the same time is not high.

The exam "test plan" that defines the fundamental principles, tasks, and elements of knowledge that you need to know has been made public. If you want to know what you have to know in order to pass, you can print out the test plan from the web site of the California Board for Professional Engineers and Land Surveyors. What you will find out is that you need to know everything in this book and then some. I didn't include the actual test plan in this book because it's too scary-looking. You cannot learn anything from it, other than that the scope of the exam is huge. Besides, the test plan is vulnerable to the "change without notice" syndrome, anyway. "You's just pays your monies, and you's just takes your chances."

NOMENCLATURE

Unless defined otherwise in the text, the following symbols are used in this book. Consistent units are presented. In some cases (such as modulus of elasticity and drift), however, it is customary to report values in smaller units (lbf/in^2 and in).

Symbol	Term	Units U.S.	Units SI
a	acceleration	ft/sec^2	m/s^2
a	link beam distance	ft	m
a_p	in-structure component amplification factor	–	–
A	amplitude	ft	m
A	area	ft^2	m^2
A_B	ground floor area	ft^2	m^2
A_c	combined effective area of shear wall	ft^2	m^2
A_e	minimum cross-sectional area	ft^2	m^2
A_x	torsional amplification factor	–	–
ARS	acceleration response spectrum	–	–
b	link beam distance	ft	m
b	parallel wall length	ft	m
B	damping coefficient	lbf-sec/ft	N·s/m
B	seismic parameter	–	–
C	chord force	lbf	N
C	numerical coefficient	–	–
C	seismic parameter	mi^{-2}	km^{-2}
C_a	seismic response coefficient	–	–
C_e	exposure and gust factor coefficient	–	–
C_q	pressure coefficient	–	–
C_s	base shear coefficient	–	–
C_t	numerical coefficient		

Symbol	Term	Units U.S.	Units SI
C_v	seismic response coefficient	–	–
d	depth	ft	m
d_i	thickness of layer i	ft	m
d_s	total thickness of cohesionless soil layers	ft	m
D	column depth	ft	m
D	dead load	lbf	N
D	fault slip	ft	m
D_e	length of a shear wall	ft	m
e	eccentricity	ft	m
E	earthquake load	lbf	N
E	energy released	ft-lbf	J
E	modulus of elasticity	lbf/ft^2	Pa
E_h	horizontal earthquake load	lbf	N
E_m	estimated maximum earthquake load	lbf	N
E_v	vertical earthquake load	lbf	N
EPA	effective ground acceleration	ft/sec^2	m/s^2
f	frequency	Hz	Hz
f	stress	lbf/ft^2	Pa
f'_m	masonry compressive strength	lbf/ft^2	Pa
F	force	lbf	N
F	story shear	lbf	N
F_p	design seismic forces on a part of the structure	lbf	N
F_{px}	design seismic forces on a diaphragm	lbf	N
F_t	concentrated force at the top of a structure	lbf	N
g	acceleration of gravity	ft/sec^2	m/s^2

Symbol	Term	U.S.	SI
g_c	gravitational constant (32.2)	ft-lbm/lbf-sec²	n.a.
G	shear modulus	lbf/ft²	Pa
h	height	ft	m
h	story height	ft	m
h_n	height above base to level n	ft	m
h_r	structure roof elevation with respect to grade	ft	m
h_x	element or component attachment elevation with respect to grade	ft	m
H	horizontal force	lbf	N
H	story height	ft	m
I	moment of inertia	ft⁴	m⁴
I	seismic importance factor	–	–
I_p	seismic importance factor for an element or component	–	–
I_w	wind importance factor	–	–
J	polar moment of inertia	ft⁴	m⁴
k	stiffness (spring constant)	lbf/ft	N/m
ℓ_d	development length	ft	m
L	length	ft	m
L	live load	lbf	N
m	mass	lbm	kg
M	moment	ft-lbf	N·m
M	Richter magnitude	–	–
n	cycle number	–	–
n	exponent	–	–
N	number of earthquakes	–	–
N	standard penetration resistance of soil layer	blows/ft	blows/m
N_a	near-source factor	–	–
N_{CH}	standard penetration resistance of cohesionless soil layer	blows/ft	blows/m
N_v	near-source factor	–	–
P	design wind pressure	lbf/ft²	Pa
P	magnitude of forcing function	lbf	N
P	sum of dead and live loads	lbf	N
PGA	peak ground acceleration	ft/sec²	m/s²
PI	plasticity index	–	–
q_s	wind stagnation pressure	lbf/ft²	Pa

Symbol	Term	U.S.	SI
r	radius or moment arm	ft	m
$r_{\max}$	maximum element story shear ratio	ft⁻¹	m⁻¹
R	electrical resistance	ohms	ohms
R	relative rigidity	–	–
R	response modification factor	–	–
R_p	component response modification factor	–	–
R_s	snow load reduction factor	lbf/ft²	Pa
s	distance	ft	m
S	snow load	lbf/ft²	Pa
S_a	spectral acceleration	ft/sec²	m/s²
S_d	spectral displacement	ft	m
S_u	average undrained shear strength	lbf/ft²	Pa
S_{v}	spectral velocity	ft/sec	m/s
SR	slip rate	ft/yr	m/yr
t	thickness	ft	m
t	time	sec	s
T	fundamental period of vibration	sec	s
U	strain energy (per unit volume)	ft-lbf/ft³	J/m³
U	ultimate capacity	lbf	N
v	shear stress	lbf/ft²	Pa
v	velocity	ft/sec	m/s
v_{si}	shear wave velocity in layer i	ft/sec	m/s
V	base shear	lbf	N
w	load per unit length	lbf/ft	N/m
w_i	weight of level i	lbf	N
w_{mc}	moisture content	–	–
W	load due to wind pressure	lbf	N
W	nail and spike withdrawal design values	lbf/in	N/m
W	weight	lbf	N
W_p	weight of an element or component	lbf	N
x	position or excursion	ft	m
y	height over which wind acts	ft	m
Y	number of years	–	–
Z	ductility reduction factor	–	–
Z	lateral design force	lbf	N
Z	seismic zone factor	–	–

Greek Symbol	Term	U.S.	SI
		Units	
β	magnification factor	–	–
Γ	participation factor	–	–
δ	decay decrement	–	–
δ	displacement	ft	m
δ_i	horizontal displacement at level i	in	mm
Δ	drift	in	mm
Δ_M	maximum inelastic response displacement	in	mm
Δ_{MT}	total inelastic separation	ft	m
Δ_S	design level response displacement	in	mm
ϵ	strain	–	–
θ	stability coefficient	–	–
λ	Lame's constant	–	–
μ	ductility factor	–	–
ν	Poisson's ratio	–	–
ξ	damping ratio	–	–
ρ	redundance/reliabilty factor	–	–
ρ	reinforcement ratio	–	–
ρ	rock/soil density	lbm/ft^3	kg/m^3
σ	stress	lbf/ft^2	Pa
τ	short period of time	sec	s
ϕ	mode shape factor	–	–
ω	angular frequency	rad/sec	rad/s
Ω_O	seismic amplification factor	–	–

Subscripts	
0	initial or calibration
a	allowable
ave	average
b	bending
c.r.	center of rigidity
c.m.	center of mass
d	damped
d	displacement
E	elastic strain
H	hysteresis
i	inertial or level number
j	mode number
n	cycle number
p	part or portion
P	P-wave
R	resilience
st	static
sx	between levels $x-1$ and x
S	S-wave
t	top or total
T	toughness
v	velocity
x	level number
x	with respect to x-axis
y	with respect to y-axis

1

BASIC SEISMOLOGY

1-1 THE NATURE OF EARTHQUAKES

An earthquake is an oscillatory, sometimes violent movement of the ground's surface that follows a release of energy in the Earth's crust. This energy can be generated by a sudden dislocation of segments of the crust, a volcanic eruption, or a man-made explosion. Most of the destructive earthquakes, however, are caused by dislocations of the crust.

When subjected to geologic forces from plate tectonics, the crust initially strains (i.e., bends and shears) elastically. For pure axial loading, Hooke's law gives the stress that accompanies this strain.

$$\sigma = E\epsilon \quad \text{[axial loading]} \qquad [1.1]$$

As rock is stressed, it stores *strain energy*, U. The elastic strain energy per unit volume for pure axial loading is[1]

$$U = \frac{\sigma\epsilon}{2} \quad \text{[axial loading]} \qquad [1.2]$$

When the stress exceeds the ultimate strength of the rocks, the rocks break and quickly move (i.e, they "snap") into new positions. In the process of breaking, the strain energy is released and *seismic waves* are generated. This is the basic description of the *elastic rebound theory* of earthquake generation.[2]

---

[1]When the rock is stressed in shear, an analogous term for shear strain energy can be written. Both energy forms can be present simultaneously.

[2]The *dilatational source theory* explains that earthquakes are produced from the explosion, sudden vaporization, or implosion of underground material. However, this theory is no longer favored as a description of the source of earthquakes. Other sources, such as Wiegel (1970), cover this theory in greater detail.

These waves travel from the source of the earthquake (known as the *hypocenter* or *focus*) to more distant locations along the surface of and through the Earth. The wave velocities depend on the nature of the waves and the material through which the waves travel. (See Sec. 1-14.) Some of the vibrations are of high enough frequency to be audible, while others are of very low frequency with periods of many seconds and thus are inaudible.

A modern theory may explain how some earthquakes are triggered. Geologists know that pumping fluids into the ground under high pressure can trigger earthquakes. Now there is evidence from the gas-producing regions in France that removing fluids from pores deep in the earth can also trigger earthquakes. Oil and gas are the main fluids of concern; pumping water from aquifers close to the surface is probably not as likely to result in an earthquake. The theory states that the reservoir shrinks when the gas or oil is removed, but the rocks surrounding the reservoir do not. This results in stresses in the earth that later are released in an earthquake.

1-2 EARTHQUAKE TERMINOLOGY

The *epicenter* of an earthquake is the point on the Earth's surface directly above the *focus* (also known as the *hypocenter*). The location of an earthquake is commonly described by the geographic position of its epicenter and its focal depth. The *focal depth* of an earthquake is the depth from the Earth's surface to the focus. These terms are illustrated in Fig. 1.1.

Earthquakes with focal depths of less than approximately 40 mi (60 km) are classified as *shallow earthquakes*. Very shallow earthquakes are caused by the fracturing of brittle rock in the crust or by internal strain energy that overcomes the friction locking opposite sides of a fault. California earthquakes are typically shallow.[3] *Intermediate earthquakes*, whose causes are not fully understood, have focal depths ranging from 40 to 190 mi (60 to 300 km). *Deep earthquakes* may have focal depths of up to 450 mi (700 km).

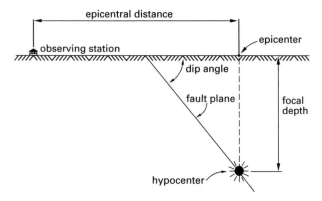

Figure 1.1   Earthquake Terminology

The slip propagates from the hypocenter along the fault with a velocity up to that of the outward-radiating seismic shear wave front—about 1.8 mi/sec (3 km/s)—until the entire affected segment is in motion. (See Sec. 1-14 for a description of shear waves.)

1-3 GLOBAL SEISMICITY

Most earthquakes occur in areas bordering the Pacific Ocean.[4] This circum-Pacific belt, nicknamed the *ring of fire*, includes the Pacific coasts of North America and South America, the Aleutian Islands, Japan, Southeast Asia, and Australia. The reason for such a concentration is explained by plate tectonics theory. (See Sec. 1-5.)

The United States has experienced less destruction than other countries located in this earthquake zone. This is partly due to the country's relatively young age and attention to earthquake-resistant construction methods, but millions of Americans now live in potential earthquake areas. Large parts of the western United States are known to be vulnerable to damage from earthquakes.[5] Nowhere is this more true than in California.

---

[3]Since there is no deep subduction zone (Fig. 1.3), earthquakes in California typically occur at depths of less than 10 mi (15 km).

[4]The other major concentration of earthquakes is in a much smaller east-west belt that runs between Asia and the Mediterranean.

[5]It is interesting that the largest earthquakes on the North American continent in the history of the United States occurred in the east (the 1811 and 1812 New Madrid, Missouri, and the 1886 Charleston, South Carolina, earthquakes, the latter of which had an estimated Richter magnitude in excess of 8.2). However, earthquakes in these regions are much less frequent than earthquakes in the western United States.

Nuclear reactors, dams, schools, hospitals, and high-rise buildings are planned and built in locations of high seismic hazard. This has created an urgent need for greater attention to the mitigation of earthquake-induced damage.

1-4 CONTINENTAL DRIFT

It has been known since the early 1900s that the continents are moving relative to one another, movement known as *continental drift*. In fact, fossilized records of past climates (the subject of the field of *paleoclimatology*) indicate that the continents have been moving slowly about the globe for hundreds of millions of years. For example, the same 300-million-year-old fossilized deposits are found in India and in the Arctic.

The theory of continental drift was reasonably established during the 1930s but was not universally accepted. In the 1950s, the emerging science of *paleomagnetism* provided new supporting evidence of continental drift. Many rocks, such as volcanic rock solidified from molten lava, contain tiny grains of magnetic minerals such as *magnetite*. When these minerals are formed, they retain the magnetic orientation of the Earth's magnetic field at the time of their formation. The magnetic orientations of rocks suggest the same ancient locations of the continents suggested by paleoclimatology and other geologic criteria.

An enormous amount of geophysical data was gathered during the 1950s and 1960s, particularly from oceanographic research vessels such as the *Omar Challenger*. A system of interconnecting submarine ridges, called *mid-ocean ridges*, was discovered circling the Earth. Such ridges are located approximately midway between continents that are moving apart (e.g., between Africa and South America). It is now recognized that new oceanic crust is being formed at the ridges and is added to the plates moving apart. This is known as *seafloor spreading*.

Great submarine trenches were also located, particularly along the convex oceanic sides of the volcanic arcs that make up the Pacific ring of fire. Inclined zones of earthquakes dip down from these trenches to as deep as 450 mi (700 km) into the mantle beneath and behind the volcanic arcs. Oceanic crust is formed at spreading ridges behind the moving plates.

The ocean crust, known to consist of alternating belts of highly and weakly magnetic oceanic crust material, represents magnetic records as new crust forms in the gaps behind separating plates. The symmetrical belts record the ambient magnetism on opposite sides of the newly formed ridges.

The crust is destroyed at the same rate elsewhere as oceanic plates dip down at the trenches and slide deep into the mantle along the seismic zones. However, there is a global balance between crust formation and destruction. The formation of plate material in the Atlantic Ocean is compensated by absorption of plate material, primarily in the Pacific Ocean.

1-5 PLATE TECTONICS

Most earthquakes are a manifestation of the fragmentation of the Earth's outer shell (known as the *lithosphere*) into various large and small plates. (The academic field that studies plate motion is known as *plate tectonics*.) There are seven very large plates, each consisting of both oceanic and continental portions. There are also a dozen or more small plates, not all of which are shown in Fig. 1.2.

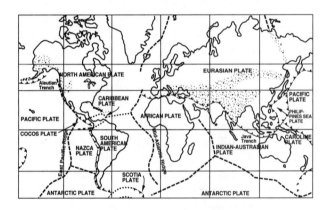

Figure 1.2   Lithosphere Plates

Each plate is approximately 50 to 60 mi (80 to 100 km) thick and has thick and thin parts. The thinner part deforms by elastic bending and brittle breakage. The thicker part yields plastically. Beneath the plate is a viscous layer on which the entire plate slides. The plates themselves tend to be internally rigid, interacting only at the edges.

These plates move relative to each other with steady velocities that approach 0.4 ft/yr (0.13 m/a).[6] Although plate velocities are slow by human standards, they are extremely rapid geologically. For example, a motion of 0.15 ft/yr (0.05 m/a) adds up to 30 mi (50 km) in only 1 million years. Some plate motions have been continuous for 100 million years.

Depending on location, the plates can be moving apart, colliding slowly to build mountain ranges, or slipping laterally past or sliding over and under one another.

1-6 SUBMARINE RIDGES

Where plates are pulling apart, particularly along the system of submarine ridges, hot material from the deeper mantle wells up to fill the gap. Some of the mantle material appears as lava in volcanic material. Most solidifies beneath the surface, forming a submarine ridge. The ridge is high relative to the ocean bottom because the mantle material is hot and, hence, low in density.

As the plates move apart, the ridge material gradually cools and contracts and its surface sinks. Ridges generally form step-like alterations in height perpendicular to the direction of plate motion. Strike-slip faults form parallel to the direction of plate motion. (See Sec. 1-10.)

1-7 SUBMARINE TRENCHES

Where plates converge, one dips down and slides beneath the other in a process known as *subduction*. Generally, an oceanic plate slides, or subducts, beneath a continental plate (as is happening along the west coast of South America) or beneath another oceanic plate (as is happening along the east side of the Philippine Sea plate). A trench is formed where the subducting plate dips down. The sediment from the ocean floor is scraped off against the front edge of the top plate. This is illustrated in Fig. 1.3.[7]

Far back under the top plate, inclined zones of earthquakes reach down into the mantle. The average depth of these zones is approximately 80 mi (125 km), but the zones can approach 450 mi (700 km) in depth. The hypocenters of earthquakes in these zones indicate the trajectory of the subducted plate.

A belt of volcanoes typically occurs above this earthquake zone, roughly paralleling the plate edges. (The Pacific Ocean coastal region of South America is typical of such an area.) Rock melting, which ultimately produces the volcanoes, starts when water combined in the crystalline structures of various minerals, or otherwise trapped, is removed by the increase in pressure on the subducted plate. The water loss lowers the net energy required to melt the remaining rock.

---

[6]There are grounds for suggesting that the African plate may be fixed relative to the deep mantle. If so, it is the only major plate that is fixed.

[7]Details and dimensions are those for western Java and the Java trench system, but other systems are similar.

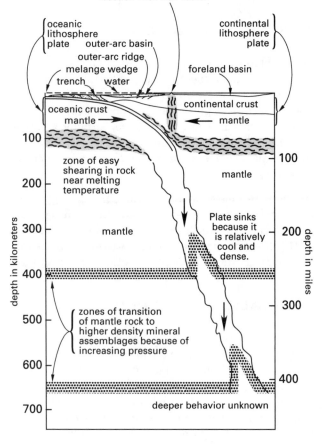

Figure 1.3   Zone of Subduction

(i.e., an aspect of the dilatational theory). However, recent research seems to indicate the former explanation is more appropriate than the latter.

Only a fraction of the energy released in an earthquake actually appears in seismic waves. Most of the released strain energy is reabsorbed locally by the moving, deforming, and heating of the rock. The fraction absorbed increases irregularly with increasing size of earthquakes. Minor earthquakes generally do not represent a sufficient release of energy to dissipate the strain energy and prevent great earthquakes, although a slow creep along a fault can provide a partial release. A great earthquake, however, does not necessarily release all of the strain energy either.

Great earthquakes occur primarily along convergent (subducting) plate boundaries.[9] Submerged ridges (where plates are spreading apart) are so hot at relatively shallow depths that the solid rock above them cannot store enough elastic strain energy to produce great earthquakes. The infrequent large earthquakes that do occur in these ridge systems are mostly on the longer strike-slip faults. (See Sec. 1-10.)

1-9 SEISMIC SEA WAVES

When the seafloor suddenly rises up during a great earthquake, water also rises with it and then rushes away to find a level surface. If the floor drops, water rushes in. An enormous mass of water is suddenly set in motion, and a complex sloshing back and forth between continents continues for many hours. The result is a train of surface-water waves, each of which is known as a *seismic sea wave* (also known as a *tidal wave*), or, in Japanese, a *tsunami*. The most pronounced sudden changes in seafloor depth, and hence the greatest sea waves, result from shallow subduction-zone earthquakes.

As with any surface wave or surge wave, the velocity of a tsunami depends primarily on the ocean depth.[10] In deep ocean, waves travel at about 500 mi/hr (800 km/h). The waves at sea may be an hour apart and perhaps only 1 ft (0.3 m) in height. Combined with a wave period of 5 to 60 min, they are virtually undetectable. As a wave approaches land, however, the wave velocity decreases due to increased friction with

1-8 EARTHQUAKE ENERGY RELEASE[8]

Shallow earthquakes represent sudden slippages and are accompanied by a release of elastic strain energy stored in the rock over a long period. It is not totally clear whether deep mantle subduction-zone earthquakes are accompanied by similar elastic releases (i.e., the elastic rebound theory) or are merely abrupt contractions of part of the subducting plateinto rock of higher density

---

[8]See Sec. 2-5 also.

---

[9]The great 1906 San Francisco earthquake, however, was not a subduction-zone earthquake.

[10]This is greatly simplified. There has been much research on the effect of depth and other aspects of tsunami generation and propagation. A good survey of the subject is contained in Murty (1977).

the increasingly shallow seafloor. As the wave velocity decreases, the wave height increases.

Where seafloor topography and orientation are optimal for tsunami formation (where there is a gently sloping seafloor and where the slope is parallel to wave direction), the wave can form a wall of water more than 50 ft (15 m) in height. Such a wave can cause enormous destruction when it rushes onto shore. Nearby coastal points, where the bottom configuration is much different (i.e., more abrupt in depth change), may see the same wave pass as only a rapid surge and withdrawal of water.

Only normal (dip-slip) and thrust (reverse) faults produce tsunamis. The greater the depth of water, the larger the energy content of the tsunami.

1-10 FAULTS

A fault is a fracture in the Earth's crust along which two blocks slip relative to each other. One crustal block may move horizontally in one direction while the opposite block moves horizontally in the opposite direction. Alternatively, one block may move upward while the other moves downward.

One of the ways movement along faults can occur is by sudden displacement, or *slip*, of the crust or rock along a fault. During the 1906 San Francisco earthquake, the ground was displaced as much as 21 ft (6.5 m) in northern California along the San Andreas Fault. By comparison, the 1989 Loma Prieta earthquake had a maximum displacement of approximately 6 ft (2 m).

Most of the faults in California are vertical or near-vertical breaks. Movement along these breaks is predominantly horizontal in the northerly or northwesterly direction.[11] With *right-lateral movement* (such as the movement of earthquakes in the San Andreas system), a block on the opposite side of the fault (relative to an observer) moves to the right. Conversely, the block moves to the left in a *left-lateral fault*. Lateral movement is produced by *strike-slip (wrench) faults.*

A fault in which the movement is vertical is called a *dip-slip fault.* In a *normal fault*, the hanging wall moves down relative to the foot wall. In a *reverse fault*, also known as a *thrust fault*, the hanging wall moves up relative to the foot wall.

Along many faults movement is both horizontal and vertical. Such faults are named by combining the names

of each kind of movement. For example, Fig. 1.4 shows a left-lateral normal fault. The term *oblique fault* is also used.

A few *reverse faults* have been active in California. The planes of such faults are inclined to the Earth's surface. The rocks above the fault plane have been thrust upward and over the rocks below the fault plane. The Arvin-Tehachapi earthquake of 1952 was caused by the White Wolf reverse fault. The San Fernando earthquake of 1971 was caused by a sudden rupture along a reverse fault at the foot of the San Gabriel Mountains.

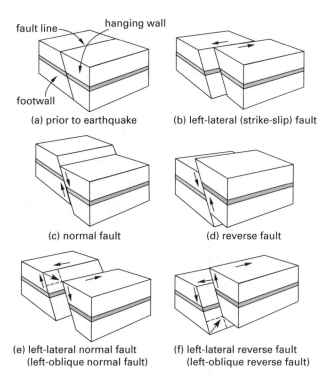

(a) prior to earthquake

(b) left-lateral (strike-slip) fault

(c) normal fault

(d) reverse fault

(e) left-lateral normal fault
(left-oblique normal fault)

(f) left-lateral reverse fault
(left-oblique reverse fault)

Figure 1.4 Types of Faults

1-11 CALIFORNIA FAULTS

The most earthquake-prone areas in the United States are those that are adjacent to the San Andreas Fault system of California, as well as the fault system that separates the Sierra Nevada from the Great Basin. Many of the individual faults of these major systems are known to have been active during the past 200 years. Others are believed to have been active since the end of the last great ice advance about 10,000 years ago.

---

[11]Notable exceptions are the Garlock and Big Pine left-lateral faults, which trend westerly.

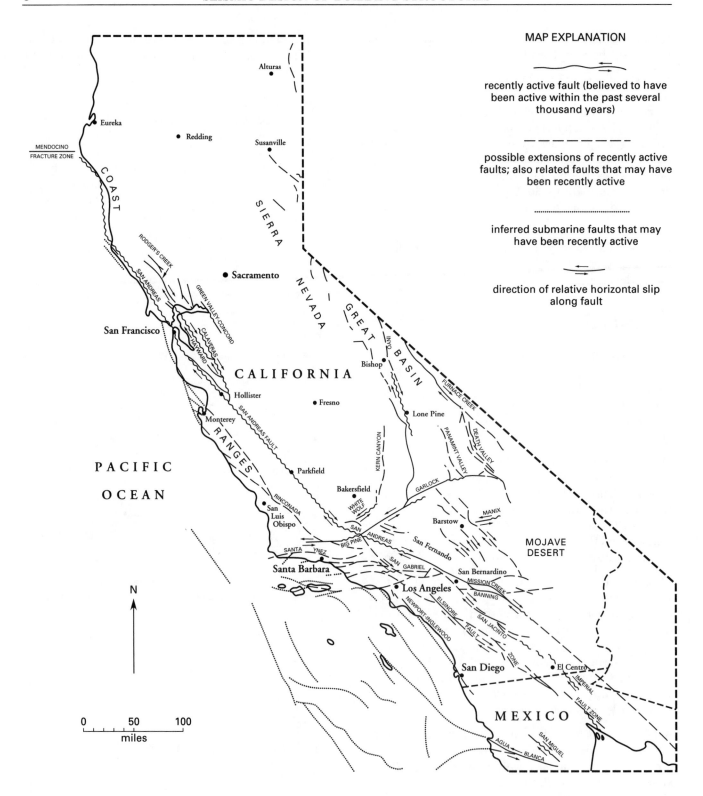

Figure 1.5   Active California Faults

During the past 200 years, many of the faults shown in Fig. 1.5 have experienced either sudden slip or slow creep. Activity of other faults, however, can only be inferred from geologic and topographic relations that indicate the faults have been active during the past several thousand years. Such activity suggests that these faults could slip or creep again. (See Sec. 1-13.)

Earthquakes in California are relatively shallow and are clearly related to movement along active faults. Many California earthquakes have produced surface rupture.

1-12 SAN ANDREAS FAULT

The San Andreas Fault is the major fault of a network that cuts through rocks of the California coastal region. This right-lateral fault is a huge fracture more than 600 mi (950 km) long. It extends almost vertically into the Earth to a depth of at least 20 mi (30 km). In detail, it is a complex zone of crushed and broken rock from only a few feet wide to a mile wide. Many smaller faults branch from and join the San Andreas Fault.

A linear trough in the surface of the Earth reveals the presence of the San Andreas Fault over much of its length. From the air, the linear arrangements of lakes, bays, and valleys is apparent. On the ground, the fault zone can be recognized by long, straight escarpments, narrow ridges, and small, undrained ponds formed by the settling of small areas of rock. However, people on the ground usually do not realize when they are on or near the fault.

Geologists who have studied the fault between Los Angeles and San Francisco have suggested that the total accumulated displacement along the fault may be as much as 350 mi (550 km). Similarly, geological study of a segment of the fault between the Tejon Pass and the Salton Sea has revealed geologically similar terrains on opposite sides of the fault separated by 150 mi (250 km). This indicates that the separation is a result of movement along the San Andreas and branching San Gabriel faults.

Since 1934, earthquake activity along the San Andreas Fault system has been concentrated in three areas: (1) an off-shore area at the northernmost tip of the fault known as the *Mendocino fracture zone*, (2) the area along the fault between San Francisco and Parkfield, and (3) the southernmost fault section roughly bounded by Los Angeles and the border with Mexico. Creep, slip, and moderate earthquakes have occurred on a regular basis in these areas.

The two zones between these three active areas have had almost no earthquakes or known slip since the great earthquakes of 1857 in the southern segment and 1906 in the northern segment. This implies that these two zones of the San Andreas Fault system are temporarily locked and that strain energy is building. The lack of seismic activity in the locked sections could mean that the sections are subject to less frequent but larger fault movements and, correspondingly, more severe earthquakes.

1-13 CREEP

In addition to fault rupture, a second type of fault movement known as *creep* can occur. Creep is characterized by continuous or intermittent movement without noticeable earthquakes. Fault creep occurring on portions of the Hayward, Calaveras, and San Andreas faults has produced cumulative offsets ranging from mere millimeters to almost 1 ft (0.3 m) in curbs, streets, and railroad tracks.

The offsets observed seem consistent with the creep rate measured. Precise surveying shows a slow drift approaching 2 in/yr (5 cm/yr) in some places along the San Andreas Fault. Over 350 mi (550 km) of offset has occurred during the past 100 million years.

1-14 SEISMIC WAVES

Seismic waves are of three types: compression, shear, and surface waves. Compression and shear waves travel from the hypocenter through the Earth's interior to distant points on the surface. Only compression waves, however, can pass through the Earth's molten core. Because *compression waves* (also known as *longitudinal waves*) travel at great speeds (19,000 ft/sec, or 5800 m/s, in granite) and ordinarily reach the surface first, they are known as *P-waves* (for "primary waves").[12] P-wave velocity is given by Eq. 1.3.

$$v_P = \sqrt{\frac{(\lambda + 2G)g_c}{\rho}} \quad \text{[U.S.]} \quad [1.3(a)]$$

$$v_P = \sqrt{\frac{\lambda + 2G}{\rho}} \quad \text{[SI]} \quad [1.3(b)]$$

In Eq. 1.3, λ is Lame's constant,

$$\lambda = \frac{G(E - 2G)}{3G - E} = \frac{\nu E}{(1 + \nu)(1 - 2\nu)}$$

---

[12]The term *dilatation* is used to describe negative compression (i.e., the "expansion" of rock from its normal density). (See Fig. 1.6.)

Shear waves (also known as *transverse waves*) do not travel as rapidly (10,000 ft/sec, or 3000 m/s, in granite) through the Earth's crust and mantle as do compression waves. Because they ordinarily reach the surface later, they are known as *S-waves* (for "secondary waves"). Instead of affecting material directly behind or ahead of their lines of travel, shear waves displace material at right angles to their path. Equation 1.4 gives the velocity of S-waves. While S-waves travel more slowly than P-waves, they transmit more energy and cause the majority of damage to structures.

$$v_S = \sqrt{\frac{Gg_c}{\rho}} \quad \text{[U.S.]} \quad\quad [1.4(a)]$$

$$v_S = \sqrt{\frac{G}{\rho}} \quad \text{[SI]} \quad\quad [1.4(b)]$$

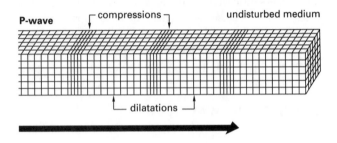

P-wave

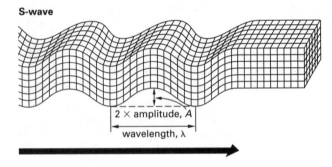

S-wave

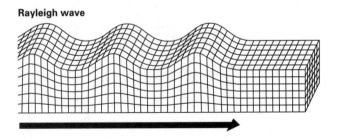

Rayleigh wave

Figure 1.6 Types of Seismic Waves

Surface waves, also known as *R-waves* (for "Rayleigh waves") or *L-waves* (for "Love waves"), may or may not form. They arrive after the primary and secondary waves. In granite, R-waves move at approximately 9000 ft/sec (2700 m/s).

1-15 LOCATING THE EPICENTER

The first indication of an earthquake will often be a sharp "thud" signaling the arrival of the compression wave front. This will be followed by the shear waves and then the ground roll caused by the surface waves. The times separating the arrivals of the compression and shear waves at various seismometer stations can be used to locate the epicenter's position and depth.

The distance, s, from a seismometer to the epicenter can be determined from the wave velocities and the observed time between the arrival of the compression (P-) and shear (S-) waves.[13]

$$t_S - t_P = \left(\frac{1}{v_S} - \frac{1}{v_P}\right)s \quad\quad [1.5]$$

The epicenter and hypocenter correspond to the locations of initial fault slip but do not necessarily coincide with the center of energy release. For small and medium earthquakes (i.e., Richter magnitude $M < 6$), the points of initial fault slip and energy release are relatively close. For larger earthquakes, however, hundreds of kilometers can separate the two.

---

[13]It is not always possible to accurately determine the difference in arrival times from the seismometer record.

2

EARTHQUAKE CHARACTERISTICS

2-1 INTENSITY SCALE

The *intensity* of an earthquake (not to be confused with *magnitude*, see Sec. 2-3) is based on the damage and other observed effects on people, buildings, and other features. Intensity varies from place to place within the disturbed region. An earthquake in a densely populated area that results in many deaths and considerable damage may have the same magnitude as a shock in a remote area that does nothing more than frighten the wildlife. Large magnitude earthquakes that occur beneath the oceans may not even be felt by humans.

An intensity scale consists of a series of responses, such as people awakening, furniture moving, and chimneys being damaged. Although numerous intensity scales have been developed, the scale encountered most often in the United States is the *Modified Mercalli Intensity scale*, developed in 1931 by the American seismologists Harry Wood and Frank Neumann.[1]

The Modified Mercalli scale consists of 12 increasing levels of intensity (expressed as Roman numerals following the initials MM) that range from imperceptible shaking to catastrophic destruction. The lower numbers of the intensity scale generally are based on the manner in which the earthquake is felt by people. The higher numbers are based on observed structural damage. The numerals do not have a mathematical basis and therefore are more meaningful to nontechnical people than to those in technical fields.

---

[1] The original Mercalli scale was developed in 1902 by the Italian seismologist and volcanologist of the same name. The *Rossi-Forel scale* (its ten values are used in Fig. 2.1 to describe the 1906 San Francisco earthquake) was developed in the 1880s.

Table 2.1
Modified Mercalli Intensity Scale

intensity	observed effects of earthquake
I	Not felt except by very few under especially favorable conditions.
II	Felt only by a few persons at rest, especially by those on upper floors of buildings. Delicately suspended objects may swing.
III	Felt quite noticeably by persons indoors, especially in upper floors of buildings. Many people do not recognize it as an earthquake. Standing vehicles may rock slightly. Vibrations similar to the passing of a truck. Duration estimated.
IV	During the day, felt indoors by many, outdoors by a few. At night, some awakened. Dishes, windows, doors disturbed; walls make cracking sound. Sensation like heavy truck striking building. Standing vehicles rock noticeably.
V	Felt by nearly everyone; many awakened. Some dishes, windows broken. Unstable objects overturned. Pendulum clocks may stop.
VI	Felt by all, many frightened. Some heavy furniture moved. A few instances of fallen plaster. Damage slight.
VII	Damage negligible in buildings of good design and construction; slight to moderate in well-built ordinary structures; considerable damage in poorly built structures. Some chimneys broken.

(continued on next page)

Table 2.1 (continued)

intensity	observed effects of earthquake
VIII	Damage slight in specially designed structures; considerable damage in ordinary substantial buildings, with partial collapse. Damage great in poorly built structures. Fallen chimneys, factory stacks, columns, monuments, walls. Heavy furniture overturned.
IX	Damage considerable in specially designed structures; well-designed frame structures thrown out of plumb. Damage great in substantial buildings, with partial collapse. Buildings shifted off foundations.
X	Some well-built wooden structures destroyed; most masonry and frame structures with foundations destroyed. Rails bent.
XI	Few, if any, masonry structures remain standing. Bridges destroyed. Rails bent greatly.
XII	Damage total. Lines of sight and level are distorted. Objects thrown into air.

2-2 ISOSEISMAL MAPS

It is possible to compile a map of earthquake intensity over a region. Data for such an *isoseismal map* can be obtained by observation or, in some cases, by questions answered by residents after an earthquake.

2-3 RICHTER MAGNITUDE SCALE

In 1935, Charles F. Richter of the California Institute of Technology developed the Richter magnitude scale to measure earthquake strength. The magnitude, M, of an earthquake is determined from the logarithm to base ten of the amplitude recorded by a seismometer. (See Secs. 2-4 and 2-5.) Adjustments are included in the magnitude to compensate for the variation in the distance between the various seismometers and the epicenter. Because the Richter magnitude is a logarithmic scale, each whole number increase in magnitude represents a ten-fold increase in measured amplitude.

Richter magnitude is expressed in whole numbers and decimal fractions. For example, a magnitude of 5.3 might correspond to a *moderate earthquake*. A *strong earthquake* might be rated at 7.3. *Great earthquakes* have magnitudes above 7.5. Earthquakes with magnitudes of 2.0 or less are known as *microearthquakes*. While recorded on seismometers, microearthquakes are rarely felt by people.

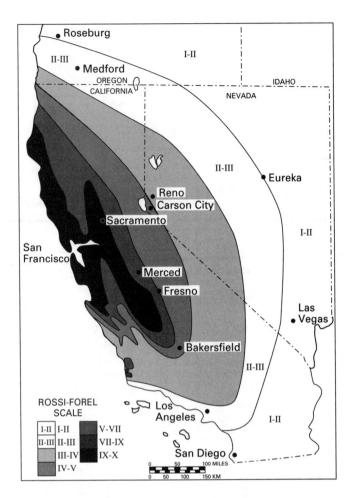

Figure 2.1 Isoseismal Map of 1906 San Francisco Earthquake (based on the Rossi-Forel Intensity Scale)

Source: *Earthquake Engineering*, Robert L. Wiegel, ed., copyright © 1970 by Prentice-Hall, Inc. Redrawn from map 23 in *Report of the State Earthquake Investigation Committee*, Atla, Carnegie Institution of Washington, publication 87 (1908).

Several thousand seismic events with magnitudes of approximately 4.5 or greater occur each year and are strong enough to be recorded by seismometers all over the world. Earthquakes of this size and below have little potential to cause structural damage. Great earthquakes, such as the 1906 San Francisco earthquake and the 1964 Alaskan earthquake, occur, on the average, once each year.

The magnitude of an earthquake depends on the length and breadth of the *fault slip*, as well as on the amount of slip. The largest examples of fault slip recorded in California accompanied the earthquakes of 1857, 1872, and 1906—all of which had estimated magnitudes over 8.0 on the Richter scale.

Although the Richter scale has no lower or upper limit (i.e., it is "open ended"), the largest known shocks have had magnitudes in the 8.7 to 8.9 range.[2] The actual factor limiting energy release—and hence Richter magnitude—is the strength of the rocks in the Earth's crust.[3]

Because of the physical limitations of the faults and crust in the area, earthquakes larger than 8.5 in southern California are considered to be highly improbable.

2-4 RICHTER MAGNITUDE CALCULATION

The Richter magnitude, M, is calculated from the maximum amplitude, A, of the seismometer trace, as illustrated in Fig. 2.2. A_0 is the seismometer reading produced by an earthquake of standard size (i.e., a *calibration earthquake*). Generally, A_0 is 3.94×10^{-5} in (0.001 mm).

$$M = \log_{10} \left(\frac{A}{A_0} \right) \qquad [2.1]$$

Equation 2.1 assumes that a distance of 62 mi (100 km) separates the seismometer and the epicenter. For other distances, the nomograph of Fig. 2.3 and the following procedure can be used to calculate the magnitude. Due to the lack of reliable information on the nature of the Earth between the observation point and the earthquake epicenter, an error of 5 to 20 mi (10 to 40 km) in locating the epicenter is not unrealistic.

step 1: Determine the time between the arrival of the P- and S-waves.

step 2: Determine the maximum amplitude of oscillation.

step 3: Connect the arrival time difference on the left scale and the amplitude on the right scale with a straight line.

step 4: Read the Richter magnitude on the center scale.

step 5: Read the distance separating the seismometer and the epicenter from the left scale.

Whereas one seismometer can determine the approximate distance to the epicenter, it takes three seismometers to determine and verify the location of the epicenter.

---

[2]Depending on the sensitivity of the seismograph, earthquakes with negative Richter magnitudes can occur.

[3]It is said that Richter magnitudes much higher than 9 would correspond to an energy release sufficient to destroy the Earth itself. While this may theoretically be true for much larger Richter magnitudes (due to the logarithmic nature of the measurement), the limited strength of the rock itself serves to ensure that such a doomsday earthquake will never occur.

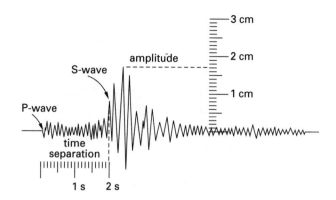

Figure 2.2 Typical Seismometer Amplitude Trace

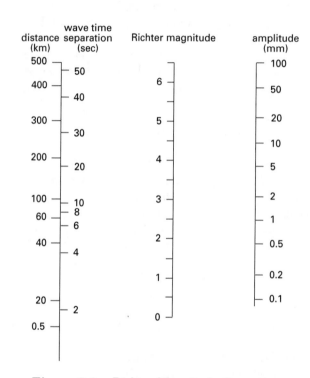

Figure 2.3 Richter Magnitude Correction Nomograph

2-5 ENERGY RELEASE AND MAGNITUDE CORRELATION

Once the Richter magnitude, M, is known, an approximate relationship can be used to calculate the energy, E, radiated. Most of the relationships are of the form of Eq. 2.2.

$$\log_{10} E = \log_{10} E_0 + aM \qquad [2.2]$$

In 1956, Gutenberg and Richter determined the approximate correlation to be as given in Eq. 2.3. E is the energy in ergs. (See App. A for conversions to other units.) Although there have been other relationships developed, Eq. 2.3 has been verified against data from underground explosions and is the primary correlation cited.

$$\log_{10}E = 11.8 + 1.5M \qquad [2.3]$$

The radiated energy is less than the total energy released by the earthquake. The difference goes into heat generation and other nonelastic effects, which are not included in Eq. 2.3. Little is known about the amount of total energy release.

The fact that a fault zone has experienced an earthquake offers no assurance that enough stress has been relieved to prevent another earthquake. As indicated by the logarithmic relationship between seismic energy and Richter magnitude, a small earthquake (of magnitude 5, for example) would radiate approximately only 1/32 of the energy of an earthquake just one magnitude larger (of magnitude 6, for example). Thus it would take 32 small earthquakes to release the same energy as an earthquake one magnitude larger.

2-6 LENGTH OF ACTIVE FAULT

Equation 2.4 correlates the Richter magnitude, M, with the approximate total fault length, L in kilometers, involved in an earthquake. Such correlations are very site-dependent, and even then, there is considerable scatter in such data. Equation 2.4 should be considered only representative of the general (approximate) form of the correlation.

$$\log_{10}L = 1.02M - 5.77 \qquad [2.4]$$

2-7 LENGTH OF FAULT SLIP

Equation 2.5 (as derived by King and Knopoff in 1968) correlates the Richter magnitude, M, and the fault length, L (in meters), with the approximate length of vertical or horizontal fault slip, D (for *displacement*) in meters.[4] As with Eq. 2.4, this correlation should be considered representative of the general relationship.

$$\log_{10}(LD^2 \times 10^6) = 1.90M - 2.65 \qquad [2.5]$$

2-8 PEAK GROUND ACCELERATION

The *peak* (maximum) *ground acceleration*, PGA, is easily measured by a seismometer (see Sec. 2-14) or accelerometer (see Sec. 2-15) and is one of the most

important characteristics of an earthquake.[5] The PGA can be given in various units, including ft/sec², in/sec², or m/s². However, it is most common to specify the PGA in "g's" (i.e., as a fraction or percent of gravitational acceleration).

$$\mathrm{PGA} = \frac{a_{\mathrm{ft/sec^2}}}{32.2} \times 100\% \qquad \text{[U.S.]} \quad [2.6(a)]$$

$$= \frac{a_{\mathrm{in/sec^2}}}{386} \times 100\% \qquad \text{[U.S.]} \quad [2.6(b)]$$

$$= \frac{a_{\mathrm{m/s^2}}}{9.81} \times 100\% \qquad \text{[SI]} \quad [2.6(c)]$$

Significant ground accelerations in California include 1.25 g (1971 San Fernando earthquake, Pacoima dam site), 0.50 g (1966 Parkfield earthquake), 0.65 g (1989 Loma Prieta earthquake), and 1.85 g (1992 Cape Mendocino earthquake).

Equation 2.7 (as determined by Gutenberg and Richter in 1956) is one of many approximate relationships between the Richter magnitude, M, and the PGA at the epicenter. Of course, the ground acceleration (in rock) will decrease as the distance from the epicenter increases, and for this reason, relationships called *attenuation equations* have been developed. (See Sec. 2-13.)

Blume's 1965 equation (Eq. 2.7) for California earthquakes depends on the epicentral distance (R' in kilometers), the local depth (h in kilometers), and a specific site factor (b).

$$\mathrm{PGA}_{\mathrm{gravities}} = \frac{y_o}{1 + \left(\dfrac{R'}{h}\right)^2}$$

$$\log y_o = -(b+3) + 0.81M - 0.027M^2 \qquad [2.7]$$

Attenuation equations are very site dependent. Since Eq. 2.7 was developed, newer studies have resulted in better correlations in different formats and for many different locations, but they are all based on limited data. Such studies regularly result in revisions of the seismic provisions of building codes.

Table 2.2 is a commonly cited correlation between magnitude, PGA, and duration of *strong-phase shaking* (see Sec. 2-14) in the vicinity of the epicenter of California earthquakes.[6] The values of acceleration in the table

---

[4]King, Chi-Yu, and L. Knopoff, "Stress Drop in Earthquakes," *Bulletin of the Seismological Society of America*, 58 (1968): 249.

[5]It is possible for an earthquake to exceed the range of the accelerometer or seismometer, in which case the PGA will not be recorded.

[6]This table gives the impression of high correlation even though the correlation is actually low. For example, the 1989 Loma Prieta earthquake magnitude was approximately 7.1, but the peak ground acceleration was 0.65 g.

are somewhat on the high side. Ground acceleration in observed earthquakes usually has been lower.

Table 2.2
Approximate Peak Ground Acceleration and Duration of Strong-Phase Shaking (California Earthquakes)

magnitude	maximum acceleration (g)	duration (sec)
5.0	0.09	2
5.5	0.15	6
6.0	0.22	12
6.5	0.29	18
7.0	0.37	24
7.5	0.45	30
8.0	0.50	34
8.5	0.50	37

2-9 CORRELATION OF INTENSITY, MAGNITUDE, AND ACCELERATION WITH DAMAGE

Although there are some empirical relationships, no exact correlations of intensity, magnitude, and acceleration with damage are possible since many factors contribute to seismic behavior and structural performance. For example, seismic damage depends on the care that was taken at the time of building design and construction. Buildings in villages in undeveloped countries fare much worse than high-rise buildings in developed countries in earthquakes of equal magnitudes. This damage causes a corresponding lack of correlation between intensity and magnitude.

However, within a geographical region with consistent design and construction methods, fairly good correlation exists between structural performance and ground acceleration, because the Mercalli intensity scale is based specifically on observed damage.

Table 2.3
Approximate Relationship Between Mercalli Intensity and Peak Ground Acceleration

MMI	PGA (g)
IV	0.03 and below
V	0.03–0.08
VI	0.08–0.15
VII	0.15–0.25
VIII	0.25–0.45
IX	0.45–0.60
X	0.60–0.80
XI	0.80–0.90
XII	0.90 and above

2-10 VERTICAL ACCELERATION

The shear (transverse) waves are at right angles to the compression (longitudinal) waves. (See Sec. 1-14.) Since there is nothing constraining the shear waves to a horizontal direction, it is not surprising that the S-wave (shear wave) can be broken down into horizontal and vertical components. When necessary, these are identified as SH-waves and SV-waves for horizontal and vertical shear waves, respectively.

Vertical ground acceleration is known to occur in almost all earthquakes. The peak vertical acceleration is usually approximately one-third of the peak horizontal acceleration, but often reaches a ratio of two-thirds. Combined with resonance site effects, vertical forces can become substantial. Furthermore, forces from all three coordinate directions combine into a resultant force that can easily exceed the yield (and, sometimes, the ultimate) strength of a member.

The current UBC-97 seismic design code is generally based on horizontal acceleration alone [Sec. 1630.1.1]. This practice is justified by assuming that structures with horizontal seismic resistance will automatically have adequate vertical seismic resistance. One of the reasons this assumption has been accepted is that factors of safety should have been applied during the building design to ensure that a member is able to withstand a force equal to one gravity downward.

Experience has shown, however, that disregarding details to resist vertical forces can be a serious problem. Columns and walls in compression, cantilever beams, overhangs, and prestressed concrete structures that have not been designed according to specific seismic provisions are particularly susceptible to damage by vertical accelerations because they have little factor of safety against upward vertical acceleration. *Transfer girders*, horizontal members that support exterior perimeter columns in *tube buildings* (see Sec. 5-18), are definitely sensitive to vertical acceleration. The UBC-97 covers these special cases in Sec. 1630.11 and, for dynamic analysis, in Secs. 1631.2, Item 5, and 1631.5.5. (See Sec. 6-37.)

2-11 PROBABILITY OF OCCURRENCE[7]

The probability that an earthquake of magnitude M or greater will occur in a specific region in any given

---

[7]Wiegel (1970) contains a more complete presentation of this subject.

year is given approximately by Eq. 2.8.[8] B is a seismic parameter that has been approximately determined as 2.1 for the entire state of California and 0.48 for 100,000 mi^2 (260 000 km^2) of southern California. While Eq. 2.8 does not place any upper bound on M, the probability of exceeding 8.5 is effectively zero.

$$p\{M\} = e^{-M/B} \qquad [2.8]$$

The expected number of earthquakes having magnitude greater than M during Y years is given by Eq. 2.9. Equation 2.9 is known as a *recurrence formula*. For northern California, $C = 76.7 \text{ yr}^{-1}$ and $B = 0.847$, approximately. For the San Francisco area, $C = 19,700 \text{ yr}^{-1}$ and $B = 0.463$, approximately.

$$N = CYe^{-M/B} \qquad [2.9]$$

2-12 FREQUENCY OF OCCURRENCE

For a specific area, an equation for the expected number, N, of earthquakes of a given magnitude, M, per year will be of the form

$$\log_{10}N = a - bM \qquad [2.10]$$

For the south-central segment of the San Andreas fault, a and b have values of 3.3 and 0.88, respectively. Taking the entire world as a whole, the approximate relationship (up to approximately $M = 8.2$) is

$$\log_{10}N = 7.7 - 0.9M \qquad [2.11]$$

Table 2.4 gives the expected number of earthquakes of any given magnitude per 100 years in California. (The table does not give the frequency over any particular location in the state.) Table 2.4 cannot be derived exactly from Eq. 2.11 because adjustments have been made to account for California's increased seismicity.

Table 2.4

Approximate Expected Frequency of Occurrence of Earthquakes (per 100 years)

magnitude	number
4.75–5.25	250
5.25–5.75	140
5.75–6.25	78
6.25–6.75	40
6.75–7.25	19
7.25–7.75	7.6
7.75–8.25	2.1
8.25–8.75	0.6

---

[8]The form of Eq. 2.8 is easily derived from a Poisson distribution, which is commonly used to calculate the probability of an infrequent event.

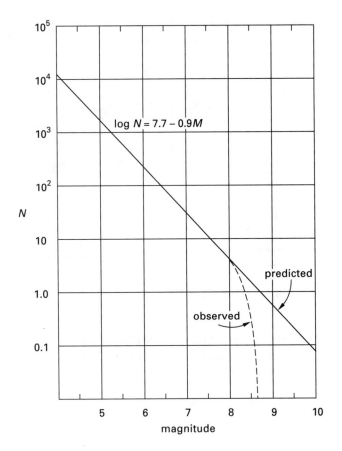

Figure 2.4   Expected Number of Earthquakes per Year (World)

2-13 ATTENUATION OF GROUND MOTION

Ground motion at a site is related to the seismic energy received at that site, and when the propagation path is through rock, the amount of energy decreases the farther a site is from the epicenter.[9] This decrease is known as *attenuation*. Some of the factors affecting attenuation include path line, path length, focal depth, geological formations, properties of the crustal rock, and orientation of the fault.

Unfortunately, the geology and local conditions affect the actual values so much that little more than

---

[9]While the overall wave energy attenuates when the transmission path is through rock, the damage at a site does not necessarily decrease with distance. There are other factors that can concentrate the energy that reaches the site, as was proved by the Mexico City (1985) and Loma Prieta (1989) earthquakes. This is analogous to the decrease in the intensity of sunlight with distance from the sun. While the light may be diffused when it reaches the Earth's surface, it can be sufficiently concentrated with a lens to kindle a fire.

generalizations such as the following are possible about the rate of attenuation.[10]

◇ Intensity generally decreases with distance from the epicenter.

◇ There is little attenuation in the vicinity of the epicenter.

◇ Higher-frequency components of the seismic wave attenuate faster than slower components do.

2-14 SEISMOMETER

Seismic waves (see Sec. 1-14) travel through the Earth and are recorded on seismometers. A *seismometer* is the detecting and recording part of a larger apparatus known as a *seismograph*. Seismometers are pendulum-type devices that are mounted on the ground and measure the displacement of the ground with respect to a stationary reference point. Since a seismometer usually records motion in only one orthogonal direction, three seismometers are needed to record all components of ground motion. Figure 2.2 illustrates a typical seismometer trace, known as a *seismogram*. Appendix F is the actual seismogram of the 1940 El Centro earthquake.

Notice that while seismic activity usually continues for some time after the start of the earthquake, the major movement occurs in a concentrated period known as the *strong phase*. The longer the earthquake shakes, the more seismic energy is absorbed by buildings; thus, the duration of strong-phase shaking greatly affects the damage inflicted.[11] The assumed duration of strong shaking for the current design earthquake in the UBC-97 for zone 4 is 10 to 20 sec.

Seismometers record the varying amplitude of ground oscillations beneath the instrument. Sensitive seismometers greatly magnify these ground motions and can detect strong earthquakes occurring anywhere in the world. The time, location, and magnitude of an earthquake can be determined from the data recorded by seismometer stations.

Since a seismometer is a spring-mass-dashpot device, it will magnify or distort earthquakes with frequencies in certain ranges. The ratio of actual damping to critical damping can be changed to minimize such distortion. Good seismometer design calls for a damping ratio of between 0.6 and 0.7 with a natural period of vibration smaller than the smallest period to be measured.[12] (See Sec. 4-7 for damping ratio.)

2-15 ACCELEROMETER

An *accelerometer (accelerograph)* is a seismometer mounted in buildings for the purpose of recording large accelerations.[13] For this reason, they are also known as *strong motion seismometers*. The large swings accelerometers record typically exceed the scale limits of most seismometers. An accelerometer located in a building does not run continually. It is triggered by a P-wave (see Sec. 1-14) and runs for a fixed period of time.

Buildings located in seismic zones 3 and 4 over ten stories in height or over six stories and total floor areas of 60,000 ft^2 (5574 m^2) or more are required by the UBC-97 [Sec. 1649, App. Chap. 16, Division II] to have at least three approved recording accelerographs.

2-16 OTHER SEISMIC INSTRUMENTS

A *tiltmeter* installed in the ground works on the same principle as a carpenter's level. The slightest movement of a bubble floating in a spherical dome is electronically detected to reveal tilting of the ground.

The strain (deformation) of rock under pressure can be measured by a *magnetometer*. Such strain changes the magnetic permeability of the rock, resulting in a local change in the magnetic field of the Earth.

Strain gauges measure how much the earth deforms. *Dilatometers* measure the earth's dilations. A dilatometer is a closed, fluid-filled tube approximately 10 ft (3 m)

---

[10]Numerous attenuation relationships have been published, particularly for sites and faults in California. However, there is little similarity between the correlations. Nevertheless, for sites located on firm soil, the attenuation laws are quite useful for predicting expected ground motions from future earthquakes.

[11]For example, the 1985 Chile earthquake (magnitude 7.8) had almost 80 sec of strong ground motion. There were approximately 60 sec of strong ground motion in the 1985 Mexico earthquake (magnitude 8.1). Both of these earthquakes resulted in significant loss of life and destruction. By comparison, the strong-phase motion of the 1940 El Centro earthquake (magnitude 7.1) lasted only 10 sec, and the 1989 Loma Prieta earthquake (magnitude 7.1) had a mere 5 sec. There is great debate and no consensus on whether or not long-duration earthquakes can occur in California.

[12]These design principles are incorporated into the Wood-Anderson seismometer, which has a damping ratio of approximately 0.8 (almost critical) and a natural period of 0.8 sec (i.e., a natural frequency of 1.25 Hz).

[13]The distinction between the terms *seismometer* and *accelerometer* is not always made. However, it is important to recognize that seismometers typically run continually and record displacement, while accelerometers are triggered by P-waves and record acceleration.

long that is buried in the ground. Changes in the earth's "squeeze" are detected and measured by a pressure sensor or gauge at the top of the tube.

Scintillation counters are installed in wells to measure the amount of radioactive radon gas in the water. Minute amounts of radon are released into well water by rocks under stress.

Changes in the resistance of rock can be measured by a *resistivity gauge* and are indications of density and water content changes. Both density and water content change during periods of fluctuating stress.

A *creepmeter* measures minute gradual movement along a fault. In the past, such a meter relied on a wire stretched across a fault. Movement of the fault increased the tension in the wire. Current creepmeters use laser technology.

A *gravimeter* responds to variations in the local force of gravity. Such variations are the result of changes in underground rock density.

A *laser* can measure the round-trip travel time of a light beam between two points. When the relative positioning of the two points changes as a direct result of an earthquake, the travel time also changes.

2-17 EARTHQUAKE PREDICTION

While reliable long-term earthquake prediction remains elusive, short-term predictions based on the observation or measurement of various precursors (*premonitory signs*) seem possible. Most of the measuring devices mentioned in Secs. 2-14 through 2-16 can be adapted to instantaneous reporting. It may be possible to correlate sudden and unexpected changes in behavior (i.e., creep rate, tilt, accumulating strain, elevation, fluid pressure, seismic wave speed, electrical conductivity, and magnetic susceptibility) with the probability of an impending earthquake. Thus far, however, no reliable indicators have been found.

One proposed precursor, as related by the *seismic bay hypothesis*, may be the large-scale volume of the rock itself. As rock masses along a fault develop stresses, they crack and increase in volume. The volume increase is not consistently detectable, but in some cases, the volume increase can be observed.[14]

Another precursor is a decrease in the ratio of velocities of the two major seismic wave types. Normally, the ratio of velocities of the P- and S-waves is approximately

1.7, but this ratio decreases as the fault rock breaks. After a time, groundwater seeps in and fills the fissures (forming a "seismic bay"), and the velocity ratio returns to 1.7. According to the hypothesis, an earthquake can then occur.

Another possible precursor may be the emission of ultra-low-frequency (ULF) radio waves before an earthquake.[15] Additional research in this area is required.

2-18 OTHER EARTHQUAKE CHARACTERISTICS

In addition to the peak ground acceleration, two other characteristics that contribute significantly to the effects of an earthquake are its duration of strong shaking (motion) and frequency content.[16] Roughly speaking, the longer the duration of strong shaking, the greater the energy that can be imparted to a structure. Since various parts of a structure can absorb only a limited amount of elastic strain energy, a longer earthquake has a greater chance of driving structural performance into inelastic behavior.

The shaking in the 1989 Loma Prieta earthquake lasted approximately 10 to 15 sec. It is generally believed that a longer duration (i.e., another 20 sec) would have resulted in significantly greater damage. The UBC-97 provisions are intended to accommodate earthquakes with durations of 10 to 30 sec.

The effects of resonance on all types of machinery and structures are well known. Basically, if a regular disturbing force is applied at the same frequency as the natural frequency (see Sec. 4-5), the oscillation of the structure can be greatly magnified. In such cases, the effects of damping are minimal. While earthquakes are never as regular as a sinusoidal waveform, there is usually a predominant waveform that is roughly regular.[17]

---

[14]The *Palmdale bulge* is thought to be an example of this volume increase.

[15]Ultra-low-frequency radio waves (below 5 Hz) were received from the vicinity of the epicenter in the October 1989 Loma Prieta earthquake. However, a direct cause-and-effect relationship has yet to be proved.

[16]Note that this characteristic is defined as the duration of strong shaking, not the overall duration of the earthquake.

[17]Fourier analysis can be used to separate the dominant frequencies (i.e., the frequencies at which most seismic energy arrives at a site) of a specific earthquake. However, this is rarely done for most seismic design projects.

3

EFFECTS OF EARTHQUAKES ON STRUCTURES

3-1 SEISMIC DAMAGE

Structural damage due to an earthquake is not solely a function of the earthquake ground motion. The primary factors affecting the extent of damage are

⋄ *earthquake characteristics*, such as (a) peak ground acceleration, (b) duration of strong shaking, (c) frequency content, and (d) length of fault rupture

⋄ *site characteristics*, such as (a) distance between the epicenter and structure, (b) geology between the epicenter and structure, (c) soil conditions at the site, and (d) natural period of the site

⋄ *structural characteristics*, such as (a) natural period and damping of the structure, (b) age and construction method of the structure, and (c) seismic provisions (i.e., detailing) included in the design

3-2 SEISMIC RISK ZONES

In order to design a structure to withstand the effects of an earthquake, it is necessary to determine the expected earthquake magnitude. While extensive mathematical models could be developed for each location, seismic codes have evolved a simplified model based on *seismic zones*. The seismic zone map referenced by the 1997 Uniform Building Code is shown in Fig. 3.1.[1]

There are several methods of evaluating the significance of the seismic risk zones. One method is to correlate the zones with the approximate accelerations and magnitudes, as Table 3.1 does.

---

[1]The map shown in Fig. 3.1 is subject to change with different versions of the seismic code.

Table 3.1
Approximate Code Maximum Zone Acceleration and Magnitude

zone	maximum acceleration	maximum magnitude
0	0.04 g	4.3
1	0.075 g	4.7
2A	0.15 g	5.5
2B	0.2 g	5.9
3 (not near a great fault)	0.30 g	6.6
4 (near a great fault)	0.40 g	7.2

Another interpretation of the significance of the zones is to correlate them to the effects of an earthquake and the Modified Mercalli intensity scale as in Table 3.2.

Table 3.2
Effects of an Earthquake by Zone

zone	effect
0	no damage
1	minor damage corresponding to MM intensities V and VI; distant earthquakes may damage structures with fundamental periods greater than 1.0 sec
2	moderate damage corresponding to MM intensity VII
3	major damage corresponding to MM intensity VIII
4	major damage corresponding to MM intensity VIII and higher

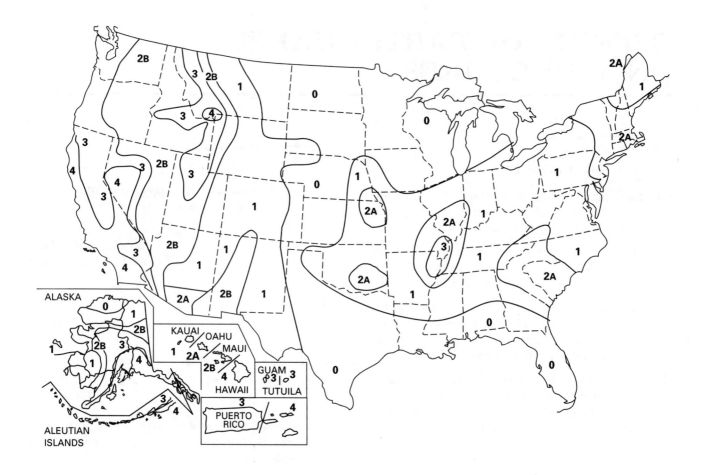

Figure 3.1 Seismic Risk Zones

Reproduced from the 1997 edition of the *Uniform Building Code*™, copyright © 1997, with the permission of the publisher, the International Conference of Building Officials. [UBC-97 Fig. 16-2]

3-3 RISK MICROZONES

Certain limited areas, referred to as *microzones*, consistently experience higher ground accelerations than do surrounding areas.[2] This tendency is primarily attributed to the site (i.e., soil) conditions in the microzone, as the soil profile affects the peak ground acceleration, frequency content, and duration of strong motion. Inasmuch as seismic damage is at least partially related to ground acceleration, knowledge of such microzonification is essential. The *microzonification* concept, however, has not been explicitly incorporated in the Uniform Building Code, although it is implicit in the design process.

Mexico City has incorporated microzones into its rebuilding plan following the devastating earthquake in 1985. Considering the significant variations in damage in different areas of the San Francisco Bay Area during the Loma Prieta earthquake, microzonification has also been widely proposed. (For example, unreinforced buildings in Chinatown were not damaged, while similarly constructed buildings in the Marina district were. Portions of the Cypress structure on clay collapsed, but other portions built on firmer soil did not.)

3-4 PROBABLE AND MAXIMUM CONSIDERED EARTHQUAKE GROUND MOTIONS

A *maximum probable earthquake* ground motion at a site is the largest earthquake shaking that has a significant probability of occurring within the lifetime of a structure due to earthquakes from all sources. Since the potential losses of property and life are very great, the probability does not have to be very large to be significant.

The *maximum considered earthquake* ground motion at a site is the maximum possible earthquake ground motion based on the mapped spectral response acceleration, modified by site soil profile effects. The maximum considered earthquake is difficult to evaluate.

3-5 EFFECTIVE PEAK GROUND ACCELERATION

The *effective peak ground acceleration*, EPA (or A_a as it is used in some documents), is nominally equal to the maximum ground acceleration associated with the design earthquake, but is more of a contrived design parameter set by code than a feature of an actual earthquake.[3] As a code provision, the EPA depends on the region and corresponds numerically in gravities to the Z coefficient (see Sec. 6-12) in the UBC-97.

As explained in the 1990 Blue Book commentary, the EPA is derived from the log-tripartite graph given in ATC 3-06 scaled downward from the spectral acceleration by dividing it by a spectral amplification factor. (See Sec. 3-9 for spectral acceleration.) The value of the spectral amplification factor depends on the amount of damping present in the building and the probability of the earthquake's occurrence. For example, for 5% damping and a hazard level (probability of occurrence) of 10% in a 50-year period, the spectral amplification factor is 2.5.

The method of deriving the EPA is subject to continuing study and analysis.

3-6 SITE PERIOD

The *site (soil) period* is now recognized as a significant factor contributing to structural damage.[4] When a site has a natural frequency of vibration that corresponds to the predominant earthquake frequency, site movement can be greatly magnified. This is known as *resonance*. (See Sec. 4-13.) Thus, the buildings can experience ground motion much greater than would be predicted from only the seismic energy release.

Determining the actual site period is no easy matter. Since the site period can be computed precisely from widely available formulas and still be grossly inaccurate, such determinations are best left to experts familiar with the area.

Other soil characteristics, including density, bearing strength, moisture content, compressibility (i.e., tendency to settle), and sensitivity (i.e., tendency to liquefy), are additional factors not addressed by the

---

[2] For example, in the 1989 Loma Prieta earthquake, peak ground accelerations in San Francisco did not generally exceed 0.09 g. However, a 31-story instrumented office building in Emeryville experienced a horizontal acceleration of 0.26 g at the ground. Peak accelerations at the Bay Bridge are believed to have ranged between 0.22 g and 0.33 g. The Golden Gate Bridge experienced 0.24 g, while 0.33 g was recorded at the San Francisco Airport. Similar ground accelerations were recorded near the collapsed Interstate 880 Cypress structure.

---

[3] The terms *effective peak ground acceleration* (EPA) and *effective peak velocity* (EPV) were originally defined in the commentary of ATC 3-06.

[4] The resonance-induced damages of the 1985 Mexico City earthquake and the 1989 Loma Prieta earthquake are prime examples.

seismic code. These factors, nevertheless, must be considered in structural design.

3-7 SOIL LIQUEFACTION

Liquefaction occurs in soils, particularly in soils of saturated cohesionless particles such as sand, and is a sudden drop in shear strength. This is experienced as a drop in bearing capacity. In effect, the soil turns into a liquid, allowing everything it previously supported to sink. It is not necessary for the soil to be located on a cliff or other escarpment for liquefaction to occur. Perfectly flat soil layers can become major mud puddles if the conditions are right.[5]

Continued cycles of reversed shear in saturated sand can cause *pore water pressure* to increase, which in turn decreases the *effective stress* and *shear strength*.[6] When the shear strength drops to zero, the sand liquefies.

Conditions most likely to contribute to or indicate a potential for liquefaction include (1) a lightly loaded sand layer within 49 to 66 ft (15 to 20 m) of the surface, (2) uniform particles of medium size, (3) a saturated condition below a water table, and (4) a low penetration-test value.

3-8 BUILDING PERIOD

When a lightly damped building is displaced laterally by an earthquake, wind, or other force, it will oscillate back and forth with a regular *period*. (This building period should not be confused with the site period mentioned in Sec. 3-6 or with the period of the earthquake.)

The natural period of modern buildings can seldom, if ever, be calculated from simple vibrational theory. Five other methods, listed below, can be used when knowledge of a building period is needed.

1. Analytical models based on finite element analysis (FEA) and other modeling techniques can be used.

2. A scale model of the building can be constructed and the natural period extrapolated from measurements on the model. (This is seldom done, however.)

3. If the building has been constructed, actual measurements can be taken.

4. Empirical relations (such as are incorporated in the UBC-97 Sec. 1630.2.2) can be used. (See Sec. 6-26.)

5. Rayleigh's method can be used.

3-9 SPECTRAL CHARACTERISTICS

Despite some inherent regularity, earthquake seismograms are quite "noisy." It is difficult to determine how a building behaves at all times during an earthquake consisting of many random pulses. It is also unnecessary in many cases to know the entire time-history response of the building, since the maximum seismic force on (and, hence, damage in) a structure depends partially on the effective peak acceleration experienced, not on lower accelerations that might have occurred during the earthquake.

The maximum acceleration[7] that is experienced by a single-degree-of-freedom vibratory system (see Sec. 4-2) is known as the *spectral acceleration*, S_a.[8] Similarly, the maximum displacement and velocity are known as the *spectral displacement*, S_d, and *spectral velocity*, S_v, respectively.

3-10 SEISMIC FORCE

The theoretical maximum seismic force (referred to as "base shear" in the UBC), V, on a structure of mass m (weight W) is given by Newton's second law ($F = ma$).[9]

---

[5]Dramatic examples of liquefaction occurred in the 1964 earthquakes in Alaska and Niigata, Japan.

[6]These are standard terms used in soils and foundations handbooks.

[7]The maximum building acceleration should not be confused with the effective ground acceleration. The building acceleration is typically higher than the ground acceleration. The ratio of building to ground acceleration depends on the building period, a concept that is discussed elsewhere in this book. For infinitely stiff buildings (with zero natural periods), the ratio is 1. The spectral acceleration from typical California design earthquakes (i.e., those used as a basis in establishing the UBC-97 provisions) for a 10% damped building located on rocks or other firm soil is approximately 2.0 to 2.5 times the peak ground acceleration. (See Sec. 3-5.)

[8]Another name is *spectral pseudo acceleration*—"pseudo" because the value does not correspond exactly to the maximum acceleration.

[9]The total building dead load (dead weight) is used in the calculation of the base shear. This practice should not be confused with calculations of diaphragm force, which omit half of the ground floor dead load. Except for a warehouse, no live load is included. While these may seem like arbitrary provisions, the UBC-97 is specific in including and excluding certain fractions of the building weight and live load. (See Sec. 6-29.)

$$V = \frac{mS_a}{g_c} = \frac{WS_a}{g} \qquad \text{[U.S.]} \quad \text{[3.1(a)]}$$

$$V = mS_a = \frac{WS_a}{g} \qquad \text{[SI]} \quad \text{[3.1(b)]}$$

Design *base shear*, as defined in the UBC-97, takes on the form of $V = CS_aW$.

3-11 RELATIONSHIP BETWEEN SPECTRAL VALUES

Equation 3.2 indicates that the spectral displacement, velocity, and acceleration can be derived from one another if the natural frequency (in rad/sec) of vibration, ω, is known.[10] Equation 3.2 is exact for the case of an undamped, single-degree-of-freedom system in simple harmonic motion but is approximate otherwise (i.e., is approximate with damping and for multiple-degree-of-freedom systems). (See Sec. 4-2.)

$$|S_d| = \left|\frac{S_v}{\omega}\right| = \left|\frac{S_a}{\omega^2}\right| \qquad \text{[undamped SDOF]} \qquad \text{[3.2]}$$

---

[10]Equation 3.2 is easily derived for the case of sinusoidal oscillation. Starting with a sinusoidal position equation, $x(t) = A\sin\omega t$, the first derivative (i.e., the velocity equation) is $v(t) = \omega A\cos\omega t$, whose maximum amplitude is ω multiplied by the amplitude of the position function. The maximum acceleration amplitude is similarly determined.

4
VIBRATION THEORY

4-1 TWO APPROACHES TO SEISMIC DESIGN

There are two greatly different approaches to seismic design, both of which are "correct" in their own ways. In a *dynamic analysis*, the overall building and story stiffnesses and rigidities are calculated. (See Sec. 4-4.) A specific design earthquake, including magnitude and loading history, is selected and applied to a mathematical model (consisting of lumped masses, damping, and spring stiffness) of the building. The solution may rest heavily on vibrational theory, finite element analysis, and other advanced structural techniques requiring computer analysis. The response of the system (including the displacement and acceleration functions) is calculated and used to determine the forces in each member as a function of time. This method is now almost always used for critical structures such as dams and power plants.

There are a number of factors that can render the dynamic approach inappropriate. The building itself may be too simple or too standardized to warrant the rigorous approach of the design analysis. Conversely, the building may be too complex and have too many degrees of freedom to model mathematically. Also, in the initial design phases, the member sizes and locations may not be known, making it difficult to estimate stiffnesses and rigidities.[1] The dynamic approach is inappropriate, too, when the design earthquake is not known. Additionally, the analysis may be beyond the financial or computational abilities of the engineering firm performing the design. And, finally, unless the building is particularly

irregular as defined in UBC-97 Sec. 1629.5.3, there may be no code requirement to perform a dynamic analysis.

The alternative to a dynamic analysis is a *static analysis*. The *equivalent lateral (seismic) force* is calculated as simply some fraction of the dead weight. Chapter 16 of the UBC-97 codifies this analysis so that there is no need to know the design earthquake.

In the chapters and sections that follow, these two methods are at times discussed separately and, at other times, aspects of each method are combined. Although considerably different in approach, the static method is based on engineering logic that can, in many cases, be traced back to vibration theory.

4-2 SIMPLE HARMONIC MOTION[2]

Ideal vibrational systems that consist of springs and masses and that are not acted upon by external disturbing forces (after an initial displacement) are known as *simple harmonic oscillators*. During steady-state motion, such oscillators move in a repetitive sinusoidal pattern known as *simple harmonic motion*. Simple harmonic motion is characterized by the absence of a continued disturbing force and a lack of frictional damping.

Examples of simple harmonic oscillators are a mass hanging on an ideal spring (Fig. 4.1(a)), a pendulum on a frictionless pivot, and a slab supported on two massless cantilever springs (Fig. 4.1(b)).

---

[1]Although there are some real design programs, most "design" programs are actually analysis programs that require the user to input information about the locations and characteristics of the structural members.

[2]There is no need actually to develop the differential equations of oscillatory motion for a building. However, this section introduces some of the terms and concepts related to structural dynamics.

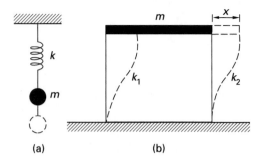

Figure 4.1   Simple Harmonic Oscillator

The number of variables needed to define the position of all parts of a system is known as its *degrees of freedom.* If the oscillator is constrained to move in one dimension only, or alternatively, if one linear or angular variable is sufficient to describe the position of the oscillator, the system is known as a *single-degree-of-freedom* (SDOF) *system.* The moving mass in an SDOF system is usually concentrated at one point and is known as a *lumped mass.*

Oscillation of the SDOF system shown in Fig. 4.2 is initiated by displacing and releasing the mass. The displacement, x, is measured from the equilibrium position. Once the system has been displaced and released, no further external force acts on it. Because there is no friction once it is set in motion, the mass remains in motion indefinitely.

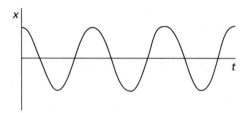

Figure 4.2   Time Response of an Undamped Simple Harmonic Oscillator

4-3 STIFFNESS AND FLEXIBILITY

When a force, F, acts on an ideal linear spring, *Hooke's law* predicts the magnitude of the spring deflection, x. In Eq. 4.1, k is the *stiffness* or *spring constant* in lbf/ft (N/m). The stiffness is the force that must be applied in order to deflect the spring a distance of one unit.

$$F = kx \quad \text{[Hooke's law]} \quad [4.1]$$

Referring to the mass-spring system shown in Fig. 4.1(a), the spring is undeflected until the mass is

attached to it. After the mass is attached, the spring will deflect an amount known as the *static deflection,* x_{st}.

$$W = \frac{mg}{g_c} = kx_{\text{st}} \quad \text{[U.S.]} \quad [4.2(a)]$$

$$W = mg = kx_{\text{st}} \quad \text{[SI]} \quad [4.2(b)]$$

The stiffness, k, of a beam can be calculated as the ratio of applied force to deflection from the beam deflection tables that are typically in every mechanics of materials textbook. Table 4.1 summarizes some of these terms.

Compliance (flexibility) is the reciprocal of stiffness. It is the deflection obtained when a unit force is applied. Therefore, its units are ft/lbf (m/N).

4-4 RIGIDITY

Strictly speaking, *rigidity*, R, is the reciprocal of deflection. In buildings where all members consist of the same material and all walls have the same thickness (for example, a masonry-walled building or an all-concrete building), the deflection is traditionally calculated with arbitrary values of applied force, modulus of elasticity, and wall thickness. This is permitted when distributing the applied lateral loads to vertical members because the load "taken" by each member is proportional to the member's *relative rigidity.*

$$R = \frac{1}{x_{\text{inches}}} \quad \text{[U.S.]} \quad [4.3(a)]$$

$$R = \frac{25.4}{x_{\text{mm}}} \quad \text{[SI]} \quad [4.3(b)]$$

Both moment and shear contribute to the deflection experienced by a vertical member (e.g., a shear wall).[3] Consider the wall shown in Fig. 4.3(a). This wall is fixed at the top and bottom and bends in double curvature since the top and bottom must remain vertical. Such a wall is known as a *fixed pier.* The deflection due to both shear and flexure of a fixed pier is given by Eq. 4.4.

$$x_{\text{fixed}} = \frac{Fh^3}{12EI} + \frac{1.2Fh}{AG} \quad \text{[fixed pier]} \quad [4.4]$$

$$A = td \quad [4.5]$$

$$I = \frac{td^3}{12} \quad [4.6]$$

[3]Common beam deflection equations, such as those presented in Table 4.1, usually disregard the effect of shear. However, shear contributes to deflection when the ratio of height to depth is low. In general, shear deflection should not be neglected, unless beam spans are long.

Table 4.1
Deflection and Stiffness for Various Systems
(Due to Bending Moment Alone)

System	Maximum Deflection (x)	Stiffness (k)
	$\dfrac{Fh}{AE}$	$\dfrac{AE}{h}$
	$\dfrac{Fh^3}{3EI}$	$\dfrac{3EI}{h^3}$
	$\dfrac{Fh^3}{12EI}$	$\dfrac{12EI}{h^3}$
	$\dfrac{wL^4}{8EI}$	$\dfrac{8EI}{L^3}$
	$\dfrac{Fh^3}{12E(I_1+I_2)}$	$\dfrac{12E(I_1+I_2)}{h^3}$
	$\dfrac{FL^3}{48EI}$	$\dfrac{48EI}{L^3}$
(w is load per unit length)	$\dfrac{5wL^4}{384EI}$	$\dfrac{384EI}{5L^3}$
	$\dfrac{FL^3}{192EI}$	$\dfrac{192EI}{L^3}$
(w is load per unit length)	$\dfrac{wL^4}{384EI}$	$\dfrac{384EI}{L^3}$

A wall that is fixed at the bottom but free to rotate at the top bends in simple curvature and is known as a *cantilever pier*. The deflection of a cantilever wall due to both effects is

$$x_{\text{cantilever}} = \frac{Fh^3}{3EI} + \frac{1.2Fh}{AG} \quad \text{[cantilever pier]} \quad [4.7]$$

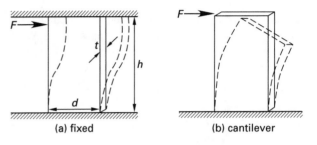

Figure 4.3 Fixed and Cantilever Piers

For concrete, $E \approx 3 \times 10^6$ psi $(2.1 \times 10^7$ kPa) and $G \approx 0.4E$. For masonry, $E \approx 1 \times 10^6$ psi $(6.9 \times 10^6$ kPa) and $G \approx 0.4E$. For steel, $E \approx 3 \times 10^7$ psi $(2.1 \times 10^8$ kPa) and $G \approx 1.2 \times 10^7$ psi $(8.3 \times 10^7$ kPa). For wood, $E \approx 1 \times 10^6$ psi to 1.8×10^6 psi $(6.9 \times 10^6$ kPa to 12×10^6 kPa). However, since the shear that is distributed to each vertical member (i.e., each pier) is in proportion to the relative rigidity and does not depend on the actual rigidity, the deflections can be calculated with arbitrary values of total shear, F, and wall thickness, t. Equations 4.8 and 4.9 use $F = 100{,}000$ lbf $(445\,000$ N), $t = 1.0$ in $(25$ mm), $E = 1 \times 10^6$ psi $(6.9 \times 10^6$ kPa), and arbitrary units.

$$R_{\text{fixed}} = \frac{1}{(0.1)\left(\dfrac{h}{d}\right)^3 + (0.3)\left(\dfrac{h}{d}\right)} \quad [4.8]$$

$$R_{\text{cantilever}} = \frac{1}{(0.4)\left(\dfrac{h}{d}\right)^3 + (0.3)\left(\dfrac{h}{d}\right)} \quad [4.9]$$

4-5 NATURAL PERIOD AND FREQUENCY

The time for a complete cycle of oscillation of an SDOF system is known as the *natural period*, T, usually expressed in seconds. The reciprocal of natural period is the *linear natural frequency*, f, usually called *natural frequency* or just *frequency*, and is expressed in Hz (i.e., cycles per second). It is important to distinguish between the natural frequency of a system (building, oscillator, etc.) and the frequency of an applied force. The natural frequency, f, in Eq. 4.10 has nothing to do with an external force.

$$f = \frac{1}{T} \quad [4.10]$$

The natural frequency can also be expressed in radians per second (rad/sec), in which case it is known as the *circular frequency*, *angular natural frequency*, or just *angular frequency*, ω.

$$\omega = 2\pi f = \frac{2\pi}{T} \qquad [4.11]$$

It is easy to derive the natural frequency for the case of a simple harmonic oscillator.[4] For a mass on a spring,

$$\omega = \sqrt{\frac{kg_c}{m}} = \sqrt{\frac{kg}{W}} \qquad \text{[U.S.]} \quad [4.12(a)]$$

$$\omega = \sqrt{\frac{k}{m}} \qquad \text{[SI]} \quad [4.12(b)]$$

Substituting k from Hooke's law (Eq. 4.1) and recognizing that the mass, m, can be calculated from the weight, W, an expression is derived for the natural frequency in terms of the static deflection, x_{st}, calculated in Sec. 4-3.

$$\omega = \frac{2\pi}{T} = \sqrt{\frac{Fg_c}{x_{st}m}} = \sqrt{\frac{Fg}{x_{st}W}} \qquad \text{[U.S.]} \quad [4.13(a)]$$

$$\omega = \frac{2\pi}{T} = \sqrt{\frac{F}{x_{st}m}} \qquad \text{[SI]} \quad [4.13(b)]$$

Since Eq. 4.13 can be used to calculate the natural period, it is tempting to substitute the maximum allowable code drift (i.e., 2.5% of the total building height; see Sec. 6-40) for the static deflection in order to calculate the natural building period.[5] Such a substitution would require no structural analysis at all but implies that the building will have maximum flexibility permitted by the code. The problem with this approach is that it assumes the maximum allowable drift to be the same for all zones, although the flexibility actually depends on the zone since flexibility is affected by the building's seismic resistance. Thus, while the lateral forces on the building differ, the maximum drift and, thus, the period, do not. Obviously, the building period cannot be calculated in this way.

---

[4]This is done in virtually every physics, dynamics, and earthquake book, but not here.

[5]The UBC-97 permits the drift limits to be exceeded when it is demonstrated that greater drift can be tolerated by both structural elements and nonstructural elements.

Example 4.1

A 5-lbm (2.27-kg) mass hangs from two ideal springs as shown. (Assume the block is constrained so that it does not rotate.) What is the natural period of vibration?

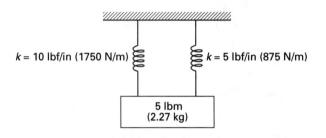

Customary U.S. Solution

Both springs must deflect in order for the mass to move. The total composite spring constant is

$$k_t = k_1 + k_2 = \left(5\,\frac{\text{lbf}}{\text{in}} + 10\,\frac{\text{lbf}}{\text{in}}\right)\left(12\,\frac{\text{in}}{\text{ft}}\right)$$
$$= 180\,\text{lbf/ft}$$

From Eqs. 4.11 and 4.12, the natural period is

$$T = 2\pi\sqrt{\frac{m}{g_c k}}$$
$$= 2\pi\sqrt{\frac{5\,\text{lbm}}{\left(32.2\,\frac{\text{ft-lbm}}{\text{lbf-sec}^2}\right)\left(180\,\frac{\text{lbf}}{\text{ft}}\right)}}$$
$$= 0.185\,\text{sec}$$

SI Solution

Both springs must deflect in order for the mass to move. The total composite spring constant is

$$k_t = k_1 + k_2 = 1750\,\frac{\text{N}}{\text{m}} + 875\,\frac{\text{N}}{\text{m}}$$
$$= 2625\,\text{N/m}$$

From Eqs. 4.11 and 4.12, the natural period for vertical translation is

$$T = 2\pi\sqrt{\frac{m}{k}} = 2\pi\sqrt{\frac{2.27\,\text{kg}}{2625\,\frac{\text{N}}{\text{m}}}}$$
$$= 0.185\,\text{s}$$

Example 4.2

A small water tank is supported on a slender column as shown. Neglecting the weight of the column, calculate the natural period of vibration.

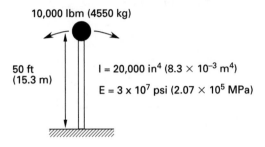

10,000 lbm (4550 kg)

50 ft (15.3 m)

$I = 20{,}000 \text{ in}^4 \ (8.3 \times 10^{-3} \text{ m}^4)$

$E = 3 \times 10^7 \text{ psi} \ (2.07 \times 10^5 \text{ MPa})$

Customary U.S. Solution

Consider the water tower to be a cantilever beam. The stiffness is the force required to deflect the tank 1 ft laterally.

From Table 4.1,

$$k = \frac{3EI}{h^3} = \frac{(3)\left(3 \times 10^7 \, \dfrac{\text{lbf}}{\text{in}^2}\right)(20{,}000 \text{ in}^4)}{(50 \text{ ft})^3 \left(12 \, \dfrac{\text{in}}{\text{ft}}\right)^2}$$

$$= 1 \times 10^5 \text{ lbf/ft}$$

From Eqs. 4.11 and 4.12, the period is

$$T = 2\pi\sqrt{\frac{m}{g_c k}}$$

$$= 2\pi \sqrt{\frac{10{,}000 \text{ lbm}}{\left(32.2 \, \dfrac{\text{ft-lbm}}{\text{lbf-sec}^2}\right)\left(1 \times 10^5 \, \dfrac{\text{lbf}}{\text{ft}}\right)}}$$

$$= 0.35 \text{ sec}$$

SI Solution

Consider the water tower to be a cantilever beam. The stiffness is the force required to deflect the tank 1 m laterally.

From Table 4.1,

$$k = \frac{3EI}{h^3}$$

$$= \frac{(3)(2.07 \times 10^5 \text{ MPa})\left(10^6 \, \dfrac{\text{Pa}}{\text{MPa}}\right)(8.3 \times 10^{-3} \text{ m}^4)}{(15.3 \text{ m})^3}$$

$$= 1.44 \times 10^6 \text{ N/m}$$

From Eqs. 4.11 and 4.12, the period is

$$T = 2\pi\sqrt{\frac{m}{k}} = 2\pi \sqrt{\frac{4550 \text{ kg}}{1.44 \times 10^6 \, \dfrac{\text{N}}{\text{m}}}}$$

$$= 0.35 \text{ s}$$

4-6 DAMPING

Damping is the dissipation of energy from an oscillating system, primarily through friction. The kinetic energy is transformed into heat. All structures have their own unique ways of dissipating kinetic energy, and in certain designs, mechanical systems known as *dampers* (see Sec. 14-4) can be installed to increase the damping rate.[6]

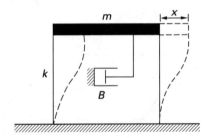

Figure 4.4   Oscillator with Damping

There are several sources of damping. *External viscous damping* is caused by the structure moving through surrounding air (or water, in some cases). It is generally small in comparison to other sources of damping. *Internal viscous damping*, commonly the only type of damping actually modeled, is related to the viscosity of the structural material. It is proportional to velocity. (See Eq. 4.14.) *Body-friction damping*, also known as *Coulomb friction*, results from friction between members in contact. It includes friction at connection points. Sections of opposed cracked masonry walls rubbing back and forth against one another are very effective body-friction dampers. Another source of damping, *radiation damping*, occurs as a structure vibrates and becomes a source of energy itself. Some of the energy is reradiated

---

[6]A damper is similar in design to a shock absorber and is often depicted as a plunger moving through a pot of viscous fluid. In modeling, dampers are also known as *dashpots*, although this term is more common among mechanical engineers.

through the foundation back into the ground. Finally, *hysteresis damping* occurs when the structure yields during reversals of the load. (See Sec. 5-7.)

For internal viscous damping, the frictional damping force opposing motion is given by Eq. 4.14. The exponent n in Eq. 4.14 is usually taken as 1.0 for slow-moving systems and 2.0 for fast-moving systems. However, even these values are idealizations. The coefficient B in Eq. 4.14 is known as the *damping coefficient*.

$$F_{\text{damping}} = Bv^n \qquad [4.14]$$

4-7 DAMPING RATIO

An oscillating system with a small amount of damping will continue to oscillate, although the amplitude of the oscillations will decay. Many cycles and a long time may elapse before the system eventually reaches the motionless equilibrium position. This type of system is known as an *underdamped system*.

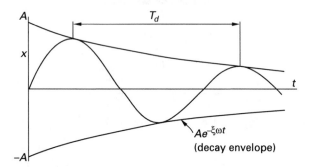

Figure 4.5   Underdamped Motion
(Moderate Damping)

Conversely, a system may have a large amount of damping. When displaced, such an *overdamped system* seems to "hang in space," taking an extremely long time to return to the motionless equilibrium position.[7]

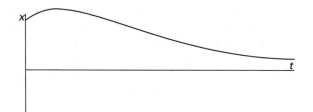

Figure 4.6   Overdamped Motion

---

[7]An example of an overdamped system is a door with a slow-closing device that will not permit the door to slam shut. Instead, the door approaches the fully closed position slowly.

Both the underdamped and overdamped cases bring the system back to the equilibrium position only after a long time. There is one particular amount of damping, known as *critical damping*, that brings the system to equilibrium in a minimum time without oscillation. In this case, the damping coefficient, B, is known as the *critical damping coefficient*, B_{critical}.

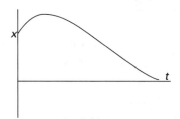

Figure 4.7   Critically Damped Motion

Most systems are not critically damped. The ratio of the actual damping coefficient to the critical damping coefficient is known as the *damping ratio*, ξ.

$$\xi = \frac{B}{B_{\text{critical}}} \qquad [4.15]$$

4-8 DECAY ENVELOPE

For small and moderate amounts of damping (i.e., the underdamped case), the oscillation will be bounded by a *decay envelope* as was illustrated in Fig. 4.5. The equation of the decay envelope is given by Eq. 4.16.

$$x = Ae^{-\xi\omega t} \qquad [4.16]$$

The ratio of one cycle's amplitude to the subsequent cycle's amplitude is the *decay decrement*. The natural logarithm of the decay decrement is the *logarithmic decrement*, δ.

$$\delta = \ln\left(\frac{x_n}{x_{n+1}}\right) = \frac{2\pi\xi}{\sqrt{1-\xi^2}} \qquad [4.17]$$

4-9 DAMPING RATIO OF BUILDINGS

The exact damping ratio, ξ, of an actual structure is difficult to determine. Furthermore, the damping ratio increases during large swings. Available data on actual structures suggest the values given in Table 4.2. There is little evidence to support damping ratios in real structures that exceed 15%.

Table 4.2
Typical Damping Ratios

type of construction	ξ

steel frame
    welded connections
    flexible walls
	0.02

steel frame
    welded connections
    normal floors
    exterior cladding
	0.05

steel frame
    bolted connections
    normal floors
    exterior cladding
	0.10

concrete frame flexible internal walls	0.05

concrete frame
    flexible internal walls
    exterior cladding
	0.07

concrete frame
    concrete or masonry
      shear walls
	0.10

concrete or masonry shear wall	0.10
wood frame and shear wall	0.15

Although the damping ratio is essentially constant for a given building, the damping ratio of a particular building type or construction material appears to depend on the natural period of the building. Buildings with natural periods of less than 1.0 sec may have damping ratios two to three times higher than buildings with similar construction but natural periods greater than 1.0 sec. While generalizations that do not consider all factors are possible, it appears that the building's damping ratio, period, and construction method are all related.

4-10 DAMPED PERIOD OF VIBRATION

The period of oscillation of a system will be slightly greater with damping than without it, since the damping slows down the movement. Equations 4.18 and 4.19 give the damped frequency and period. Most buildings have only small amounts of damping. Therefore, the damped and undamped periods are almost identical.

$$\omega_d = \omega\sqrt{1 - \xi^2} \qquad [4.18]$$

$$T_d = \frac{2\pi}{\omega_d} \qquad [4.19]$$

4-11 FORCED SYSTEMS

A *forced system* is an oscillatory system that is supplied energy on a regular, irregular, or random basis. The force that supplies the energy is known as a *forcing function*. Forcing functions can be constant (i.e., a *step function*), applied and quickly removed (i.e., an *impulse function*), sinusoidal, or random.

An example of a regularly forced system is a flexible floor supporting an out-of-balance motor. When turning, the motor will generate a force at a frequency proportional to the motor's rotational speed. An example of a randomly forced system is a structure acted on by wind or seismic forces. In the latter case, there is little or no regularity to the applied forces.

It is not significant whether a lateral force (e.g., seismic force or wind) is applied to a building directly or whether the base moves out from under the building (e.g., as in an earthquake). In the latter case, the equivalent lateral force is an inertial force, but it is just as effective at displacing the building relative to its base as any direct force is.

The system response (i.e., the behavior of a building) to a force depends on the nature of the forcing function. Unfortunately, earthquakes are never simple sinusoids and buildings have more than a single degree of freedom (see Sec. 4-16), so the determination of system response is time consuming and complex. Computers and numerical techniques, however, greatly simplify the analysis.[8]

4-12 MAGNIFICATION FACTOR

It is not difficult to show that when a sinusoidal forcing function with the form $F(t) = P\sin\omega_f t$ is applied to a system with stiffness k, the steady-state response will be of the form of Eq. 4.20.

$$x(t) = \beta\left(\frac{P}{k}\right)\sin\omega_f t \qquad [4.20]$$

In Eq. 4.20, P/k is the *static deflection*, x_{static} (see Sec. 4-3), that is experienced if a constant force P is applied to the system. β is a dynamic *magnification factor* that depends on all other characteristics of the system.[9]

---

[8]It may not always be a simple matter, however, to interpret the results of the analysis.

[9]The dynamic magnification factor depends on the natural and forcing frequencies, the mass in motion, and the amount of damping (or, alternatively, on the damping ratio). Formulas for calculating the magnification factor for damped and undamped cases are given in virtually every textbook covering vibration theory.

4-13 RESONANCE

For a given system, the dynamic magnification factor, β, can be less than or greater than unity, depending on the ratio of the natural and forcing frequencies. Figure 4.8 illustrates how the magnification factor varies for different frequency ratios. At one point, corresponding to where the forcing function frequency equals the natural frequency of the system, the magnification factor is very large (theoretically infinite for undamped systems). Such a condition is known as *resonance*. The ratio ω_f/ω must be greater than $\sqrt{2}$ for β to drop below $|1.0|$.

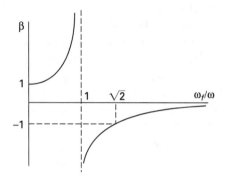

Figure 4.8   Undamped Magnification Factor

The 1985 Mexico City magnitude 8.1 earthquake occurred on September 19, with a 7.5 aftershock occurring the next day. Approximately 400 buildings were destroyed, and 700 were damaged. The death toll was over 5000. The earthquake consisted of (approximately) twenty 0.18 g pulses coming every 2 sec (the natural period of the ground). This coincided with the period for buildings in the 7 to 20 story range. The resulting resonance-related yielding was the primary cause of structural failure. Quality of construction was not a major factor in the widespread destruction.

Resonance is now considered a prime factor in the collapse of the Oakland Interstate 880 Cypress structure during the October 17, 1989, Loma Prieta earthquake. The structure had a natural frequency of 2 to 4 Hz, which coincided with the 3 to 5 Hz natural period of the deep mud that underlaid piles that supported portions of the freeway that collapsed. The depth of the mud and the length of the piles varied between 20 to 80 ft. The natural period also varied. Portions of the freeway built on harder alluvial sediments remained standing.

Although the Cypress structure was built to the standards of its time, it was poorly designed, and it is now recognized that use of nonductile reinforced concrete joints and bents with only three hinges, and inadequate confinement of the structure made its failure predictable.

4-14 IMPULSE RESPONSE—UNDAMPED SYSTEM

Seismic energy is applied to a structure in a nonregular manner. While a Fourier analysis can be used to analyze the structure response, it is also possible to break the irregular seismic loading into a series of short-duration rectangular impulses. An *impulse* is a force, F, that is applied over a duration, dt, that is much less than the natural period, T, of the structure. The product $F\,dt$ is 1.0 for a *unit impulse*. Therefore, a study of the response, $x(t)$, of a system to an impulse is of great interest.[10]

Equation 4.21 indicates that the same response will be achieved from all short-duration impulses (sine, rectangular, square, triangular, random, etc.) that have the same value of $\int F\,dt$. Notice that the response is sinusoidal even though the loading is not.

$$x(t) \approx \frac{g}{W\omega} \int (F\,dt)\sin\omega t \text{ [undamped]} \quad \text{[U.S.]} \quad \text{[4.21(a)]}$$

$$x(t) \approx \frac{1}{m\omega} \int (F\,dt)\sin\omega t \text{ [undamped]} \quad \text{[SI]} \quad \text{[4.21(b)]}$$

4-15 DUHAMEL'S INTEGRAL FOR AN UNDAMPED SYSTEM

If an undamped structure is acted upon by an irregular force of any duration, the loading can be treated as a series of impulses. The response in this case is given by Eq. 4.22, known as *Duhamel's integral*. Equation 4.22 is the application of superposition to a series of pulses, each ending at time τ.

$$x(t) = \frac{g}{W\omega} \int_0^t F(\tau)\sin\omega(t-\tau)d\tau \quad \text{[U.S.]} \quad \text{[4.22(a)]}$$
$$\text{[undamped]}$$

$$x(t) = \frac{1}{m\omega} \int_0^t F(\tau)\sin\omega(t-\tau)d\tau \quad \text{[SI]} \quad \text{[4.22(b)]}$$
$$\text{[undamped]}$$

Several numerical methods can be used to evaluate the integral in Eq. 4.22. However, when the ground motion is not known in advance, such an integration is not possible. Since earthquake motions are both nonregular and generally unexpected, it is usually acceptable to

---

[10]An impulse loading can occur from a projectile impact, bomb blast, sudden wind gust, or short-duration seismic tremor.

work with a maximum value of acceleration (or velocity or displacement). This is the principle behind the spectral values discussed in Sec. 3-9. From Eqs. 3.1 and 3.2, the total force (i.e., the *base shear*) on the structure is

$$F_{\max} = \frac{m a_{\max}}{g_c} \approx \frac{m v_{\max} \omega}{g_c} \qquad \text{[U.S.]} \quad [4.23(a)]$$

$$F_{\max} = m a_{\max} \approx m v_{\max} \omega \qquad \text{[SI]} \quad [4.23(b)]$$

◇   ◇   ◇   ◇   ◇   ◇   ◇

Example 4.3

A mass of 2×10^6 lbm (9.1×10^5 kg) is supported on two vertical members with lateral stiffnesses of 25,000 lbf/in (4.4×10^6 N/m) each. The columns have no mass and are fixed at both ends. The lateral forcing function consists of a ramp up to 50,000 lbf (220 kN) taking 0.08 sec, a uniform loading for 0.08 sec, and a ramp down to zero taking 0.08 sec. Use Duhamel's integral to determine the response (displacement) as a function of time.

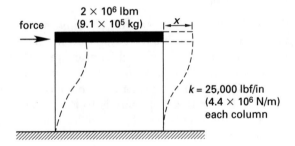

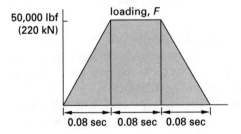

Customary U.S. Solution

The total combined stiffness, k_t, of the two columns is

$$k_t = (2)\left(25{,}000\,\frac{\text{lbf}}{\text{in}}\right) = 50{,}000\,\text{lbf/in}$$

From Eqs. 4.11 and 4.12, the natural period is

$$T = 2\pi \sqrt{\frac{m}{g_c k}}$$

$$= 2\pi \sqrt{\frac{2 \times 10^6\,\text{lbm}}{\left(32.2\,\frac{\text{ft-lbm}}{\text{lbf-sec}^2}\right)\left(50{,}000\,\frac{\text{lbf}}{\text{in}}\right)\left(12\,\frac{\text{in}}{\text{ft}}\right)}}$$

$$= 2.02\,\text{sec}$$

Since the total period over which the loading is applied is much less than the period (3×0.08 sec < 2.02 sec), the loading can be considered an impulse.

The natural frequency is

$$\omega = \frac{2\pi}{T} = \frac{2\pi}{2.02\,\text{sec}} = 3.11\,\text{rad/sec}$$

The total impulse is

$$\int F\,dt = \left(\frac{1}{2}\right)(0.08\,\text{sec})(50{,}000\,\text{lbf})$$
$$+ (0.08\,\text{sec})(50{,}000\,\text{lbf})$$
$$+ \left(\frac{1}{2}\right)(0.08\,\text{sec})(50{,}000\,\text{lbf})$$
$$= 8000\,\text{lbf-sec}$$

From Duhamel's integral (Eq. 4.22), the response is

$$x(t) = \int \frac{F\,dt}{m\omega}\sin\omega t$$

$$= \left(\frac{\dfrac{8000\,\text{lbf-sec}}{(2 \times 10^6\,\text{lbm})\left(3.11\,\dfrac{\text{rad}}{\text{sec}}\right)}}{\left(32.2\,\dfrac{\text{ft-lbm}}{\text{lbf-sec}^2}\right)\left(12\,\dfrac{\text{in}}{\text{ft}}\right)}\right)\sin 3.11t$$

$$= (0.50\,\text{in})\sin 3.11t$$

SI Solution

The total combined stiffness, k_t, of the two columns is

$$k_t = (2)\left(4.4 \times 10^6\,\frac{\text{N}}{\text{m}}\right) = 8.8 \times 10^6\,\text{N/m}$$

From Eqs. 4.11 and 4.12, the natural period is

$$T = 2\pi\sqrt{\frac{m}{k}} = 2\pi\sqrt{\frac{9.1 \times 10^5\,\text{kg}}{8.8 \times 10^6\,\dfrac{\text{N}}{\text{m}}}} = 2.02\,\text{s}$$

Since the total period over which the loading is applied is much less than the period $(3 \times 0.08 \text{ s} < 2.02 \text{ s})$, the loading can be considered an impulse.

The natural frequency is

$$\omega = \frac{2\pi}{T} = \frac{2\pi}{2.02 \text{ s}} = 3.11 \text{ rad/s}$$

The total impulse is

$$\int F \, dt = \left(\frac{1}{2}\right)(0.08 \text{ s})(220 \times 10^3 \text{ N})$$
$$+ (0.08 \text{ s})(220 \times 10^3 \text{ N})$$
$$+ \left(\frac{1}{2}\right)(0.08 \text{ s})(220 \times 10^3 \text{ N})$$
$$= 3.52 \times 10^4 \text{ N·s}$$

From Duhamel's integral (Eq. 4.22), the response is

$$x(t) = \int \frac{F \, dt}{m\omega} \sin \omega t$$
$$= \left(\frac{3.52 \times 10^4 \text{ N·s}}{(9.1 \times 10^5 \text{ kg})\left(3.11 \, \dfrac{\text{rad}}{\text{s}}\right)}\right) \sin 3.11t$$
$$= (1.24 \times 10^{-2} \text{ m}) \sin 3.11t$$

◇   ◇   ◇   ◇   ◇   ◇   ◇

4-16 MULTIPLE-DEGREE-OF-FREEDOM SYSTEMS

A system with several lumped masses, such as a building with multiple concrete floors supported by steel columns, whose positions are independent of one another is a *multiple-degree-of-freedom* (MDOF) *system*.

An MDOF system has as many ways of oscillating as there are lumped masses. These "ways" are known as *modes*. Each mode has its own characteristic *mode shape* and natural frequency of vibration, each being some multiple of the previous mode's frequency. The mode with the longest period is known as the *first* or *fundamental mode*. Higher modes have higher frequencies (smaller periods), and the periods decrease rapidly

from the fundamental mode.[11] Typical mode shapes of an MDOF system with three lumped masses are shown in Fig. 4.9.

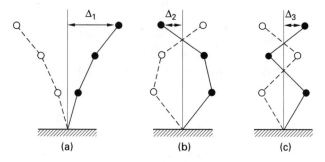

Figure 4.9   Typical Mode Shapes for a Three-Degree-of-Freedom System

4-17 RESPONSE OF MDOF SYSTEMS

Each modal frequency results in a specific mode shape, as illustrated in Fig. 4.9. However, an earthquake contains waveforms with varied frequency content; therefore, all of the modes may be present simultaneously in an earthquake. This makes it difficult to determine the building's response.

Since MDOF response can be determined as the superposition of many SDOF responses, matrix analysis (on a computer) can be used to evaluate MDOF systems based on the equivalent SDOF performance. As with SDOF systems, considerable simplification can be achieved by limiting the analysis to the maximum deflections. However, even this simplification requires a probabilistic analysis because the modal maxima do not occur at the same time, nor do they necessarily have the same sign.

Various approximation formulas are used to combine the modal maxima. The sum-of-the-squares approximation is commonly quoted.[12] If the maximum displacements, Δ_i, are known for the first n modes for some particular point (e.g., the top story), Eq. 4.24 usually gives a conservative estimate of the total displacement.

$$\Delta_t = \sqrt{\sum \Delta_i^2} \qquad [4.24]$$

---

[11]For example, for a typical high-rise building with a uniform plan view and a moment-resisting frame, the decrease is in the order of 1, 1/3, 1/5, 1/7, 1/9, and so on.

[12]This is an easy computational approximation. Whether or not it is an accurate approximation is beyond the scope of this book.

The method of combining modal responses by taking the square root of the sum of the squares is referred to in the UBC-97 as the *SRSS method* [Sec. 1633.1]. An alternative to SRSS is the *Complete Quadratic Combination* (CQC) method described in the Commentary to the 1990 SEAOC Blue Book.

Theoretically, all mode shapes must be included in the summation, but, in practice, most of the vibration energy goes into the first three to six modes, and higher modes can be disregarded. (With the use of a computer, however, there is no need to stop with such a small number of modes.) Since the lower modes dominate, the response spectra for MDOF systems are similar to those of SDOF systems. (See Sec. 5-1.) For short periods (e.g., less than 1 sec), the MDOF response is usually slightly less than for first-mode SDOF systems. For periods exceeding 1 sec, the response usually slightly exceeds SDOF response.[13]

The UBC-97 requires that all significant modes be included [Sec. 1631.4.1]. This can be accomplished by making sure that for all modes considered, at least 90% of the mass of the structure is included in the calculation of response for the horizontal direction being investigated [Sec. 1631.5.2].

◇ ◇ ◇ ◇ ◇ ◇ ◇

Example 4.4

Determine the three modal frequencies for the MDOF system shown.

$k_3 = 100$
③ $m_3 = 0.5$
② $m_2 = 1$
$k_2 = 100$
① $m_1 = 1$
$k_1 = 100$

Solution

Let x_1, x_2, and x_3 be the displacements—measured with respect to the equilibrium position—of masses 1, 2, and 3, respectively. Then, neglecting the inertial (ma) force, the spring forces on each mass are

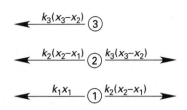

The free bodies shown are not in equilibrium. (This is particularly evident for mass 3.) According to D'Alembert's *principle of dynamic equilibrium*, an inertial force resisting motion must be added. This inertial force is

$$F_{\text{inertial}} = ma$$

However, from Eq. 3.2, the acceleration is approximately $\omega^2 x$.

$$F_{\text{inertial}} \approx m\omega^2 x$$

Therefore, the equilibrium equations for the three masses are found by adding the inertial force to the spring forces and then combining coefficients for the three displacements.

mass 1: $\left(m_1\omega^2 - (k_1 + k_2)\right)x_1 + k_2 x_2 = 0$

mass 2: $k_2 x_1 + \left(m_2\omega^2 - (k_2 + k_3)\right)x_2 + k_3 x_3 = 0$

mass 3: $k_3 x_2 + (m_3\omega^2 - k_3)x_3 = 0$

The masses and stiffnesses are known. Writing the three equilibrium equations in matrix form,

$$\begin{bmatrix} \omega^2 - 200 & 100 & 0 \\ 100 & \omega^2 - 200 & 100 \\ 0 & 100 & 0.5\omega^2 - 100 \end{bmatrix} \begin{bmatrix} x_1 \\ x_2 \\ x_3 \end{bmatrix} = \begin{bmatrix} 0 \\ 0 \\ 0 \end{bmatrix}$$

Disregarding the trivial solution, the coefficient matrix must have a determinant of zero. Setting the determinant equation to zero results in the following equation.

$$\omega^6 - (600)\omega^4 + (90{,}000)\omega^2 - 2{,}000{,}000 = 0$$

Being a cubic, this equation has three roots. Each root is a modal frequency.

$$\omega_1 = 5.18 \text{ rad/sec}$$
$$\omega_2 = 14.14 \text{ rad/sec}$$
$$\omega_3 = 19.32 \text{ rad/sec}$$

◇ ◇ ◇ ◇ ◇ ◇ ◇

---

[13]This generalization is highly dependent on the response spectrum and the soil type at the site.

4-18 MODE SHAPE FACTORS

The *mode shape factors*, ϕ, are relative numbers that represent the ratios of each of the story deflections (from the equilibrium position) to some common basis, usually the deflection of the first or last story. Since mode shape factors are relative, they can usually be determined by initially assuming a value of one of the deflections.

$$\phi_i = \frac{x_i}{x_1} \qquad [4.25]$$

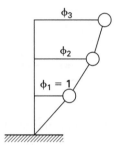

Figure 4.10   Mode Shape Factors

In some cases, the mode shape factors are normalized by dividing by $\sqrt{\sum m_i \phi_i^2}$. Then Eq. 4.26 will be valid.

$$\sum m_i \phi_i^2 = 1 \qquad [4.26]$$

Example 4.5

Find the normalized first mode shape for the system in Ex. 4.4.

Solution

The first equilibrium equation is

$$(m_1\omega^2 - (k_1 + k_2))x_1 + k_2 x_2 = 0$$
$$((1)(5.18)^2 - (100 + 100))x_1 + 100x_2 = 0$$
$$-173.2x_1 + 100x_2 = 0$$

Since the mode shape factors are relative, let $x_1 = 1$. (This will result in an unnormalized mode shape.) Then, $x_2 = 1.732$.

Similarly, the equilibrium equation for mass 3 is

$$k_3 x_2 + (m_3\omega^2 - k_3)x_3 = 0$$
$$(100)(1.732) + ((0.5)(5.18)^2 - 100)x_3 = 0$$
$$x_3 = 2$$

The unnormalized mode shape is

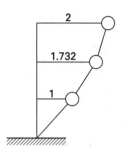

$$\sqrt{\sum m_i \phi_i^2} = \sqrt{(1)(1)^2 + (1)(1.732)^2 + (0.5)(2)^2}$$
$$= \sqrt{6}$$

Dividing each of the unnormalized mode shape factors by $\sqrt{6}$ results in the following mode shape.

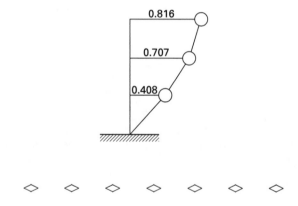

4-19 RAYLEIGH METHOD

Examples 4.4 and 4.5 show the significant computational burden of performing a full dynamic analysis for even a simple MDOF system. While the computer is an ideal tool for doing this, there may be some situations in which such an analysis is unnecessary or inappropriate. (See Sec. 6-33.)

For such situations, it may be possible to use one of several iterative procedures, most of which are variations of the *Rayleigh method*. This method starts by assuming a mode shape. (Even poor initial assumptions converge rapidly to the correct answer.) Then, the maximum kinetic energy is set equal to the maximum potential energy. Eventually, the mode shape is calculated and used as the starting point for the subsequent iteration.

The *Stodola method*, consisting of the following steps, is one such iterative process.[14]

step 1: Assume a mode shape. That is, assume a deflection, x, for each mass. (A good starting point is the shape taken by the structure when it is turned 90 degrees and acted upon by gravity.)

step 2: Compute the inertial forces for each mass from Eq. 4.27.

$$F_{\text{inertial}} \approx m\omega^2 x \qquad [4.27]$$

step 3: Compute the spring forces on each mass as the sum of the inertial forces acting on the springs.

step 4: Compute the spring deflections.

step 5: Calculate the mode deflections from the spring deflections. Repeat from step 1 as required.

Example 4.6

Use one iteration of the Stodola method to determine the mode shape of the system in Ex. 4.4.

Solution

step 1: Assume the following mode shape.

$$x_3 = 2.0$$
$$x_2 = 1.5$$
$$x_1 = 1$$
$$x_0 = 0 \text{ (ground)}$$

step 2: The inertial forces are given by Eq. 4.27.

$$F_{i3} = (0.5)(\omega^2)(2) = \omega^2$$
$$F_{i2} = (1)(\omega^2)(1.5) = 1.5\omega^2$$
$$F_{i1} = (1)(\omega^2)(1) = \omega^2$$

step 3: The spring forces are

$$F_{s3} = F_{i3} = \omega^2$$
$$F_{s2} = F_{i2} + F_{i3} = 2.5\omega^2$$
$$F_{s1} = F_{i1} + F_{i2} + F_{i3} = 3.5\omega^2$$

---

[14]The *Holzer method* is another iterative procedure; it is not discussed in this book. See Wakabayashi (1986) and other structural engineering analysis books.

step 4: The spring deflections are

$$x_{s3} = \frac{F_{s3}}{k_3} = 0.01\omega^2$$
$$x_{s2} = \frac{F_{s2}}{k_2} = 0.025\omega^2$$
$$x_{s1} = \frac{F_{s1}}{k_1} = 0.035\omega^2$$

step 5: Dividing by ω^2, the new relative mode deflections are

$$x_1 = x_{s1} = 0.035$$
$$x_2 = x_{s1} + x_{s2} = 0.060$$
$$x_3 = x_{s1} + x_{s2} + x_{s3} = 0.070$$

Dividing by 0.035, the mode shape is

$$x_3 = 2.00$$
$$x_2 = 1.71$$
$$x_1 = 1$$

These values can be used to repeat the procedure. Eventually, the values from steps 1 and 5 will agree.

4-20 PARTICIPATION FACTOR

The *participation factor*, Γ_j, is the fraction of the total building mass that acts in any particular mode, j. It can be used to calculate the *story drift*, x. (See also Sec. 4-19.) The denominator of Eq. 4.28 is the same as Eq. 4.26 and will be equal to 1.0 if normalized mode shape factors are used. (If the mode shape factors are normalized, the denominator is not needed.) Weight, W, can be substituted for mass, m, in Eq. 4.28.

$$\Gamma_j = \frac{\sum m_i \phi_{ij}}{\sum m_i \phi_{ij}^2} \qquad [4.28]$$

$$x = \Gamma S_d \phi \qquad [4.29]$$

The participation can also be used to calculate the *floor force*, F_x, that acts at story x (i.e., the force that acts at that level) and the cumulative *story shear*, V_x, that acts at that level and above. This can be done in two ways, one method derived from Hooke's law and using the spring constant, and the other method derived from Newton's law and using the mass. (Section 6-35 describes the UBC-97 method of distributing the base shear to the stories.)

$$F_x = \frac{\Gamma W S_a \phi}{g} = \Gamma k S_d \phi \qquad \text{[U.S.]} \quad [4.30(a)]$$

$$F_x = \Gamma m S_a \phi = \Gamma k S_d \phi \qquad \text{[SI]} \quad [4.30(b)]$$

Example 4.7

Determine the drifts, story shears, and total base shear for the structure in Ex. 4.5. Assume the spectral displacement and acceleration are 4 (arbitrary units) and 0.28 g (108 in/sec^2), respectively. Assume consistent units are used.

Solution

First, calculate the participation factor from the normalized mode shape factors determined in Ex. 4.5. Since normalized values are used, the denominator has a value of 1.0 and is not needed.

$$\Gamma = \sum m_i \phi_i$$
$$= (1)(0.408) + (1)(0.707) + (0.5)(0.816)$$
$$= 1.523$$

Use Eq. 4.29 to calculate the total drifts.

$$x_1 = \Gamma S_d \phi = (1.523)(4)(0.408) = 2.49$$
$$x_2 = (1.523)(4)(0.707) = 4.31$$
$$x_3 = (1.523)(4)(0.816) = 4.97$$

The story drifts are relative to the floors below.

$$x_{3-2} = 4.97 - 4.31 = 0.66$$
$$x_{2-1} = 4.31 - 2.49 = 1.82$$
$$x_{1-\text{ground}} = 2.49 - 0 = 2.49$$

Calculate the story shears from the story drifts. Each of the lateral stiffnesses was 100.

$$V_1 = kx_1 = (100)(2.49) = 249$$
$$V_2 = (100)(1.82) = 182$$
$$V_3 = (100)(0.66) = 66$$

The floor forces can be calculated from the story shears or the participation factors.

$$F_1 = V_1 - V_2 = 249 - 182 = 67$$
$$F_2 = V_2 - V_3 = 182 - 66 = 116$$
$$F_3 = V_3 = 66$$

Alternatively,

$$F_1 = \Gamma m S_a \phi = (1.523)(1)(108)(0.408) = 67.1$$
$$F_2 = (1.523)(1)(108)(0.707) = 116.3$$
$$F_3 = (1.523)(0.5)(108)(0.816) = 67.1$$

The base shear is the sum of the floor forces.

$$V = F_1 + F_2 + F_3$$
$$= 67.1 + 116.3 + 67.1$$
$$= 250.5$$

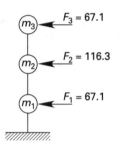

concentrated floor forces

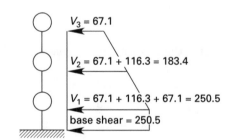

cumulative loading: story
shears and base shear

5

RESPONSE OF STRUCTURES

5-1 ELASTIC RESPONSE SPECTRA

The response of a building to earthquake ground motion depends on the dynamic characteristics of the building. Specifically, the natural period (Sec. 3-8) and the damping ratio (Sec. 4-7) affect building response more than do other factors. For a given damping ratio, ξ, a curve known as a *response spectrum of spectral acceleration*, S_a, can be drawn that plots the maximum acceleration response of an elastic single-degree-of-freedom system for a given ground motion against the natural period of the system. The response spectrum for a particular earthquake can be used to determine the theoretical maximum acceleration response of the building by entering the plot with the natural period and damping ratio of the building.

There will always be a region on the response spectrum where the acceleration is highest. This occurs where the natural building period coincides with the predominant earthquake period—when the building is in resonance with the earthquake. For California earthquakes, the peak usually occurs in the 0.2 to 0.5 sec period range.[1] Theoretically, infinite resonant response (i.e., an infinite magnification factor) is possible, though it is highly unlikely since all real structures are damped.[2]

It seems intuitively logical that a building with large amounts of internal damping will resist acceleration

(i.e., motion) to a greater extent than will a similar building with no damping. Such behavior is actually observed as spectral acceleration decreases because damping increases, although the effect of damping at lower periods is slight (since the natural periods of undamped and lightly damped structures are essentially the same). A family of curves (i.e., *response spectra*) for an actual earthquake for various damping ratios is illustrated in Fig. 5.1. Similar response spectra can be developed for *spectral velocity* and *spectral displacement*.

The spectra shown in Fig. 5.1 are for *elastic response* to an earthquake. That is, the structures used to develop the curves moved and swayed during the earthquake, but there was no yielding. For that reason, the curves are known as *elastic response spectra*.

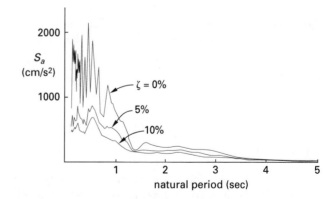

Figure 5.1 Typical Elastic Response Spectra (1940 El Centro Earthquake in N-S Direction)

---

[1] This is not always the case, as shown by the Loma Prieta earthquake.

[2] A properly designed and constructed building seldom experiences true resonance. Planned or unplanned yielding occurs before true resonant response is achieved, and this yielding damps out the resonance.

5-2 IDEALIZED RESPONSE SPECTRA

The response spectra derived from the behavior of one SDOF system in one particular earthquake are usually quite jagged, as shown in Fig. 5.1. It is not possible to use such a historical record for design, since it is unlikely that an earthquake matching the original earthquake in duration, magnitude, or time history will occur. Also, even if the design earthquake was completely specified, the significant variation in spectral values over small period ranges would require an unreasonable accuracy in the determination of the building period. To get around these problems, a smoothed average *design response spectrum* based on the envelopes of performance of several earthquakes is developed.

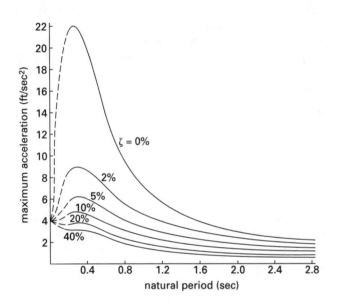

Figure 5.2　Average Elastic Design Response Spectra (based on the 1940 El Centro Earthquake) [multiplier = 1]

Example 5.1

The primary support for an industrial drill press with a mass of 100,000 lbm (45 000 kg) is the structural steel bent shown. The beam-column and base connections are rigid. The horizontal beam has a mass of 119 lbm/ft (160 kg/m), neglecting the weight of the vertical supports. The system has 5% damping. Determine the elastic response (i.e., base shear) for a 1940 El Centro earthquake in the north-south direction. Use average design spectra.

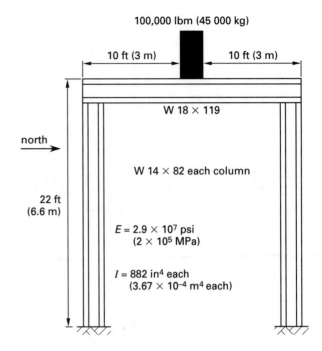

Customary U.S. Solution

The total mass, m, of the moving system is

$$m = 100{,}000 \text{ lbm} + (20 \text{ ft})\left(119 \frac{\text{lbm}}{\text{ft}}\right)$$
$$= 102{,}380 \text{ lbm}$$

From Table 4.1, the combined stiffness, k_t, of the two vertical supports is

$$k_t = (2)\left(\frac{12EI}{h^3}\right)$$
$$= \frac{(2)(12)\left(2.9 \times 10^7 \frac{\text{lbf}}{\text{in}^2}\right)(882 \text{ in}^4)}{(22 \text{ ft})^3 \left(12 \frac{\text{in}}{\text{ft}}\right)^2}$$
$$= 4 \times 10^5 \text{ lbf/ft}$$

From Eqs. 4.11 and 4.12, the natural period of vibration, T, is

$$T = 2\pi\sqrt{\frac{m}{g_c k}}$$
$$= 2\pi\sqrt{\frac{102{,}380 \text{ lbm}}{\left(32.2 \frac{\text{ft-lbm}}{\text{lbf-sec}^2}\right)\left(4 \times 10^5 \frac{\text{lbf}}{\text{ft}}\right)}}$$
$$= 0.56 \text{ sec}$$

From Fig. 5.2, the spectral acceleration for this period and 5% damping is $S_a = 5.5$ ft/sec^2. From Eq. 3.1, the base shear is

$$V = \frac{mS_a}{g_c} = \frac{(102{,}380 \text{ lbm}) \left(5.5 \dfrac{\text{ft}}{\text{sec}^2}\right)}{32.2 \dfrac{\text{ft-lbm}}{\text{lbf-sec}^2}}$$

$$= 1.75 \times 10^4 \text{ lbf}$$

SI Solution

The total mass, m, of the moving system is

$$m = 45\,000 \text{ kg} + (6 \text{ m}) \left(160 \frac{\text{kg}}{\text{m}}\right) = 45\,960 \text{ kg}$$

From Table 4.1, the combined stiffness, k_t, of the two vertical supports is

$$k_t = (2) \left(\frac{12EI}{h^3}\right)$$

$$= \frac{(2)(12)(2 \times 10^5 \text{ MPa}) \left(10^6 \dfrac{\text{Pa}}{\text{MPa}}\right)(3.67 \times 10^{-4} \text{ m}^4)}{(6.6 \text{ m})^3}$$

$$= 6.13 \times 10^6 \text{ N/m}$$

From Eqs. 4.11 and 4.12, the natural period of vibration, T, is

$$T = 2\pi \sqrt{\frac{m}{k}} = 2\pi \sqrt{\frac{45\,960 \text{ kg}}{6.13 \times 10^6 \dfrac{\text{N}}{\text{m}}}}$$

$$= 0.54 \text{ s}$$

From Fig. 5.2, the spectral acceleration for this period and 5% damping is approximately $S_a = 1.65$ m/s^2. From Eq. 3.1, the base shear is

$$V = mS_a = (45\,960 \text{ kg}) \left(1.65 \frac{\text{m}}{\text{s}^2}\right)$$

$$= 7.58 \times 10^4 \text{ N}$$

◇   ◇   ◇   ◇   ◇   ◇   ◇

5-3 RESPONSE SPECTRA FOR OTHER EARTHQUAKES

The design response spectra in Fig. 5.2, although normalized and averaged over several earthquakes, are adjusted for an earthquake of a specific magnitude and peak ground acceleration. Based on historical data and probability studies, the recurrence interval for an earthquake of that magnitude can be determined. For example, an earthquake of the 1940 El Centro magnitude is

expected at that site, on the average, every 70 years. However, smaller earthquakes will be experienced more frequently than every 70 years, and larger earthquakes will be experienced less frequently than every 70 years.

In order to apply the average design response spectra to other earthquakes, they are simply scaled upward or downward for larger and smaller earthquakes, respectively. For example, Table 5.1 gives the scale factor (to be used to scale Fig. 5.2 downward) for other recurrence intervals at the El Centro site.

Table 5.1
Scale Factors for Other Recurrence Intervals
(based on elastic response to the
1940 El Centro Earthquake)

recurrence interval (yr)	scale factor
2	2.77
20	1.83
32	1.50
70	1.00

5-4 LOG TRIPARTITE GRAPH

Since spectral acceleration, velocity, and displacement for linear elastic response are all related (see Eq. 3.2), all three spectral quantities can be shown by a single curve on a graph with three different scales. Such a graph is done on a logarithmic scale and is known as a *log tripartite plot*. Both elastic and inelastic (see Sec. 5-9) tripartite plots are widely in use. However, for inelastic response, the spectral acceleration, velocity, and displacement cannot be represented by a single curve on the tripartite plot.

Tripartite plots, both elastic and inelastic, can differ in how the axes are arranged. Figure 5.3 illustrates two common arrangements for presenting the information.

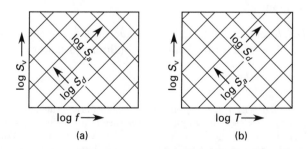

Figure 5.3   Two Types of Log Tripartite Plots

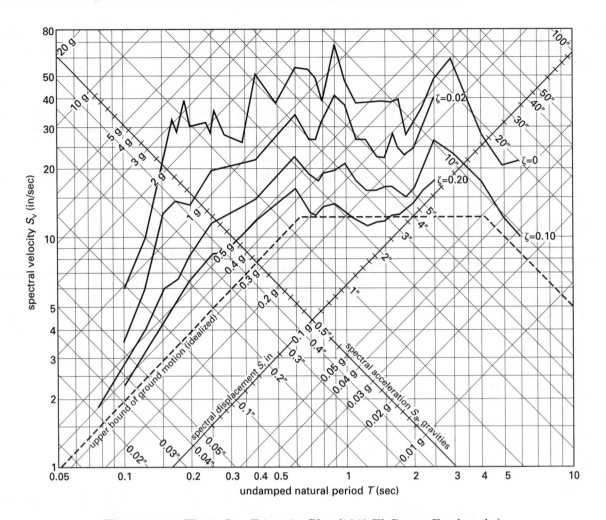

Figure 5.4   Elastic Log Tripartite Plot (1940 El Centro Earthquake)

Reprinted from *Design of Multistory Reinforced Concrete Buildings for Earthquake Motions*, by John A. Blume,
Nathan M. Newmark, and Leo H. Corning, 1961, with permission from the Portland Cement Association, Skokie, IL.

5-5 DUCTILITY

The expected magnitude of seismic loads and the nature of building codes make it necessary to accept some yielding during large earthquakes.[3] The design provisions in modern seismic codes could not create a purely elastic response during a large earthquake; in any case, building a structure with such a response would not be economical.

Displacement ductility (or just *ductility*) is the capability of a structural member or building to distort and

yield without collapsing. During an earthquake, a ductile structure can dissipate large amounts of seismic energy after local yielding of connections, joints, and other members has begun.

The actual ductility of a joint or structural member is specified by its *ductility factor*, μ. There are a number of definitions of the ductility factor, all of which represent the ratio of some property at failure (i.e., fracture) to that same property at yielding. For example, the ductility factor may be specified in terms of energy absorption, as in Eq. 5.1.

$$\mu = \frac{U_{\text{fracture}}}{U_{\text{yield}}} \qquad [5.1]$$

In addition to the definition based on the ratio of energies, there are definitions of the ductility factor based on ratios of linear strain and angular strain (rotation). These definitions are not interchangeable, although they

---

[3]The high seismic loading expected in California and the high cost of a totally elastic design make it necessary to accept some yielding. Therefore, the building is designed to withstand a smaller effective peak acceleration (see Sec. 3-5) without yielding, thereby ensuring yielding when a larger ground acceleration is experienced.

are related.[4] Generally, however, the basic concept (i.e., the ratio of some failure property to the same yield property) is all that is needed to explain the significance of a ductile structure.

The minimum assumed ductility (based on strain or deformation) of building structures with good connections and good redundancy that are designed to modern seismic codes is 2.2. (Ductility of bridge structures is much less.) Desirable levels vary, although it is best to have large values of the ductility factor—4 to 6 for concrete frames and 6 to 8.5 for steel frames. In order to achieve these levels of ductility in the structure overall, the structural members themselves must have special detailing with inelastic deformation in mind.

5-6 STRAIN ENERGY AND DUCTILITY FACTOR

The area under the stress-strain curve represents the *strain energy* absorbed, U. The maximum energy that can be absorbed without yielding (i.e., the area under the curve up to the yield point) is known as the *modulus of resilience*, U_R. The maximum energy that can be absorbed without failure is the *modulus of toughness (rupture)*, U_T. One definition of the ductility factor, μ, can be calculated from the ratio of these two quantities.

$$\mu = \frac{U_T}{U_R} \qquad [5.2]$$

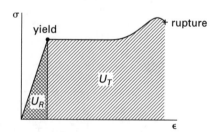

Figure 5.5 Strain Energy

5-7 HYSTERESIS

Hysteresis (hysteretic) damping is the dissipation of part of the energy input when a structure is subjected to load

---

[4]For ideal (linear) elastoplastic systems, the ductility based on energy absorption, μ_U, can be calculated from the ductility based on strain, μ_ϵ, as

$$\mu_U = 2\mu_\epsilon - 1$$

This means that if the ductility, as calculated from linear strain, is 4 to 6, the ductility will be 7 to 11 when calculated from Eq. 5.1.

reversals in the *inelastic* range. Such dissipation occurs in the structure itself as well as in the soil around the foundation and, therefore, depends on the nature of the building, foundation, and soil. The energy lost per cycle, U_H, is the area within the *hysteresis loop*, as shown in Fig. 5.6. Hysteresis losses are unaffected by the velocity of the structure.

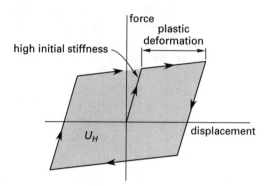

Figure 5.6 Hysteresis Loop

Inasmuch as it is difficult to evaluate hysteresis losses, hysteresis damping in seismic studies is sometimes accounted for by defining an equivalent internal viscous damping. Such an approximation works reasonably well in some cases, but the validity deteriorates as the deflection increases.

5-8 LARGE DUCTILITY SWINGS

The effect of *reversal deformation* after a few cycles of very high ductility is significant. Tests and actual experience indicate that even modern structures can fail after only a few deformation reversals if the strain is well into the inelastic region.[5] It is particularly easy to show that threaded rods are susceptible to such failure.

5-9 INELASTIC RESPONSE SPECTRA

The total seismic energy, U, received by a building structure in an earthquake is stored or dissipated in four primary ways. Some of the energy is stored as elastic strain energy (U_E); some is converted into kinetic energy (U_K); some is dissipated as hysteretic or plastic losses (U_H); and some is lost due to frictional

---

[5]This type of failure was first predicted in a controversial paper written by Vitelmo V. Bertero and Egor Popov in the 1960s. Such failures were actually observed in the 1964 Alaska and 1971 San Fernando earthquakes.

and damping effects (U_ξ). In the simplest models, the sum of these four terms equals the total energy input.

Particularly when the building is stressed in the elastic region, input energy is dissipated relatively slowly, primarily because of internal friction (i.e., damping) of the structure converting the kinetic energy into heat. However, it takes much more energy to plastically deform parts of the structure. Since the amount of input energy is limited to what was received and the frictional losses are approximately constant, an increase in this energy of deformation is accompanied by a decrease in the kinetic energy of oscillation. Thus, each yielding connection, every broken column, and every sheared pin dissipates a finite amount of kinetic energy. Therefore, a building's amplitude of oscillation number of oscillation cycles decrease as major portions of the building yield.[6]

The *inelastic design response spectra* (IDRS) show what the acceleration will be when some of the seismic energy is removed inelastically. It is appropriate to consider the inelastic effects when the response of a building to a major earthquake is being determined. The inelastic response spectra are usually derived from the elastic response spectra.

There are several well-known methods of obtaining the inelastic response spectra from the elastic response spectra, but few of them are suitable for manual calculations. Perhaps the quickest and easiest, though not necessarily the most rigorous, method is simply to scale the elastic curves downward by some function of the ductility factor.

$$S_{a,\text{inelastic}} = \frac{S_{a,\text{elastic}}}{\text{factor}} \quad [5.3]$$

The "factor" in Eq. 5.3 depends on the period. For extremely small periods (i.e., frequencies greater than approximately 33 Hz), there is no reduction at all. For periods greater than approximately 0.5 to 1.5 sec (i.e., frequencies less than 2 Hz), the ductility factor (based on strain), μ_ϵ, itself can be used as the reduction factor. For intermediate periods (33 Hz $> f > 2$ Hz), the reduction factor is approximately $\sqrt{\mu_U} = \sqrt{2\mu_\epsilon - 1}$.

In converting an elastic response spectrum to an inelastic response spectrum, the ductility factor, μ_ϵ, used to calculate the reduction factor may be known as the *structure deflection ductility factor* or *design ductility factor*, μ_Δ. It is the ratio of the deflection at ultimate collapse to the deflection at first yield, measured at the roof of the structure. Estimates of this value are known

---

[6]A yielding structure experiences larger localized deformations than an elastic structure does. This is different, however, from the overall oscillation of the structure.

to be unreliable at low natural periods (i.e., high frequencies), but the simple division by μ_Δ or μ_ϵ is favored because of its simplicity.

Values of the design ductility factor in excess of 6 are not often used, as excessive damage (architectural as well as structural) would be experienced, even though larger values (up to 10 for ductile steel structures) are readily achievable.

At high periods (i.e., low frequencies), energy absorption effects dominate, and a ductility factor based on energy (rather than strain), μ_U, is more appropriate for use in determining the inelastic response spectrum. (See Ftn. 4.)

5-10 NORMALIZED DESIGN RESPONSE SPECTRA

For design purposes, the response spectrum should be representative of the characteristics of all seismic properties experienced at a specific site. The design response spectrum should be based on geologic, tectonic, seismological, and soil characteristics associated with that specific site if these are known. If not, it may be constructed according to the spectral shape presented by Fig. 5.7 (UBC-97 Fig. 16-3). Figure 5.7 is normalized with respect to the peak ground acceleration (EPA) and uses the site-specific values of *seismic response coefficients*, C_a and C_v, spectral accelerations that define the shape of the response spectrum.

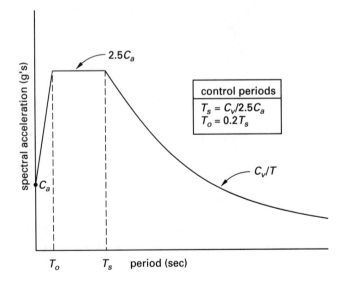

Figure 5.7 Design Response Spectrum

The seismic response coefficients C_a and C_v account for the potential amplification of the ground vibration generated at a specific site by an earthquake (seismic response throughout the spectral range). They are influenced by the regional seismicity and geology, the expected recurrence rates and maximum magnitudes of events on known faults and seismic zones, the proximity of the site to active seismic sources (i.e., faults), and the characteristics of the site soil profile.

The seismic response coefficient, C_a, corresponds to the site-dependent effective peak acceleration at grade and is controlled by the short period portion of the spectrum for structures having a fundamental period of equal to or less than $C_v/2.5C_a$. C_v corresponds to the site-dependent effective peak acceleration response at 1.0 sec period and is controlled by the longer period portion of the spectrum. UBC-97 Tables 16-Q and 16-R give values of C_a and C_v.

In seismic zone 4, the seismic response coefficients, C_a and C_v, are functions of N_v. N_v is the *near-source factor* related to both the proximity of the building or structure to known active faults with specific magnitudes and slip rates as given in UBC-97 Tables 16-T and 16-U. (See Sec. 6-18.) For sites more than 9 mi (15 km) away from major identified faults and more than 1.2 mi (2 km) away from minor faults, $N_v = 1.0$.

5-11 RESPONSE SPECTRUM FOR UBC-97-DEFINED SOIL PROFILES

Six different soil profile types are classified in the UBC-97 based on average shear wave velocity in the top 100 ft (30 m) of the soil layer: S_A, S_B, S_C, S_D, S_E, and S_F (see Sec. 6-15). The UBC-97 design response spectrum (Fig. 5.7) is site dependent by virtue of the soil-profile dependence of seismic coefficients C_a and C_v. In zone 4, the coefficients are also dependent on near-source factors N_a and N_v. For near-source factors of 1.0 , the site-specific UBC-97 response spectra are shown in Fig. 5.8.

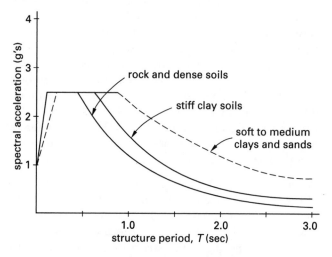

Figure 5.8   Normalized Response Spectra Shapes
$(N_v = 1.0)$

The functions plotted in the figure are easily derived by computing quantities C_a and C_v from UBC-97 Tables 16-Q and 16-R and T_o and T_s from the definitions shown in Fig. 5.7.

5-12 DRIFT

Drift (also known as *story drift*) is the lateral displacement (deflection) of one floor relative to the floor below. Story drift is shown graphically in Fig. 5.10. There are two main reasons to control drift. First, excessive movement in upper stories has strong adverse psychological and physical effects on occupants. Second, it is difficult to ensure structural and architectural integrity with large amounts of drift.[7] Excessive drift can be accompanied by large secondary bending moments and inelastic behavior. (See Sec. 5-13.) In a severe earthquake in which yielding is experienced, a modern highrise building can be expected to experience a drift of approximately 2% of its total height at the roof level.[8]

There are three components of drift: (1) column and girder bending and shear, (2) joint rotation, and (3) frame bending. The first component is sometimes referred to as *bent action*. The first two components together are referred to as *shear drift*. The third component is known as *chord drift* and *cantilever displacement*.

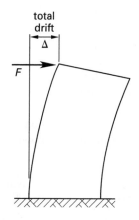

Figure 5.9   Total Drift

---

[7]*Architectural failures* are such nonstructural damage as failure of partitions, windows, and hung ceilings. In low-rise construction, damage to stairwells and elevator shafts can also be considered nonstructural. However, in high-rise construction, stairwells and elevator shafts usually constitute the most critical structural elements in the structural core.

[8]The UBC-97 limits the drift under the code-specified design lateral forces based on the fundamental period of the structure [UBC-97 Sec. 1630.10.2]. (See Sec. 6-25.) Under larger forces, the drift will be larger.

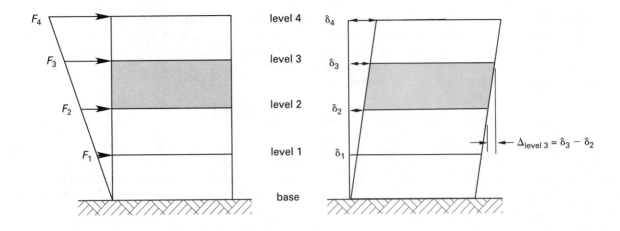

Figure 5.10   Story Drift

Table 5.2 contains generalizations about the effects of different variables on drift. The drift is proportional to the variables (raised to the powers indicated) defined in the table.

Table 5.2

Effect of Different Variables on Drift

variable	column drift	girder drift	joint drift	chord drift
story height, H	H^3	H^2	H^2	none
building height, H	none	none	none	H^3
girder length, L	none	L	none	L^3
column depth, D	D^2	none	D^{-1}	none
girder depth, D	none	D^2	D^{-1}	none
column height, H	H	none	H^2	none
shear load, F	none	none	none	F
frame length, L	none	none	none	L^{-2}

The *story drift ratio* is the story drift divided by the height (floor to floor) of the story.

The UBC-97 requires computation of seismic building drifts based on the response that occurs during the design earthquake. Simply, the drift value should correspond to the drift that would occur when the structure responds inelastically to a 10%-in-50-year earthquake.

In the earlier version of the UBC-97, drifts (Δ_w) were determined from "working stress level" lateral forces. In the UBC-97, design seismic forces are "strength level," and the corresponding design level response displacements are denoted as Δ_S. Displacements Δ_S are computed from static, elastic lateral analyses using the design seismic forces of the UBC-97 [Sec. 1630.2.1], as specified in UBC-97 Sec. 1630.9.1. The elastic, design-level deformations (Δ_S) are then converted to the actual drifts expected for the design earthquake, Δ_M, by multiplying by 0.7 times the overstrength and ductility factor R, as specified in UBC-97 Sec. 1630.9.2. The resulting drifts Δ_M are called *maximum inelastic response displacements* and include an estimated inelastic contribution to the total deformation.

With the design basic ground motion, structures experience forces larger than both the working stress and strength level design forces. The corresponding Δ_M is several times larger than either Δ_w or Δ_S. (See Sec. 6-40.)

5-13 *P*-Δ EFFECT

The column members in a structure are loaded in compression by the vertical live and dead loads. Normally, these loads are concentric with the bases of the members. When the structure is acted upon by a lateral (horizontal) seismic load, the structure becomes laterally displaced and the applied vertical loads become eccentric with respect to the bases. This results in additional forces and moments and increased story displacements. This *secondary effect* on shears, axial forces, moments, and displacements of frame members is referred to as the *P-Δ effect*.

When the total vertical load is concentric with the base of the structure, the overturning moment is referred to as the *primary moment*. The magnitude of the primary moment is Fh, where F is the lateral seismic load and h is the height of the structure. When the vertical loads become eccentric with respect to the base, the overturning moment adds an eccentric bending stress to the columns. This additional column stress is referred to as the *secondary moment*. The magnitude of the secondary moment is $P\Delta$, where P is a function of the building weight (i.e., dead load, live load, and snow load) and Δ is the drift.

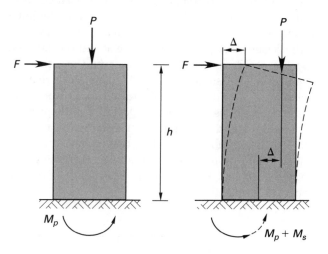

Figure 5.11 P-Δ Effects

According to the UBC-97 [Sec. 1630.1.3], the member forces and moments and the story displacements generated by P-Δ effects should be considered in the evaluation of overall structural frame stability. For this evaluation, the forces producing the displacements of Δ_S [UBC-97 Sec. 1630.9.1] should be used. (See Sec. 6-40.)

If the *overturning moment* increases faster than the *restoring moment* from the frame stiffness, the frame will be unstable. Since the vertical load is constant (i.e., is not transient as is the seismic load that causes the initial eccentricity), the column members will eventually fail and the frame will buckle. Based on the 1991 Bernal research, P-Δ has very little effect on structural response until dynamic instability is approached. Protection against instability failures is provided by wall X-bracing and thick shear walls.

Unstable frames can be inadvertently designed in non-seismic areas where only vertical loads are considered. Designing a frame to withstand large lateral (seismic) loads has the effect of limiting drift; such frames, therefore, are unlikely to experience a problem caused by P-Δ instability.

The UBC-97 [Sec. 1630.1.3] states that the P-Δ effect need not be considered in the analysis of the entire structure when: (1) the ratio of secondary moment to primary moment (i.e., *stability coefficient*) in any story is equal to or less than 0.10, and/or (2) the story drift ratio does not exceed $0.02/R$ in seismic zones 3 and 4 for all stories.

5-14 TORSIONAL SHEAR STRESS

A building's *center of mass*, C_M, (on plan view) is a point through which the base shear (i.e., the total lateral seismic force) can be assumed to act. This base shear is resisted by the vertical members at the ground level. Each such member may have a different rigidity and thus provides a different lateral resisting force in the opposite direction of the base shear. The building's *center of rigidity*, C_R, is a point through which the resultant of all the resisting forces acts.[9]

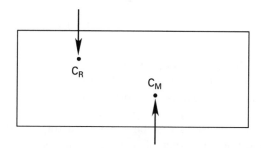

Figure 5.12 Centers of Mass and Rigidity (building plan view)

If the building's center of mass does not coincide with its center of rigidity, the building will tend to act as if it is "pinned" at its center of rigidity. It is said to be acted upon by a torsional moment, $M_{\text{torsional}}$, calculated as the product of the shear, V, and the eccentricity, e. This *eccentricity* is the distance (measured perpendicular to the direction of lateral load) between the centers of mass and rigidity.

$$M_{\text{torsional}} = Ve \qquad [5.4]$$

Equation 5.5 calculates the maximum torsional shear stress in circular members. In Eq. 5.5, J is the polar moment of inertia which, for circular cross sections, can be calculated as the sum of the moments of inertia taken with respect to the x- and y-axes.

$$v = \frac{M_{\text{torsional}}r}{J} \qquad [5.5]$$

$$J = I_x + I_y \qquad \begin{bmatrix} \text{circular cross} \\ \text{sections only} \end{bmatrix} \qquad [5.6]$$

---

[9]The implication here is that the structure has rigid diaphragms (see Sec. 7-4) between the floors so that the torsional moment can be transferred to the various resisting members distributed at that level. Structures with flexible diaphragms (see Sec. 7-6) are incapable of distributing torsional moments to vertical resisting elements. The UBC-97 [Sec. 1630.6] gives a method of determining whether or not a diaphragm can be considered flexible. Specifically, the diaphragm is flexible if the maximum lateral diaphragm deformation is more than twice the average story drift.

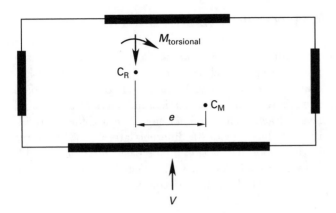

Figure 5.13 Eccentricity

In buildings, the rotation about the center of rigidity is resisted by a torsional shear force (stress) in all members.[10] This torsional force (stress) is proportional to the distance, r, from the building's center of rigidity to the resisting member. A satisfactory substitute for the polar moment of inertia, J, can be calculated from the relative rigidities of the resisting elements. If R_i is the relative rigidity of shear wall i and r_i is the distance of wall i from the center of rigidity, then the polar moment of inertia is

$$J = \sum R_i r_i^2 \qquad [5.7]$$

The units of J are somewhat ambiguous since the units used to determine the relative rigidities are not necessarily known.

The shear force, $F_{i,\text{torsion}}$, in member i due to torsion is

$$F_{i,\text{torsion}} = \frac{R_i r_i M_{\text{torsional}}}{J}$$
$$= \frac{R_i r_i M_{\text{torsional}}}{\sum R_i r_i^2} \qquad [5.8]$$

Then, the torsional shear stress in member i is found by dividing the shear force by the cross-sectional area of the member.

$$v = \frac{F_{i,\text{torsion}}}{A} \qquad [5.9]$$

Equation 5.7 shows that the contribution of a stiff element to torsional rigidity increases with the square of the distance of that element from the center of rigidity. If R is the relative rigidity of a shear wall and r is the distance of the wall from the center of rigidity, then

[10]Unlike the base shear, which is resisted only by walls parallel to the seismic force, the torsional shear is resisted by all walls and columns.

the contribution of the wall to the torsional moment of inertia is $J_w = Rr^2$. Therefore, shear walls should be located as near the building perimeter (and hence as far from the center of rigidity) as possible.

The UBC-97 [Sec. 1630.6] requires that an *accidental eccentricity* (e_a) of ±5% (based on the maximum building dimension at that level perpendicular to the direction of the seismic load) be added to the actual eccentricity, if any, in the design of all buildings, even those that are symmetrical. (Also, see Sec. 6-39.) This eccentricity is included to account for accidental errors in workmanship, uncertainties in the actual location of the centers of mass and rigidity, nonuniform distribution of dead and live loads, nonuniformities that result from subsequent building modifications, and eccentricities that develop during an earthquake after the failure of certain structural elements.

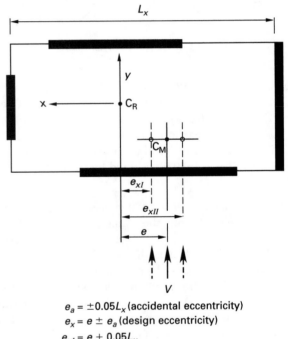

$e_a = \pm 0.05 L_x$ (accidental eccentricity)
$e_x = e \pm e_a$ (design eccentricity)
$e_{xI} = e + 0.05 L_x$
$e_{xII} = e - 0.05 L_x$

Figure 5.14 Accidental Eccentricity

Example 5.2

A crane system is modeled as a 1000-lbm (455-kg) mass attached to the end of a 5-ft (1.5-m) cantilever beam supported by a 20-in (51-cm) diameter hollow tubular vertical column. Calculate the maximum torsional shear stress for a lateral acceleration in the x-direction of 0.3 g.

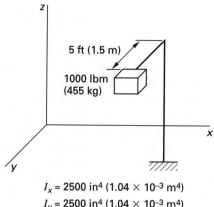

$I_x = 2500\ \text{in}^4\ (1.04 \times 10^{-3}\ \text{m}^4)$
$I_y = 2500\ \text{in}^4\ (1.04 \times 10^{-3}\ \text{m}^4)$
diameter = 20 in (0.51 m)

Customary U.S. Solution

The lateral (seismic) force is equal to the inertial force.

$$F_i = \frac{ma}{g_c} = \frac{(1000\ \text{lbm})(0.3\ \text{g})\left(32.2\ \dfrac{\text{ft}}{\text{sec}^2\text{-g}}\right)}{32.2\ \dfrac{\text{ft-lbm}}{\text{lbf-sec}^2}}$$
$$= 300\ \text{lbf}$$

From Eq. 5.4, the torsional moment is

$$M_{\text{torsional}} = Fe = (300\ \text{lbf})(5\ \text{ft})\left(12\ \frac{\text{in}}{\text{ft}}\right)$$
$$= 18{,}000\ \text{in-lbf}$$

Equation 5.6 gives the polar moment of inertia.

$$J = I_x + I_y = 2500\ \text{in}^4 + 2500\ \text{in}^4$$
$$= 5000\ \text{in}^4$$

The distance from the center of rigidity (i.e., the center of the column) to the most exterior point on the column is 10 in. The maximum torsional shear stress is given by Eq. 5.5.

$$v = \frac{M_{\text{torsional}}r}{J} = \frac{(18{,}000\ \text{in-lbf})(10\ \text{in})}{5000\ \text{in}^4}$$
$$= 36\ \text{lbf/in}^2\ (\text{psi})$$

SI Solution

The lateral (seismic) force is equal to the inertial force.

$$F_i = ma = (455\ \text{kg})(0.3\ \text{g})\left(9.81\ \frac{\text{m}}{\text{s}^2\cdot\text{g}}\right)$$
$$= 1339\ \text{N}$$

The torsional moment is

$$M_{\text{torsional}} = Fe = (1339\ \text{N})(1.5\ \text{m})$$
$$= 2009\ \text{N·m}$$

Equation 5.6 gives the polar moment of inertia.

$$J = I_x + I_y$$
$$= 1.04 \times 10^{-3}\ \text{m}^4 + 1.04 \times 10^{-3}\ \text{m}^4$$
$$= 2.08 \times 10^{-3}\ \text{m}^4$$

The distance from the center of rigidity (i.e., the center of the column) to the most exterior point on the column is 25.5 cm. The maximum torsional shear stress is given by Eq. 5.5.

$$v = \frac{M_{\text{torsional}}r}{J}$$
$$= \frac{(2009\ \text{N·m})(0.255\ \text{m})}{2.08 \times 10^{-3}\ \text{m}^4}$$
$$= 2.46 \times 10^5\ \text{Pa}$$

$\diamond \quad \diamond \quad \diamond \quad \diamond \quad \diamond \quad \diamond \quad \diamond$

5-15 NEGATIVE TORSIONAL SHEAR

The base shear causes a shear stress that acts in the same direction in all vertical base members. The torsional shear stress, however, has different signs on either side of the center of rigidity (see Fig. 5.15). On one side (i.e., where the resisting element is on the same side of the center of rigidity as the center of mass, wall 2 in the figure) the torsion increases the stress from the base shear; on the other side (wall 1 in the figure), the stress is decreased. The amount of decrease is known as *negative torsional shear*. The total lateral force is the sum of the shear force and the torsional force. Negative torsional shear should normally be neglected; that is, it should not be used to decrease the design capacity of a wall or member.

It is easy to make the error of reversing the signs of the induced stresses and adding the negative torsional stress where it should be subtracted, and vice versa. The key to avoiding this error is to always work with the stresses that *resist* the forces and moments. Thus, the stress that resists the base shear acts in a direction opposite to the base shear (i.e., opposite to the direction of ground motion). Similarly, the torsional stresses that resist the torsional moment are in the direction opposite to the applied moment.

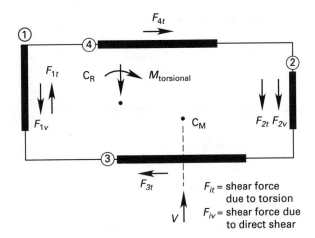

Figure 5.15　Torsional Shear

5-16 OVERTURNING MOMENT

The distribution of earthquake forces over the height of a structure causes the structure to experience overturning effects. According to the UBC-97 [Sec. 1630.8.1], every structure is to be designed to resist the overturning effects caused by seismic forces. The design overturning moment is distributed to the various resisting elements. The intent is to transfer the overturning effects on all resisting element to the foundation.

The summation of moments due to the distributed lateral forces (see Sec. 6-35) is the *overturning moment*, often given the symbol OTM. If the overturning moment is large enough, it can reverse the compression that normally exists in outer columns caused by the dead and live building loads. Because footings and concrete walls and columns can be placed in a state of tension, the overturning moment is more of a problem for concrete frame and shear wall construction (which cannot tolerate much tension) than it is for steel frame construction.

The overturning moment will increase the compressive stress in outer columns on the opposite side of the building. Such an increase must be countered by increasing the thickness of shear walls and using extra steel reinforcement in concrete columns.

Overturning moments should be calculated for each building level. The first overturning moment is the sum of all moments taken about the ground level. This moment should be used to size footings and to design the primary outer columns. The overturning moment for each subsequent floor considers only lateral forces above that floor. This moment is used to design the shear walls and other supporting structures at that floor.

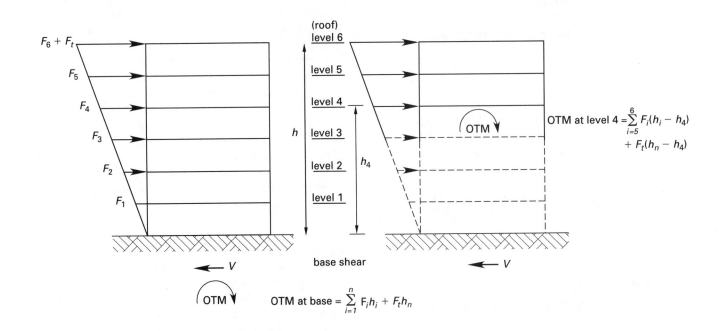

Figure 5.16　Overturning Moment

5-17 RIGID FRAME BUILDINGS

Before 1965, when the design of structural systems was still in its infancy, most tall buildings were designed as *rigid frames*.[11] In a rigid-frame building, columns and beams were welded together to create a structural grid that resisted wind and earthquake forces elastically.[12]

Such buildings were expensive to construct because they used inordinately large amounts of material, usually steel, to keep the stresses in the elastic region.[13]

5-18 HIGH-RISE BUILDINGS

The optimum design for seismic loading often conflicts with that for wind loading, a significant factor for any tall building. For an earthquake, the building needs to be flexible, even though the full flexibility might be called upon only once in 500 years. However, the full flexibility might be experienced during large windstorms, say, once every 50 years. The greater flexibility required to resist large earthquakes makes for unpleasant motions in windstorms.[14]

The rigid-frame system relies on the bending of columns and beams for its lateral stiffness. However, bending is a poor way to tap a structural member's strength compared to axial loading.[15] A *tube building* resists lateral forces in a radically different way from a rigid-frame building. The tube is like a giant box beam cantilevering out of the ground. Axial forces in the columns mainly resist the tendency to move laterally.

In order to economically design for increasing numbers of stories, different flexible structural systems were developed.[16] The most general systems are: (1) frames with bracing in the core, which creates a stiff vertical truss, good for buildings up to 30 or 40 stories; (2) framed tubes, good for up to 60 or 70 stores; and (3) diagonally braced tubes, good for up to 100 or 120 stories.

In a *pure* tube system also known as a *framed tube*, all of the lateral resistance is in the structure's exterior tube, made up of closely spaced steel columns linked by stiff deep spandrel beams. This framing is usually (but not always) located on the perimeter of the structure. In the past, framed tubes were used for tall buildings in high-wind areas and are now finding some applicability in seismic design.

---

[11]The Empire State and the Chase Manhattan Bank buildings in New York City and the Tenneco Building in Houston are rigid-frame structures.

[12]Bracing in the core (see Sec. 5-18) was not used, although it was recognized that it contributed to structural performance. Using such bracing would have been prohibitively complicated because tools for the structural analysis, such as computers and software, had not been developed. The increasing cost of land after 1965 also made it worthwhile to use costlier designs.

[13]For smaller buildings, rigid frames may be more economical in some cases, particularly where wind forces prevail or when the cost of the additional material is less than the cost of increased design and testing.

[14]This and the preceding sections are meant to document the trend in high-rise design, not to suggest using tube structures in designs for earthquake resistance.

---

[15]A measure of the "efficiency" of steel is the weight of steel per square foot of floor space for all stories. The 60-story rigid-frame Chase Manhattan Bank Building uses about 60 pounds of steel per square foot (290 kg/m^2). The 100-story John Hancock Center in Chicago uses a trussed tube structural system requiring half as much steel per unit area—about 30 pounds per square foot (145 kg/m^2).

[16]Dr. Fazlur Khan is acknowledged as being the first structural engineer to recognize the value of the alternate structural systems. One of the first (if not the first) flexible buildings was the 43-story concrete framed-tube system Chesternut Dewitt Apartment Building in Chicago that Khan designed. He also designed the One Shell Plaza building in Houston (a 50-story structure using lightweight concrete).

6

UBC SEISMIC CODE

6-1 HISTORICAL BASIS OF SEISMIC CODES IN CALIFORNIA

It is somewhat surprising that the formal study of earthquake-resistant design had to wait until considerably after the 1906 San Francisco earthquake. The only lateral force requirement placed on structures designed and constructed in San Francisco after that earthquake was a 30 lbf/ft^2 (1.44 kN/m^2) wind loading.

Only after the 1925 Santa Barbara earthquake did the California legislature direct that significant effort be expended on the study of seismology.[1]

There were 115 fatalities in the 1933 Long Beach earthquake. Inasmuch as there was widespread damage to schools in the area (caused by poor workmanship, design errors, and construction shortcuts), the *Field Act* was subsequently passed. This act gave the Division of Architecture, State Department of Public Works, responsibility for approving school designs.[2] Also, the

1933 Riley Act set minimum standards for lateral force resistance in all buildings (specifically, just 2% of the dead load). The format of the early codes was that the building had to be "strong" enough to resist a static lateral force, the *base shear*, V, of some fraction (e.g., 10% for masonry school buildings) of the weight, W.[3] The fraction was known as the *base shear coefficient*, C_s. Between 1943 and 1953, the base shear coefficient was modified several times based on the building period and/or the height of a building, but the *equivalent static force* concept remained and is in use to this day.[4,5]

$$V = C_s W \qquad [6.1]$$

Nine lives were lost in the 1940 El Centro earthquake, a 7.1 magnitude event caused by the Imperial Fault. While only approximately 10 sec in duration, a relatively high ground acceleration, 0.33 g, was observed. This earthquake was significant, not because of the widespread damage but because it was the first earthquake to occur in a heavily instrumented area. The

---

[1] Based on damage reports, the 1925 Santa Barbara earthquake is estimated at 6.3 Richter magnitude, although Richter-style seismometers had not yet been developed. Fatalities were limited, primarily because the earthquake occurred in the early morning before people were in the business district and children were in school. Widespread damage similar to that which destroyed San Francisco was averted as city engineers detected the tremors by observing fluctuations in the water pressure gauges and shut off the gas valves and electrical mains. Despite this, there were 13 fatalities and significant building damage, particularly in brick, masonry, and tile construction. Steel, wood, and properly designed reinforced concrete construction sustained little or no damage, although damage to poorly designed concrete structures occurred.

[2] The 1933 Long Beach earthquake had a Richter magnitude of 6.3, according to the newly developed seismometer. (Although the seismometer's range was exceeded, there is essentially a full record of this earthquake.) As with the 1925 Santa Barbara earthquake, the 1933 Long Beach earthquake occurred in the early morning, before children were in school. The same types

of structural failures were observed—that is, masonry and brick buildings, in particular, were damaged.

[3] This is simplifying the theory slightly, as the base shear coefficient was actually applied to the dead load and some part of the live load.

[4] The American Society of Civil Engineers (ASCE) and the Structural Engineers Association of Northern California (SEAONC) formed a committee in 1948 that recommended that the equivalent static force concept be used in San Francisco. In 1959, the Structural Engineers Association of California (SEAOC) code was expanded to a "uniform" code for all areas of the United States. This was the first "Blue Book." At this time, the type of building—frame, box, and so on—was made significant in code specifications.

[5] There are cases in which the UBC-97 [Sec. 1629.8.4] requires a dynamic analysis. (See Sec. 6-33.)

first accelerometer yielding response data on building periods was obtained.

The 1966 Parkfield earthquake on the San Andreas Fault had a relatively low magnitude of 5.5 and a very short duration, but the ground acceleration of 0.5 g was the highest observed to that date. Thus, it is apparent that magnitude and acceleration are not necessarily correlated.

A great amount of accelerometer data was obtained from the 1971 San Fernando earthquake (6.6 magnitude, San Fernando Fault zone). This earthquake was significant for two reasons: First, an unbelievable ground acceleration of 1.24 g was experienced at the Pacoima Dam site. Second, there were failures of new buildings designed with the current seismic codes.[6]

The 1979 Imperial Valley earthquake (Richter magnitude 6.6) produced the first accelerometer data from a building with extensive damage. In a building that was partially supported at the ground level by concrete columns, the period and amplitude of oscillation decreased significantly each time one of the columns failed. This is consistent with the concept that seismic energy is removed from a yielding structure.[7]

The 1989 Loma Prieta earthquake (magnitude 7.1, 62 fatalities, San Andreas Fault) was significant because of the important lessons learned about soil and site conditions. The most significant damage outside the epicenter occurred in areas where soil resonance magnified the seismic energy. Although the overall seismic energy should have been (and probably was) greatly attenuated by the large distance between the epicenter and San Francisco, the site conditions under the San Francisco Marina district and Oakland's Interstate 880 Cypress structure magnified the remaining energy.

The 1994 Northridge earthquake (magnitude 6.7) was centered in the San Fernando Valley region of Southern California. The Northridge earthquake was significant in that it was centered in a major urban area and, although of short duration, produced spectral acceleration values at many sites exceeding the UBC design spectrum. Measured peak ground accelerations routinely exceeded 0.7 g. The duration of source rupture

associated with the Northridge earthquake was 6 to 8 sec. The fault involved was the Northridge Thrust Fault (also known as the Pico Thrust Fault). Interestingly, before this earthquake, this fault was not considered a seismic danger. Because a thrust fault (blind thrust) was involved, the Northridge earthquake produced a significant vertical acceleration component, which contributed to economical losses exceeding $15 billion (the second most expensive U.S. earthquake).

One of the biggest surprises to come out of the Northridge earthquake was widespread damage to welded moment resisting steel structures, favored in earthquake country because of their supposed ductility. Review and analysis of the damage revealed no single cause, but rather a combination of factors, including the influence of vertical acceleration, connection ductility, and weld materials and quality. The majority of damage consisted of fractures in welded beam-column connections. These connections had cracks in either top or bottom flange welds, which occasionally propagated through the column flange and/or beam web. One prime reason for the Northridge failures has been identified as lack of ductility due to overconstrained joints. Appendix L offers more details regarding performance of the steel moment frame connections in this earthquake.

The 1995 Hyogoken-Nanbu (Kobe) earthquake, which struck Hyogo, located on the south-central part of Japan, had moderate to severe ground shaking with a magnitude of 6.9. The Kobe earthquake produced a fault rupture directly through the downtown section of a city, resulting in over 5400 deaths and injuries numbering in the tens of thousands, with estimated damage costs of $150 billion.

Kobe's setting is very similar to that of the San Francisco Bay Area. Both have large strike-slip faults adjacent to a bay and engineered buildings constructed on sedimentary deposits. The city of Oakland is particularly comparable to Kobe. Similarities include the proximity to a major active fault and considerable deposits of soft soils and bay mud in the downtown section, directly under many of the engineered buildings. These building structures are mostly constructed of nonductile concrete, steel, and masonry.

One major point driven home by the Kobe earthquake is the importance of characterizing near-source ground motion in the design of earthquake-resistant structures. A severe earthquake in a major urban area can create immense social and economic losses, and the impact on the economy will be significant. The main steps in mitigating this risk is to improve the model seismic design codes and to introduce mandatory requirements for seismic rehabilitation of vulnerable buildings.

---

[6]In particular, the New Olive View Hospital and the San Fernando Veterans Administration Hospital were new structures that sustained major damage.

[7]That this event proves inelastic behavior removes seismic energy from a structure should not be used to legitimize intentional design for inelastic behavior. It would be a major flaw to design columns to lose their load-carrying ability in the way they did in this instance.

6-2 SEISMIC CODES

A *code* is a set of rules adopted by an organization empowered to enforce the code. The mere publication of a set of guidelines such as those contained in the SEAOC Blue Book does not constitute a governing law. The guidelines must be adopted by a law-making body to become legal documents.

Various regions of the United States have adopted different codes. Agencies whose codes are widely adopted are referred to as "model code agencies," such as the International Conference of Building Officials (ICBO), which publishes the Uniform Building Code (UBC). ICBO members are representatives of local, regional, and state governments who investigate and research principles concerning safety to life and property in the construction, function, and location of buildings and related structures.

The *Uniform Building Code* (UBC) is dedicated to the development of better building construction and greater safety to the public. The UBC, the most widely adopted model building code in the United States, contains the most extensive seismic provisions for structures of any code.[8] Many of the seismic provisions of the UBC have been influenced by the *Recommended Lateral Force Requirements* and commentary (commonly referred to as the *Blue Book*) published by the Seismology Committee of the Structural Engineers Association of California, SEAOC.[9] The Blue Book summarizes recommendations of the SEAOC for earthquake-resistant design of structures. The Blue Book Commentary is not reproduced in the UBC, but this commentary is invaluable in understanding the significance of the UBC code provisions.

The SEAOC Blue Book provisions have been incorporated in the UBC since approximately 1960, but other building codes were slower to include more than limited seismic provisions, probably because the true seismic risk of the regions that have adopted those codes was not recognized.[10] However, following the 1971 San Fernando earthquake, when several buildings supposedly built according to current seismic provisions experienced substantial damage, other organizations received funding to develop seismic design recommendations.

- ◇ The Applied Technology Council (ATC) published ATC 3-06 in 1978. This was a massive 500-page document intended to serve as a reference for other code writers. It has now been superseded by the NEHRP provisions.

- ◇ The Building Seismic Safety Council (BSSC), with Federal Emergency Management Agency (FEMA) funding, published the National Earthquake Hazards Reduction Program (NEHRP) provisions in 1985. The 1997 NEHRP, the latest edition, is prominently featured in the seismic sections of the UBC-97.

- ◇ The American National Standards Institute (ANSI) published its A58.1 in 1982. This document deals with determining seismic loading, but it does not address detailing.

- ◇ The American Concrete Institute (ACI) included detailing to resist seismic loads as Appendix A in the 1983 edition of ACI-318. However, determination of seismic loading is not covered.

- ◇ The American Institute of Steel Construction has developed detailing requirements for steel buildings. As with the ACI document, determining the seismic loads is not covered.

While the large number of seismic documents may seem confusing, it should be noted that all are used as "source documents" for the Blue Book and the UBC. The 1988 Blue Book drew its format for the base shear equation from the ATC document.[11] The UBC-97 derives its new strength design format from the 1997 NEHRP seismic provisions, which brings the UBC more in line with the BOCA and the SBC, both based on the NEHRP.

Adoption of the UBC is up to each municipality. Most large cities have their own specific requirements that can supersede portions of the UBC or replace it entirely. Design of buildings located in Los Angeles, for example, is governed by a city code derived from the UBC-97.

General seismic provisions such as those in the UBC may be superseded by even more stringent statutory requirements. For example, Title 24 of the California Administrative Code requires that schools and hospitals be operational after an earthquake.

---

[8]Published by the International Conference of Building Officials in Whittier, California.

[9]SEAOC, 217 Second Street, San Francisco, CA 94105.

[10]The nation's two other model building codes are written by the Building Officials and Code Administrators International (BOCA) in Country Club Hills, Illinois, and the Southern Building Code Congress International (SBCCI) in Birmingham, Alabama. Both codes significantly strengthened their seismic provisions in 1992. However, the two codes differ from the UBC in their methodology.

[11]There was no "rational" basis for the individual parameters in the earlier base shear equation. In the 1988 code, all of the factors in the numerator logically represent characteristics that contribute to increased loading. (The numerator reduces the actual numerical value to a "consensus value" of design base shear.)

Bridges in California are designed according to CAL-TRANS seismic provisions. The CALTRANS method evaluates each bridge two ways (the term "loading" refers to lateral loads, moments, and shears):

1. Transverse seismic loading plus 30% of the longitudinal seismic loading

2. Longitudinal seismic loading plus 30% of the transverse seismic loading

As with the UBC method, the CALTRANS method includes both static and dynamic analyses. The static method is used for well-balanced spans with supports that are all approximately equal in stiffness. Once determined, the maximum lateral load is applied uniformly along the bridge. The dynamic method is used when the bridge is irregular in configuration or strength.

With the static method, the earthquake design force, V, is calculated as

$$V = \frac{(\text{ARS})W}{Z} \qquad [6.2]$$

ARS is the Acceleration Response Spectrum value obtained from one of four response spectra curves provided in the code or from site-specific curves, if available. Different curves are provided for different *alluvium* (i.e, soil) depths and peak rock accelerations for the *maximum credible earthquake* expected in that area. (The maximum credible earthquake is obtained from maps published by the California Division of Mines and Geology.)

Z is the *ductility risk reduction factor*, which depends on the type of structure and its period. Z varies from slightly less than 1.0 to slightly more than 8.0. A value of 5.0 is appropriate for new bridges.

W is the dead load (i.e., the weight) of the bridge.

The design bridge period, T (in seconds), is calculated from the following formula. P is the stiffness of the superstructure (i.e., the total force that, if applied uniformly, would cause the bridge to deflect one unit distance). Consistent units must be used.

$$T = 2\pi\sqrt{\frac{W}{P}} \qquad [6.3]$$

6-3 THE INTERNATIONAL BUILDING CODE AND UBC-97

In response to technical disparities among the three sets of model codes now in use in the United States, three model code agencies—the Building Officials and Code Administrators (BOCA), the International Conference of Building Officials (ICBO), and the Southern Building Code Congress International (SBCCI)—created the International Code Council (ICC) (December 9, 1994). The ICC was established as a nonprofit organization seeking to develop a single set of comprehensive and coordinated national codes. The ICC offers a single, complete set of construction codes without regional limitations—the International Codes. The International Building Code (IBC) was first published in 2000.

The UBC-97 edition incorporates both the SEAOC Seismology Committee's recommendations for seismic design and the NEHRP-recommended seismic regulations for new buildings. The UBC has been acknowledged as the prominent code publication for earthquake design provisions over the last few decades. This new edition carries on this heritage and has already functioned as a pivotal part of the foundation for the year 2000 edition of the IBC. The UBC-97 seismic changes are designed to provide a smooth transition to the 2000 edition of the IBC, which is intended to ultimately replace it.

The new seismic provisions of the 1997 code exemplify a number of notable lessons learned from earthquakes and recent advances in other seismic resource documents, including the NEHRP. The new provisions demonstrate a considerable migration from previously elected approaches in seismic design. The provisions are as follows.

◇ New design response spectra based on the 1997 NEHRP recommended provisions

◇ New devised base shear equations utilizing new soil-profile types, seismic response coefficients, seismic source classification, and near source factors

◇ Adoption of revised structure response modification factor based on elastic ground motion response

◇ Adoption of a strength-based design approach

◇ Acceptance of more precise evaluation of the drift induced in structures in response to design ground motions

◇ Acceptance of the effects of structural redundancy-reliability and overstrength factors

◇ Introduction of new structural system classification, cantilevered-column building system to simulate buildings with discontinuity in capacity (weak story), shearwall-frame interaction system, and special truss moment frames of steel (STMF)

◇ Nonstructural components provisions consistent with the 1997 NEHRP

◇ Nonbuilding structures provisions consistent with the 1997 NEHRP

6-4 THE SURVIVABILITY DESIGN CRITERIA

While little effort is expended in trying to design buildings that will be totally elastic (i.e., experience no damage) during an earthquake, it is implicit in seismic codes that catastrophic collapse must be avoided. Based on the UBC-97 [Sec. 1626.1], the purpose of the earthquake design provisions is primarily to safeguard against major structural failures and loss of life; these provisions are not intended to limit damage or maintain function. The following three design standards constitute the implied UBC seismic *survivability* (or *life-safety*) *design criteria* [1996 Blue Book Commentary, Sec. C101.1.1 and App. A]. It is notable, however, that these criteria are not actually specified in the UBC.

◇ *There should be no damage to buildings from a minor earthquake.*

◇ *There may be some architectural (nonstructural) damage, but no structural damage during a moderate earthquake.*

◇ *There may be possible structural and nonstructural damage but no collapse during a severe earthquake.*[12] Yielding is relied upon to dissipate the damaging seismic energy. Theoretically, all structural damage will be repairable when collapse is prevented, although some buildings may be condemned and replaced for reasons of economics or convenience.

The UBC-97 provisions will not prevent structural and nonstructural damage from direct earth faulting, slides, or soil liquefaction.

6-5 EFFECTIVENESS OF SEISMIC PROVISIONS

The code provides "reasonable" but not complete assurance of the protection of life. Furthermore, the code does nothing to prevent construction on land that is subject to earth slides (of the type that occurred during the Alaskan earthquake) or liquefaction (as occurred in the Niigata, Japan, earthquake).

It is important to note that the UBC-97 seismic provisions are intended as minimum requirements. The level of protection can be increased by increasing the design lateral force, energy absorbing capacity, redundancy, and construction quality assurance.

It is also important to note that seismic design is both a science and an art that, unfortunately, must be verified in the field. Thus, the history of seismic codes has been to require design features or methods and then evaluate the effectiveness of those features in practice.

Finally, the seismic code used is not the only factor affecting the performance of a structure during an earthquake. In many cases of structural failure in modern buildings, earthquake severity, duration, soil conditions, inadequate design, poor control or material quality, and poor workmanship are found to be the major factors contributing to collapse.[13] Of course, modern seismic codes cannot be blamed when pre-1973 structures fail. (See Sec. 3-1.)

6-6 APPLICABLE SEISMIC SECTIONS IN THE UBC

General seismic provisions applicable to all structures are contained in Chap. 16, Div. IV (Earthquake Design) of the UBC-97. However, the code provisions for sizing and detailing structures appear in other chapters: Chap. 18 (in particular, Sec. 1809), foundations and retaining walls; Chap. 19 (in particular, Sec. 1921), concrete; Chap. 20, lightweight metals; Chap. 21 (in particular, Secs. 2106.1.12, 2107.1.3, and 2108.2.3.8), masonry; Chap. 22 (in particular, Divs. IV and V), steel; and Chap. 23 (in particular, Sec. 2315), wood.

A section for nonbuilding structures (see Sec. 6-46) was added in the 1980 UBC version and was expanded in the 1997 edition of the UBC. Section 1634 of the UBC, Nonbuilding Structures, covers structures such as tanks, towers, chimneys, signs, billboards, and storage racks, but not such structures as retaining walls, bridges, dams, docks, and offshore platforms. Other national codes may supersede the UBC-97 where specialized design of nonbuilding structures is required.

Other parts of the UBC-97 that are occasionally useful are the UBC standards, appendices, and tables. The UBC-97 standards are listed in Vol. 1, Chap. 35; however, some excerpts from Chap. 35 are reprinted in UBC-97 Vols. 2 and 3. Appendix 16 (Structural Forces)

---

[12]These italicized sentences are the standards of survivability as they are commonly stated, but it is understood the degree of damage is dependent on the severity of the ground shaking at the building site, not on the magnitude of the earthquake at some distant epicenter.

[13]In particular, the widespread structural failures that occurred in the 1985 Mexico City earthquake are examples of how even modern buildings can be "brought down" by these contributing factors. In fact, the failures that occurred seem to validate the need for the current UBC provisions, as the very features required by that code were often not included in the design of buildings that collapsed.

of UBC-97 Vol. 2 highlights details for snow load design, earthquake recording instrumentation, seismic zone tabulation, and seismic-isolated structures. Along with key UBC-97 tables in Chap. 16, the following UBC-97 tables are particularly useful: Tables 21-E-1 and 21-F-1 (anchor bolts in masonry), Table 23-II-B-1 (nailing schedule), Tables 23-III-B-1 and 23-III-B-2 (bolting in wood), and Tables 23-III-C-1 and 23-III-C-2 (nailing in wood).

6-7 THE NATURE OF UBC SEISMIC CODE PROVISIONS

There are two major categories of seismic provisions in the UBC-97: those that relate to proportioning the structure and those that relate to detailing elements of the structure. The methods of proportioning structural elements are *allowable stress design* (ASD) and *load and resistance factor design* (LRFD). The proportions are chosen such that the structure's ability to absorb energy (i.e., its "strength") matches the application of energy, no matter how much yielding has occurred, and such that overall stability is maintained. This requires that the lateral-force-resisting elements be roughly distributed (in plan) throughout the structure. (Thus, arbitrarily increasing the strength of one element may actually have a negative effect on the overall seismic performance.) In equation form, the ratio of energy demand to energy capacity evaluation in plan should be roughly constant.

Design details prevent premature local failure by ensuring ductile behavior and preventing local instability and failure of elements that are cyclically stressed beyond their yield points. Unlike the UBC provisions for proportioning the structure, the design details can usually be determined without evaluating the stresses, drifts, or loads.

Controlled yielding in a major earthquake is implicitly anticipated by the UBC, and, therefore, a code based on yield or ultimate strengths would be preferred. In the earlier UBC versions, the seismic design forces were based on ASD (working stress or service level stress) and not on strength. This was primarily because the vertical load-carrying systems in the majority of steel highrise structures were, until recently, based on ASD.

The ASD method is being replaced in structural steel work by the LRFD method, also known as the *ultimate strength design method*. In this method, the applied loads are multiplied by a load factor. The strength of the product must be less than the ultimate strength of the structural member multiplied by a resistance factor. The conversion from working stress to strength design

is consistent with the direction of all material code requirements in the United States and with NEHRP seismic provisions.

The American Institute of Steel Construction (AISC) published design standards methods for steel buildings in seismic zones based on the Specification for Structural Steel Buildings ASD (June 1, 1989) and the LRFD (June 15, 1992). These standards cover the design of beams, columns, concentric- and eccentric-braced frames, and moment-resisting connections, including panel zones at the intersections of beams and columns.

Note that the term "LRFD" is used in the design of steel and wood structures, whereas in the design of concrete and masonry structures, the equivalent method is known as "strength design." The strength design inclusion in the UBC-97 represents the most significant development toward consistency in national seismic requirements. For smooth transition from ASD to LRFD, the 1977 UBC provides ASD provisions in parallel with strength design. Table 6.1 lists the UBC-97 sections in which strength-based and allowable stress-based seismic provisions are presented.

Table 6.1
UBC-97 Provisions by Building Material

construction material	LRFD	ASD
steel	Chap. 22, Divs. II and IV	Chap. 22, Divs. III and V
concrete	–	Chap. 19, Div. VI
masonry	–	Chap. 21 Secs. 2106 and 2107
timber	Chap. 23, Sec. 2303, Item 5.4	Chap. 23, Div. III

6-8 WIND LOADS

Wind loading is covered in the UBC-97, Chap. 16, Div. III, Secs. 1615–1625. Specifically, the wind pressure on a structure is given by Eq. 6.4 (corresponding to UBC-97 Formula 20-1).

$$P = C_e C_q q_s I_w \qquad [6.4]$$

C_e is a coefficient (given in UBC-97 Table 16-G) accounting for combined height, exposure, and gust factors. C_q is a pressure coefficient (given in UBC-97 Table 16-H) that depends on the type of structure or portion of a structure exposed to the wind. q_s is the wind stagnation pressure (given in UBC-97 Table 16-F) at a standard elevation of 33 ft (10 m). q_s depends on wind speed [UBC-97 Sec. 1618]. I_w is the importance factor for wind loading (given in UBC-97 Table 16-K) based on the occupancy category of the structure.

The wind pressure can be assumed to act on the structure in one of two ways: The *normal force method* can be used for the design of any structure, including gabled frames. A uniform wind pressure is assumed to act normal to all exterior surface [UBC-97 Sec. 1621.2]. The *projected area method* calculates the horizontal wind force by assuming that the pressure acts upon the projected area of the structure. The projected area method can be used only if the building is less than 200 ft (61 m) high and does not contain a gabled rigid frame [UBC-97 Sec. 1621.3].

Supplemental information on reducing the risk of wind-induced damages to prescriptive masonry and conventional light-frame construction in high-wind areas (wind speed varies from 80 mph to 110 mph (129 km/h to 179 km/h)), is included in the UBC-97 appendices to Chaps. 21 and 23.

6-9 SNOW LOADS

In many locations, snow applies a significant load on structures that must be considered in design. Snow regularly causes the failure of roof systems and can cause progressive collapse of entire structures. Snow loads on roofs vary widely based on the geographic location, elevation, site exposure, and slope of the roof. Structural members must be capable of supporting snow loads, which in many cases constitute the largest design load for the roof system.

Snow on a structure's roof may result in a uniform loading condition (i.e., the same load over the entire roof) or a nonuniform loading condition (i.e., a varying load) caused by wind-induced drifting or melting and refreezing of snow. Conditions giving rise to uniform loading are the exception, in reality. Unbalanced accumulation of snow at valleys, parapets, and roof structures, and offsets in roofs of uneven configuration are typical. Compound roof systems may accumulate large unbalanced loads in valleys, particularly on the leeward side of roofs. UBC-97 App. Chap. 16, Div. I, covers additional loading caused by drifting and ice dams.

Snow loading is covered in UBC-97 Sec. 1614 and App. 16, Div. I (Secs. 1637–1648). Specifically, the design snow load (P_f) for buildings and other structures is given by Eq. 6.5 (corresponding to UBC-97 Formula 40-1-1).

$$P_f = C_e I P_g \qquad [6.5]$$

C_e is the snow exposure factor (from UBC-97 Table A-16-A). I is the snow occupancy importance factor for snow loading (from UBC-97 Table A-16-B). (The basic ground snow loads (P_g) are given in UBC-97 Tables A-16-1 to A-16-3, except where controlled by the local building official.) The roof snow load is assumed to act vertically upon the area projected upon a horizontal plane. Where roof snow loads are in excess of 20 psf (958 N/m^2), the UBC-97 allows a reduction based on roof slope (expressed in degrees). Snow load (S) reduction in lbf/ft^2 (kN/m^2) for each degree of pitch (a) over 20° is determined by Eq. 6.6 (corresponding to UBC-97 Formula 14-1).

$$R_s = \frac{S}{40} - \frac{1}{2} \qquad \text{[U.S.]} \quad [6.6(a)]$$

$$R_s = \frac{S}{40} - 0.024 \qquad \text{[SI]} \quad [6.6(b)]$$

6-10 COMBINED SEISMIC, WIND, AND SNOW LOADING

Snow loads must be considered in seismic design (according to the load combinations specified in UBC-97 Sec. 1612) because snow adds to the mass of the structure. However, the UBC-97 permits a reduction in the design snow load when it is used in combination with earthquake loads.

Although codes such as AISC, AITC, and ACI provide for the combination of various loadings (e.g., snow and wind, earthquake and wind), seismic and wind loads are distinctly different in origin. Wind loads are applied over an exterior surface of a structure, whereas seismic loads are inertial in nature.

UBC-97 provisions for ductile seismic detailing must always be met, even if the wind load is greater than the seismic load. Seismic loads generally control for heavy structures and moderate-weight structures in zone 4, but wind loads often control the design in seismic zones 0 through 3 and in zone 4 for lightweight construction. It is not necessary for the building to be designed to withstand the simultaneous action of wind and seismic loads [UBC-97 Sec. 1626.3].

There is no such thing as a "governing" load when a building is in a potential earthquake area. In some cases, the maximum expected lateral wind loading will result in larger drift (see Sec. 5-12) or larger lateral forces than will an earthquake. However, even in that instance, the design must include seismic detailing. The reason for this requirement is that the structure must be able to resist seismic loads in a ductile manner even when it can resist a larger design wind load elastically.

Simply, the intent is to avoid catastrophic failure and to provide the necessary structural integrity to resist actual seismic forces, which are potentially much higher than the design seismic forces.

6-11 SEISMIC DESIGN CRITERIA SELECTION

As described in Sec. 3-10, the base shear, V, is the total design seismic force imposed by an earthquake on the structure at its base. The base shear is the sum of all the inertial story shears. The UBC-97 calculates the base shear from the total structure weight and then apportions the base shear to the stories in accordance with dynamic theory. The design seismic forces can be determined based on the UBC-97 static lateral force procedure [Sec. 1630.2] and/or the dynamic lateral force procedure [Sec. 1631]. The dynamic force procedure is always acceptable for design of any structure. However, the UBC-97 [Sec. 1631] specifies that the minimum design seismic force must be 80 to 100% (depending on the degree of structural irregularity) of that prescribed by the static lateral force procedure.

The seismic design process involves consideration of a number of structural and site characteristics, including seismic zoning, occupancy, seismic importance factors, building fundamental period, site geology and soil characteristics and soil-profile types, seismic source classification, near-source factors, seismic ground response coefficients, response modification factor, configuration, structural system, and height. Furthermore, the UBC-97 requires that all parts of the structure be designed with adequate strength to withstand the lateral displacements induced by the design ground motion [Sec. 1627] considering the inelastic response of the structure and the inherent redundancy, overstrength, and ductility of the lateral-force-resisting system. Table 6.2 furnishes the steps needed in seismic design of a structure.

Table 6.2
Seismic Design Procedure

step	description	this book's section reference
1.	Identify appropriate structural system.	6-21
2.	Classify occupancy category of the structure.	6-13
3.	Determine the components of seismic base shear coefficient.	6-30
4.	Identify structural system limitations and irregularities.	6-24 and 6-25
5.	Select appropriate lateral-force procedure.	6-31
6.	Determine the total design base shear.	6-32, 6-34, and 6-50
7.	Distribute the design base shear over the structure's height.	6-35 and 6-36
8.	Analyze P-Δ effects for the structure.	6-43
9.	Examine overturning effects caused by earthquake forces.	6-42
10.	Evaluate torsional effects for the structure.	6-39
11.	Study story drift limitations.	6-40
12.	Consider redundancy of lateral-force-resisting system.	6-22
13.	Evaluate overstrength of lateral-force-resisting system.	6-20
14.	Design elements of the structure.	6-44
15.	Confirm seismic detailing requirements with the UBC-97.	
16.	Verify structure's continuous load path completion.	

6-12 SEISMIC ZONE FACTOR: Z

The *seismic zone factor*, Z, accounts for the amount of seismic risk present in a building's seismic zone based on the anticipated maximum ground acceleration at the site. For the United States, the UBC-97 provisions contain six different seismic zones (see Sec. 3-2 and Fig. 3.1), with zone 0 representing the least risk and zone 4 encompassing areas with a chance of severe shaking. Table 6.3 (equivalent to UBC-97 Table 16-I) gives the seismic zone factors for each of the seismic zones.[14] The zone factor of 0.4 in zone 4 implies that the maximum base (unadjusted) effective peak acceleration (EPA) is 0.4 g for the design earthquake. A geographical region experiencing an EPA > 0.3 would be classified as zone 4.

---

[14]Development of the Z values was, at times, a compromise. For example, the two values for zone 2 represent the desire of building officials in the eastern United States to maintain historical force levels, regardless of the basis on which the zone boundaries were developed.

Table 6.3
UBC-97 Seismic Zone Factors (Z)
[UBC-97 Table 16-I]

seismic zone	0	1	2A	2B	3	4
Z	0	0.075	0.15	0.2	0.3	0.4

Reproduced from the 1997 edition of the *Uniform Building Code*, copyright © 1997, with the permission of the publisher, the International Conference of Building Officials.

The values of the seismic zone factor are developed considering historical records and geological data and seismological information. These values are intended to represent the effective peak ground accelerations (see Sec. 2-8) that have only a 10% chance of being exceeded in 50 years. This corresponds to a ground motion that will be exceeded, on the average, only once in 475 years (the *recurrence interval*).

For cities and countries outside the United States, the UBC-97 [Sec. 1653, Div. III] has compiled a partial listing of seismic zones.

6-13 OCCUPANCY CATEGORIES

When earthquake disaster strikes a community, any major structural and nonstructural damage to a facility that could threaten life safety demands complete closure until adequate repair measures are taken and further evaluation should deem the facility safe for occupancy again. Certain facilities such as hospitals and police and fire stations cannot be shut down under these circumstances. Accordingly, law requires that these facilities be designed to remain operational after an earthquake.[15]

It is implicit in seismic codes that catastrophic collapse of structures must be avoided to safeguard lives, minimize economic losses, and avoid disruption in the event of an earthquake. For purposes of earthquake-resistant design, each structure is specified in one of the occupancy categories listed in UBC-97 Table 16-K (also see Table 6.4 of this book). This table consists of five occupancy categories with their functions defined. Buildings in some of these categories require special review, inspection, and construction observation.

Essential facilities are emergency facilities that must remain operational after an earthquake [UBC-97 Sec. 1627]. They include hospitals with surgery and emergency treatment facilities, fire and police stations, emergency preparedness structures (including structures

housing emergency vehicles), and government communication centers required for emergency response. *Hazardous facilities* are used to store or support dangerous toxic or explosive chemicals or substances. *Special occupancy structures* are designed to house large numbers of people—for example, places of public assembly (5000 or more people), schools (300 or more students), colleges and adult education centers (500 or more students), nursing homes, daycare centers, nurseries, and jails. All other structures that house occupancies or have functions not listed are considered *standard occupancy structures*. Examples are apartment buildings, hotels, office buildings, and wholesale or retail structures. *Miscellaneous structures* are buildings or parts of buildings classed as Group U occupancies. They include private garages, carports, sheds, factories, and agricultural buildings.

Appendix K of this book (corresponding to UBC-97 Table 16-K) defines the occupancy categories in greater detail. Design and construction review requirements for each category are specified in UBC-97 Chap. 17.

Table 6.4
Occupancy Categories and Functions of Structure
[UBC-97 Table 16-K]

occupancy category	occupancy or functions of structure
1	essential facilities
2	hazardous facilities
3	special occupancy structures
4	standard occupancy structures
5	miscellaneous structures

6-14 SEISMIC IMPORTANCE FACTOR: I

Table 16-K of the UBC-97 specifies an importance factor, I, which increases seismic design forces for critical facilities.[16,17]

---

[15]After an earthquake, based on Title 24 of the California Administrative Code, hospitals and schools must be operational. Also, the California Hospital Act requires that hospitals be fully functional and operational.

[16]Seismic importance factors, I_w (for wind design) and I_p (for nonstructural component design), are discussed in Secs. 6-8 and 6-46, respectively.

[17]Increasing base shear in design and construction also increases costs. An earthquake risk management study may be used to evaluate the seismic hazard and vulnerability of a structure. The study may rationalize amplifying seismic loads to achieve greater performance goals, such as immediate occupancy in the event of a major earthquake. James L. Witt, Director of the Federal Emergency Management Agency, said, "Mitigation saved the Anheuser-Busch facility in Los Angeles after Northridge. The Anheuser-Busch Engineering Department retrofitted the plant to conform to the L.A. seismic code—and the plant was functioning within days of the earthquake. Without those revisions—they would have sustained more than $300 million in direct and interruption losses."

Table 6.5
UBC-97 Importance Factor (I) for Earthquakes
[UBC-97 Table 16-K]

occupancy or functions of structure	seismic importance factor, I
essential facilities	1.25
hazardous facilities	1.25
special occupancy structures	1.00
standard occupancy structures	1.00
miscellaneous structures	1.00

The seismic importance factor is either 1.0 or 1.25, depending on how critical it is for the structure to survive a major earthquake with minimal damage. From Table 6.5, it is obvious that a higher importance factor, I, (1.25) is designated for essential and hazardous facilities in order to ensure that these facilities remain functional and operational after a severe earthquake. For these facilities the prescribed design base shear is increased by 25% compared to other facilities. Increasing design base shear increases the seismic safety of a structure.

6-15 SOIL-PROFILE TYPES: S_A THROUGH S_F

Soft soil may amplify earthquake ground motion. Amplification of vibrations due to unfavorable soil conditions has been strikingly illustrated in many earthquakes, such as the 1985 Mexico City earthquake and the 1989 Loma Prieta earthquake. To that effect, the UBC-97 [Sec. 1629.3] specifies that each site be assigned a soil-profile type found by a properly substantiated geotechnical investigation. Subsequently, the UBC-97 [Sec. 1636, Div. V] provides the site categorization procedure for determining soil-profile types.

A *soil horizon* is a significant layer of soil with distinct characteristics extending from the surface into relatively unaltered material. In the UBC-97, the profiles are classified into six different soil types ranging from A (hard rock) to E (soft soil), and type F, assigned to those sites that require specific evaluation [UBC-97 Sec. 1629.3.1]. New solid profile designations are found in the 1997 NEHRP publication, which are based on quantitative geological parameters and recent data.

Based on the site categorization procedure of the UBC-97 [Sec. 1636], Table 6.6 (equivalent to UBC-97 Table 16-J) provides a detailed description of the UBC-97 soil profiles. This table furnishes average shear wave velocity ($\overline{v}_s$), average field standard penetration resistance ($\overline{N}$), average standard penetration resistance for cohesionless soil layers ($\overline{N}_{CH}$), and average undrained shear strength ($\overline{s}_u$) values associated with different soil profiles S_A through S_F, where applicable.

Table 6.6
Soil Profile Classification
[UBC-97 Table 16-J]

		soil profile description		
type	soil profile	shear wave velocity ft/sec (m/s)	standard penetration blows/ft	undrained shear strength lbf/ft^2 (kPa)
S_A	hard rock	$\overline{v}_s > 5000$ ($\overline{v}_s > 1500$)	–	–
S_B	rock	$2500 < \overline{v}_s \leq 5000$ ($760 < \overline{v}_s \leq 1500$)	–	–
S_C	very dense soil and soft rock	$1200 < \overline{v}_s \leq 2500$ ($360 < \overline{v}_s \leq 760$)	$\overline{N} > 50$	$\overline{s}_u \geq 2000$ ($\overline{s}_u \geq 100$)
S_D	stiff soil	$600 < \overline{v}_s \leq 1200$ ($180 < \overline{v}_s \leq 360$)	$15 \leq \overline{N} \leq 50$	$1000 \leq \overline{s}_u \leq 2000$ ($50 \leq \overline{s}_u \leq 60$)
S_E	soft soil	$\overline{v}_s < 600$ ($\overline{v}_s < 180$)	$\overline{N} < 15$	$\overline{s}_u < 1000$ ($\overline{s}_u < 50$)
	Or, any profile > 10 ft (3048 mm) of soft clay (soil profile with PI > 20, and $w_{mc} \geq 40\%$)			$\overline{s}_u < 500$ ($\overline{s}_u < 24$)
S_F	soils requiring site-specific evaluation			

The average shear wave velocity ($\overline{v}_s$) may be measured on site or estimated by a geotechnical engineer, engineering geologist, or seismologist, according to the UBC-97 [Sec. 1636.2.1], Formula 36-1.[18] The average field standard penetration resistance ($\overline{N}$) and average standard penetration resistance for cohesionless soil layers ($\overline{N}_{CH}$) can be determined from the UBC-97 [Sec. 1636.2.2], Formulas 36-2 and 36-3.[19] Lastly, the average undrained shear strength ($\overline{s}_u$) can be obtained from the UBC-97 [Sec. 1636.2.3], Formula 36-4.[20]

Where soil properties are not known in sufficient detail, soil profile S_D can be used, not S_E. Profile S_E is used only when a geotechnical study shows it to be valid or when required by the local building official.[21] Per UBC-97 [Sec. 1636.2], soil profile S_F is used with highly sensitive clays and peat vulnerable to potential failure or collapse under seismic loading (liquefiable and collapsible weakly cemented soils), and where a site-specific evaluation should be conducted in order to properly determine appropriate seismic response coefficients.

6-16 SEISMIC SOURCE CLASSIFICATION

In addition to the seismic zone, site soil profile classification, and the importance factor for establishing site seismic hazard characteristics for each site, UBC-97 provisions require the proximity of each site to active seismic sources (i.e., faults) to be known. The most recent mapping of active faults by the United States Geological Survey or California Division of Mines and Geology are used as approved geotechnical data in identifying the types of seismic sources to be used for design of buildings or structures. For routine engineering design, the ICBO publishes a resource entitled "Maps of Known Active Fault Near-Source Zones in California," which is intended for use with the UBC-97.

There are three seismic source types that are recognized in the recent UBC-97. They range from the most active source (type A) to the least active source (type

C). Faults are classified by the maximum moment magnitude potential (M) and slip rate (SR).[22] Table 6.7 characterizes the three types of faults.

Table 6.7

Seismic Source Type and Description
[UBC-97 Table 16-U]

type	seismic source description	
A	faults that are capable of generating large magnitude occurrences and a high rate of seismic activity	
	$M \geq 7.0$	SR $\geq$ 5 mm/yr
B	faults other than types A and C	
	$M \geq 7.0$	SR $<$ 5 mm/yr
	$M < 7.0$	SR $>$ 2 mm/yr
	$M \geq 6.5$	SR $<$ 2 mm/yr
C	faults considered relatively inactive	
	$M < 6.5$	SR $\leq$ 2 mm/yr

Reproduced from the 1997 edition of the *Uniform Building Code*, copyright © 1997, with the permission of the publisher, the International Conference of Building Officials.

When determining the seismic source type, it is crucial that both maximum moment magnitude potential (M) and slip rate (SR) conditions be satisfied concurrently. In California, the majority of faults fall into the type B seismic source classification. The San Andreas Fault is one notable exception, receiving a type A seismic source classification. Most faults outside California are type C seismic source.

6-17 NEAR-SOURCE FACTORS: N_a AND N_v

The ICBO resource guide "Maps of Known Active Fault Near-Source Zones in California" can be used to determine proximity to a fault zone. The United States Geological Survey and the California Department of Conservation's Division of Mines and Geology have also mapped all known active faults in establishing the location and type of seismic sources. The shortest distance to a seismic source is the minimum distance between the site and the area defined by the vertical projection of the source on the surface (i.e., surface projection of fault plane).[23] (See Fig. 6.1.)

---

[18] For these soil types, the average shear wave velocities ($\overline{v}_s$) and the average field standard penetration resistance ($\overline{N}$) are determined based on the top 100 ft (30 480 mm) of soil profile.

[19] The standard penetration resistance ($\overline{N}$) of the soil layer should be obtained in accordance with approved nationally recognized standards.

[20] The undrained shear strength ($\overline{s}_u$) should be determined in accordance with approved nationally recognized standards, not to exceed 5000 psf (250 kPa).

[21] The S_E profile is appropriate in areas with large deposits of very soft clay that are subject to large amplifications in seismic ground motion. Buildings constructed on San Francisco Bay mud or on the Mexico City lake bed are likely candidates for the S_E soil profile.

[22] Magnitude and slip rate information for faults can be obtained from the United States Geological Survey or the California Division of Mines and Geology.

[23] The surface projection should not include portions of the source at depths of 6.2 mi (10 km) or greater.

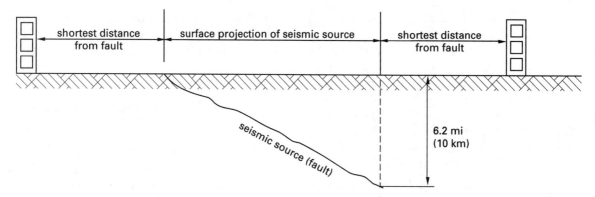

Figure 6.1   Surface Projection of a Fault Plane

In seismic zone 4 where large-magnitude earthquakes are expected, particularly severe damage to structures is likely to happen when structures are built very near or directly on top of active faults. The ground acceleration that these structures experience may be up to twice the acceleration that more distant structures experience.

The UBC-97 has adopted two *near-source factors*, N_a and N_v, to minimize this impact in seismic zone 4. These amplification factors are acceleration- (for short-period structures) and velocity- (for long-period structures) controlled factors. Near-source effects are greater for long-period structures (i.e., N_v is larger than N_a).

Table 6.8 (equivalent to UBC-97 Tables 16-S and 16-T) lists the values of the near-source factors related to both the seismic source type and the proximity of the building or structure to known faults.

Table 6.8

Near-Source Factors
[UBC-97 Tables 16-S and 16-T]

shortest distance to known seismic source

seismic source type	$\leq$ 2 km		5 km		$\geq$ 10 km		$\geq$ 15 km	
	N_a	N_v	N_a	N_v	N_a	N_v	N_a	N_v
A	1.5	2.0	1.2	1.6	1.0	1.2	1.0	1.0
B	1.3	1.6	1.0	1.2	1.0	1.0	1.0	1.0
C	1.0	1.0	1.0	1.0	1.0	1.0	1.0	1.0

(Multiply kilometers by 0.62 to obtain miles.)

Reproduced from the 1997 edition of the *Uniform Building Code*, copyright © 1997, with the permission of the publisher, the International Conference of Building Officials.

6-18 SEISMIC RESPONSE COEFFICIENTS: C_a AND C_v

In the design of a structure, the UBC-97 [Sec. 1629.4.3] requires that each structure be assigned *seismic response coefficients* C_a and C_v. These seismic response coefficients account for the increased severity of the ground motion at a specific site and are influenced by the seismic zone, proximity of the site to active seismic sources, and site soil profile characteristics. C_a functions as an acceleration-controlled coefficient for the short-period portion of the spectrum, and C_v serves as a velocity-controlled coefficient for the long-period portion of the spectrum.[24]

To determine which of these two coefficients should be used in the design of a structure, Fig. 6.2 (UBC-97 Fig. 16-3, Design Response Spectra) should be used.

In Table 6.9, in which UBC-97 Tables 16-Q and 16-R are combined, the values of these seismic response coefficients are specified per zone and soil-profile type.

To determine seismic response coefficients for soil-profile type S_F, the UBC-97 [Sec. 1629.3.1] requires that a site-specific geotechnical investigation and dynamic site response analysis be performed. For zone 4, near-source factors (N_a and N_v) are used in the determination of C_a and C_v. Based on the UBC-97 [Sec. 1629.4.2], the value of N_a should be equal to or less than 1.1 for regular structures on a site characterized with the soil-profile types either S_A, S_B, S_C, or S_D and a reliability/redundancy factor (ρ) (see Sec. 6-22) equal to 1.[25]

---

[24]The seismic response coefficient C_a represents effective peak acceleration at grade. The seismic response coefficient C_v represents acceleration response at 1.0 sec period.

[25]The redundancy factor penalizes nonredundant lateral systems that require an increase in the seismic design loads in order to reduce the magnitude of the elastic response and thus reduce the ductility demand.

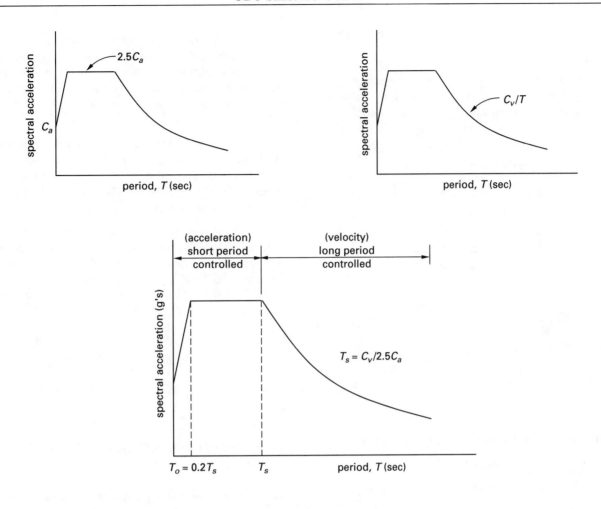

Figure 6.2   Response Spectra

Table 6.9
Seismic Response Coefficients
[UBC-97 Tables 16-Q and 16-R]

seismic zone factor, Z

soil-profile type	1		2A		2B		3		4	
	C_a	C_v	C_a	C_v	C_a	C_v	C_a	C_v	C_a	C_v
S_A	0.06	0.06	0.12	0.12	0.16	0.16	0.24	0.24	$0.32N_a$	$0.32N_v$
S_B	0.08	0.08	0.15	0.15	0.20	0.20	0.30	0.30	$0.40N_a$	$0.40N_v$
S_C	0.09	0.13	0.18	0.25	0.24	0.32	0.33	0.45	$0.40N_a$	$0.56N_v$
S_D	0.12	0.18	0.22	0.32	0.28	0.40	0.36	0.54	$0.44N_a$	$0.64N_v$
S_E	0.19	0.26	0.30	0.50	0.34	0.64	0.36	0.84	$0.36N_a$	$0.96N_v$
S_F					soil requiring site-specific evaluation					

Example 6.1

A public library building with a 6000-person capacity is being designed for the northern part of California. The building will be in the vicinity of the San Andreas Fault. The surface projection of this fault plane is as shown.

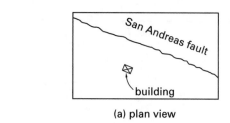

(a) plan view

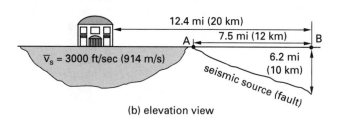

(b) elevation view

(a) What is the value of the importance factor?

(b) What is the soil-profile type?

(c) What are the values of near-source factors?

(d) What are the values of seismic response coefficients?

Solution

(a) Because of the 6000-person occupancy (more than 5000 persons) and social function of this building, this building is classified as a special occupancy structure according to the UBC-97 [Sec. 1629.2] and UBC-97 Table 16-K. According to the same table, the value of the seismic importance factor (I) for this type of occupancy is 1.0.

(b) This building is located on a soil profile that has an average shear wave velocity ($\overline{v}_s$) of 3000 ft/sec (914 m/s). Based on the site categorization procedure of the UBC-97 [Sec. 1636] and Table 6.6 (equivalent to UBC-97 Table 16-J), the soil profile corresponds to the description of soil-profile type S_B. Type S_B is described as a rock soil profile with an average shear wave velocity of 2500 ft/sec $< \overline{v}_s \leq$ 5000 ft/sec (760 m/s $< \overline{v}_s \leq$ 1500 m/s).

(c) Because the San Andreas Fault in California is an active fault that is capable of generating large-magnitude occurrences and has a high rate of seismic activity, the seismic source classification is type A.

Based on the surface projection of the fault plane (non-vertical fault), the shortest distance to the seismic source (i.e., the minimum distance between the site and the area defined by the vertical projection of the source on the surface) is 12.4 mi − 7.5 mi = 4.9 mi (20 km − 12 km = 8 km).

With seismic source type A and shortest distance of 4.9 mi (8 km) to the seismic source, the near-source factors (N_a and N_v) can be determined from Table 6.8 (equivalent to UBC-97 Tables 16-S and 16-T). The values of the near-source factors are $N_a = 1.2$ and $N_v = 1.6$.

(d) Because this site is located in seismic zone 4, in addition to the soil-profile type (S_B), near-source factors are used in the determination of seismic response coefficients (C_a and C_v). From Table 6.9 (equivalent to UBC-97 Tables 16-Q and 16-R), the seismic response coefficients are

$$C_a = 0.40N_a = (0.4)(1.2) = 0.48$$
$$C_v = 0.40N_v = (0.4)(1.6) = 0.64$$

6-19 RESPONSE MODIFICATION FACTOR: R

Building a structure to respond 100% elastically in a large-magnitude earthquake would not be economical. Therefore, the prescribed design lateral strengths are considerably lower than needed to maintain a structure in the elastic range. This reduced design strength level results in nonlinear behavior and energy absorption at displacements in excess of initial yield. Strength reductions due to nonlinear behavior are influenced by the maximum allowable displacement ductility demand, the fundamental period of the system, and the soil-profile type. Strength reductions from the elastic strength are accomplished by using a response modification factor.

The structure *response modification factor*, R, represents the inherent overstrength and global ductility capacity of structural components.[26] *Ductility* can be defined as a measure of the ability of a structural system to deform in the plastic range prior to failure. Ductile performance is important because seismic energy

---

[26]The old 1994 UBC R_w value is approximately equal to 1.4 times the new 1997 R value. This new selection for the UBC-97 is for transition from service-level to strength-level design forces. The R factor is a partly empirical, partly judgmental factor that reduces the base shear to a predetermined value. This is the only judgment factor in the base shear equation.

is dissipated through yielding of the structural components, and because it permits considerable displacements during intense earthquakes without risk to the structure's integrity and the occupants' life safety. Reductions in design forces due to inherent overstrength increase the lateral strength of the structure from the design strength to the strength that is associated with the formation of the first plastic hinge. Reductions in design forces due to global ductility capacity increase the lateral strength of the structure from the strength that can be identified with the formation of the first plastic hinge to the strength associated with the formation of a mechanism.

Because all structures are designed for strengths less than would be needed in a completely elastic structure, the value of the response modification factor (R) always exceeds 1.0. Lightly damped structures constructed of brittle materials are assigned low values of R because they cannot support deformation in excess of initial yield. Highly damped structures constructed of ductile materials are assigned higher values of R.

The level of reduction specified in the UBC-97 seismic provisions is essentially based on the analysis of the historical performance of various structural systems in strong earthquakes. The structure response modification factor is determined from the type of structural system used in design of structures, as defined for buildings in Table 6.10 (equivalent to UBC-97 Table 16-N) and for nonbuilding structures in Table 6.15 (Sec. 6-46, equivalent to UBC-97 Table 16-P). Systems with higher ductility (e.g., steel moment-resisting frames) have higher R values associated with better seismic performance expectations.

6-20 SEISMIC FORCE AMPLIFICATION FACTOR: Ω_O

The type of structural system and natural period of a structure significantly influence the structure's response to ground shaking. In the event of a severe earthquake, code-compliant structures are expected to deform beyond their elastic load-carrying capacities due to effects of system overstrength. Overloading of nonductile elements of the structure can occur if the effects of overstrength are not accounted for in design. *Overstrength* is defined as a characteristic of structures where the actual strength is greater than the design strength. The degree of overstrength depends on material type and structural system type.

A *seismic force amplification factor*, or *overstrength factor* (Ω_O), has been assigned to each identified structural system. This factor accounts for overstrength of the structure in the inelastic range.

6-21 STRUCTURAL SYSTEMS

The UBC-97 [Sec. 1629.6 and Table 16-N] recognizes seven major types of structural systems capable of resisting lateral forces. In determining the base shear and design story drift for the structural systems mentioned, UBC-97 Table 16-N provides corresponding height limitations, the appropriate response modification coefficient (R), and the seismic force amplification factor (Ω_O). Nonbuilding structures are defined in the UBC-97 [Sec. 1629.6.8]. (See Sec. 6-46.)

A. Bearing Wall Systems

A *bearing wall system* is a structural system that relies on the same elements to resist both gravity and lateral loads. By itself, the word "wall" is ambiguous because there are two main types of structural walls. (Partition walls and curtain walls are not structural walls.) A *bearing wall* is designed and constructed to resist vertical (i.e., gravity) loads. A *shear wall* is designed and constructed to resist lateral loads. A wall can be used to resist both vertical and lateral loads. A bearing wall system does not have a complete vertical load-carrying space frame. Bearing walls or bracing systems support all of the gravity loads. Lateral forces are resisted by shear walls, light bracing in bearing walls, or braced frames (where the bracing also carries lateral load).

A bearing wall system lacks redundancy and has an inadequate inelastic response capacity. Such systems do not possess a complete vertical load-carrying frame and rely on walls or braced frames to carry the vertical (gravity) and lateral (seismic) loads.[27] The distinguishing factor of these systems is that the failure of the primary seismic system also compromises the ability of the structure to support its dead and live loads.

Typical bearing wall systems are: light-framed walls with shear panels, concrete or masonry shear walls, light steel-framed bearing walls with tension-only braces, and braced frames where the bracing carries gravity loads.

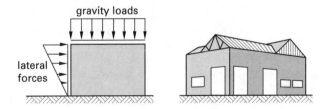

Figure 6.3   Bearing Wall System

---

[27]It is common to refer to this type of design as a *box system*.

Table 6.10
Structural Systems[1]
[UBC-97 Table 16-N]

basic structural system[2]	lateral force-resisting system description	R	Ω_O	height limit for seismic zones 3 and 4 (ft) (× 304.8 for mm)
1. bearing wall system	1. light-framed walls with shear panels			
	a. wood structural panel walls for structures three stories or less	5.5	2.8	65
	b. all other light-framed walls	4.5	2.8	65
	2. shear walls			
	a. concrete	4.5	2.8	160
	b. masonry	4.5	2.8	160
	3. light steel-framed bearing walls with tension-only bracing	2.8	2.2	65
	4. braced frames where bracing carries gravity load			
	a. steel	4.4	2.2	160
	b. concrete[3]	2.8	2.2	–
	c. heavy timber	2.8	2.2	65
2. building frame system	1. steel eccentrically braced frame (EBF)	7.0	2.8	240
	2. light-framed walls with shear panels			
	a. wood structural panel walls for structures three stories or less	6.5	2.8	65
	b. all other light-framed walls	5.0	2.8	65
	3. shear walls			
	a. concrete	5.5	2.8	240
	b. masonry	5.5	2.8	160
	4. ordinary braced frames			
	a. steel	5.6	2.2	160
	b. concrete[3]	5.6	2.2	–
	c. heavy timber	5.6	2.2	65
	5. special concentrically braced frames			
	a. steel	6.4	2.2	240
3. moment-resisting frame system	1. special moment-resisting frame (SMRF)			
	a. steel	8.5	2.8	N.L.
	b. concrete[4]	8.5	2.8	N.L.
	2. masonry moment-resisting wall frame (MMRWF)	6.5	2.8	160
	3. concrete intermediate moment-resisting frame (IMRF)[5]	5.5	2.8	–
	4. ordinary moment-resisting frame (OMRF)			
	a. steel[6]	4.5	2.8	160
	b. concrete[8]	3.5	2.8	–
	5. special truss moment frames of steel (STMF)	6.5	2.8	240
4. dual systems	1. shear walls			
	a. concrete with SMRF	8.5	2.8	N.L.
	b. concrete with steel OMRF	4.2	2.8	160
	c. concrete with concrete IMRF[5]	6.5	2.8	160
	d. masonry with SMRF	5.5	2.8	160
	e. masonry with steel OMRF	4.2	2.8	160
	f. masonry with concrete IMRF[3]	4.2	2.8	–
	g. masonry with masonry MMRWF	6.0	2.8	160
	2. steel EBF			
	a. with steel SMRF	8.5	2.8	N.L.
	b. with steel OMRF	4.2	2.8	160
	3. ordinary braced frames			
	a. steel with steel SMRF	6.5	2.8	N.L.
	b. steel with steel OMRF	4.2	2.8	160
	c. concrete with concrete SMRF[3]	6.5	2.8	–
	d. concrete with concrete IMRF[3]	4.2	2.8	–
	4. special concentrically braced frames			
	a. steel with steel SMRF	7.5	2.8	N.L.
	b. steel with steel OMRF	4.2	2.8	160
5. cantilevered column building systems	1. cantilevered column elements	2.2	2.0	35[7]
6. shear wall-frame interaction systems	1. concrete[8]	5.5	2.8	160
7. undefined systems	See UBC-97 Secs. 1629.6.7 and 1629.9.2.	–	–	–

N.L.—no limit. [1]See UBC-97 Sec. 1630.4 for combination of structural systems. [2]Basic structural systems are defined in UBC-97 Sec. 1629.6. [3]Prohibited in seismic zones 3 and 4. [4]Includes precast concrete conforming to UBC-97 Sec. 1921.2.7. [5]Prohibited in seismic zones 3 and 4, except as permitted in UBC-97 Sec. 1634.2. [6]Ordinary moment-resisting frames in seismic zone 1 meeting the requirements of UBC-97 Sec. 2214.4 may use an R value of 8.5. [7]Total height of the building including cantilevered columns. [8]Prohibited in seismic zones 2A, 2B, 3, and 4. See UBC-97 Sec. 1633.2.7.

B. Building Frame Systems

A *building frame* (*vertical load-carrying frame*) is a complete, self-contained, three-dimensional unit composed of interconnected members. *Building frame systems* use a complete space frame to carry the vertical (gravity) loads and a separate system of nonbearing shear walls or braced frames to resist the lateral (seismic) load.[28] Unlike the bearing wall system, failure of the primary lateral support system does not compromise the ability of the structure to support gravity loads.

The requirements of the "deformation compatibility" of UBC-97 [Sec. 1633.2.4] provision should be satisfied for these systems. A frame may or may not have bracing. If it does, it is known as a *braced frame*. A braced frame is a vertical truss system of interconnected members designed to resist lateral loads through the development of axial loads in the members. Braced frames can be of the concentric or eccentric types that are relatively rigid and require special detailing to ensure adequate ductile performance. Typical building frame systems are: steel eccentrically braced frames; light-framed walls with shear panels; concrete or masonry shear walls; steel, concrete, or heavy ordinary braced frames; and special steel concentrically braced frames.

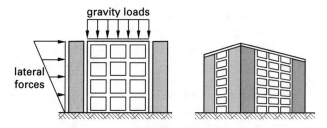

Figure 6.4 Building Frame System

C. Moment-Resisting Frames

Moment-resisting frames [UBC-97 Sec. 1629.6.4] resist forces in members and joints primarily by flexure and rely on a frame to carry both vertical and lateral loads. Lateral loads are carried primarily by flexure in the members and joints. Theoretically, joints are completely rigid. Moment-resisting frames can be constructed of concrete, masonry, or steel. There are five types of moment-resisting frames: steel and concrete special

moment-resisting frames (SMRF), masonry moment-resisting wall frames (MMRWF), concrete intermediate moment-resisting frames (IMRF), steel or concrete ordinary moment-resisting frames (OMRF), and special steel truss moment frames (STMF). These systems provide a sufficient degree of redundancy and have excellent inelastic response capacities.

Special moment-resisting frames are specially detailed to ensure ductile behavior and comply with Chap. 19 (concrete) or Chap. 22 (steel) of the UBC-97. *Intermediate moment-resisting frames*, concrete frames with less stringent requirements designed in accordance with UBC-97 Sec. 1921, cannot be used in seismic zones 3 and 4 [UBC-97 Table 16-N, Ftn. 5]. *Ordinary moment-resisting frames* are steel or concrete moment-resisting frames that do not meet the special detailing requirements for ductile behavior. Ordinary moment-resisting frames constructed of steel and concrete are restricted in use by UBC-97 Secs. 1921 and 2211.

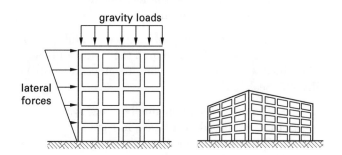

Figure 6.5 Moment-Resisting Frame System

D. Dual Systems

Dual Systems have essentially complete space frames that provide support for all vertical (gravity) loads and combine two of the previously mentioned systems to resist lateral loads. Moment-resisting frames (SMRF, IMRF, MMRWF, or steel OMRF) acting in conjunction with shear walls must be able to resist at least 25% of the design base shear independently. The two systems are designed to resist the total design base shear in proportion to their relative rigidities [UBC-97 Sec. 1629.6.5].[29]

---

[28]A *space frame* is a three-dimensional structural system, without bearing walls, consisting of interconnected members that operate as a single unit. A *frame* is a truss-like two-dimensional system with concentric or eccentric connection points in which the lateral forces are resisted by axial stresses in the members. A space frame functions with or without the aid of horizontal diaphragms or floor-bracing systems.

---

[29]Moment-resisting frames can be steel, concrete, or masonry, but concrete intermediate moment-resisting frames are prohibited in seismic zones 3 and 4, except as permitted in the UBC-97 Sec. 1634.2 [UBC-97 Table 16-N, Ftn. 5].

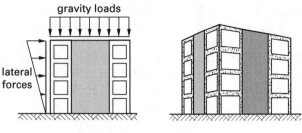

Figure 6.6 Dual System

E. Cantilevered Column Building Systems

Cantilevered column building systems have single cantilevered column elements supporting beams or framing at the top. These systems have a large portion of their mass concentrated near or at the top and are fixed at their bases. Design base shear is essentially applied at the top of the vertical base member. They are regarded as inverted pendulum-type structures since they extend from a fixed base and have zero moment restraint at the top. These systems have essentially one degree of freedom in horizontal translation.

Supporting columns or piers of inverted pendulum-type structures should be designed from the bending moment determined at the base according to procedures of the UBC-97 [Sec. 1630.2].

The cantilevered columns in this building system provide both lateral load resistance and gravity load resistance. These column elements have low redundancy and limited inelastic response capacity (energy dissipation). The potentially adverse effects that failure of the columns due to lateral forces will have on the gravity load-carrying capacity must be evaluated to determine stability. In seismic zones 3 and 4, the maximum height for these structures is 35 ft (10.7 m).

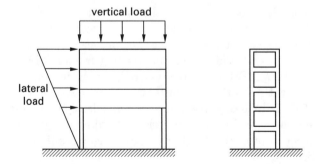

Figure 6.7 Cantilevered-Column Building System

F. Shear Wall-Frame Interaction Systems

To resist lateral forces, *shear wall-frame interaction systems* primarily use a combination of shear walls and moment frames. Building frames that are part of the lateral-force-resisting systems are required to be concrete frames. These systems are restricted to seismic zones 0 and 1 (zones of low seismicity).

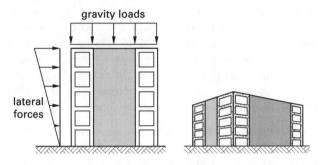

Figure 6.8 Shear Wall-Frame Interaction System

G. Undefined Systems

Undefined structural systems do not fit into any of these categories. The designer of such systems must submit a rational basis for the design force level used [UBC-97 Sec. 1629.6.7 and 1629.9.2].

6-22 RELIABILITY/REDUNDANCY FACTOR: ρ

Redundancy is an important characteristic of a structure, providing multiple paths of resistance (i.e., load paths). Higher redundancy in a structure implies better reliability. Inelastic action of a structure during a major seismic event can cause part of the structure to fail. For structures expected to experience severe inelastic demands, the lateral load-resisting system of the structure should be made as redundant as possible so that loads can be distributed to other lateral-force-resisting elements.

The *reliability/redundancy factor*, ρ, is applied as an increase in horizontal seismic forces associated with the base shear [UBC-97 Sec. 1630.1.1]. This factor effectively reduces the response modification factor, R, based on the extent of structural redundancy inherent in the design configuration of the structure and its lateral-force-resisting system. In addition to the number and distribution of vertical elements of the lateral-force-resisting system, the size of the ground floor area of the structure determines the value of ρ. The reliability/redundancy factor (ρ) value varies between 1.0 and 1.5. This factor can be calculated from Eq. 6.7 (equivalent to UBC-97 Formula 30-3).

$$\rho = 2 - \frac{20}{r_{\max}\sqrt{A_{B,\text{ft}^2}}} \qquad \text{[U.S.]} \quad \text{[6.7(a)]}$$

$$\rho = 2 - \frac{6.1}{r_{\max}\sqrt{A_{B,\text{m}^2}}} \qquad \text{[SI]} \quad \text{[6.7(b)]}$$

The UBC-97 defines the *element-story shear ratio* (r) as the ratio of the design story shear in the most heavily loaded single element divided by the total design story shear. The value of r depends on the structural system and can be taken as 1.0 in seismic zones 0, 1, and 2. r_{max} is the maximum element-story shear ratio in a given direction of loading that occurs in any of the story levels at or below two-thirds of the structure height. A_B is the ground floor area of the structure determined in ft^2 (m^2).

Observation of structural performance in Northridge, Kobe, and other large earthquakes has shown that structures with adequately redundant systems perform better than do structures with few lateral load-resisting elements. For this reason, the reliability/redundancy factor has been introduced to persuade engineers to design more highly redundant structures. In certain situations, it may be difficult to achieve a redundant design. In those cases, the magnitude of the inelastic response and the ductility demand should be reduced by increasing earthquake design loads by way of the ρ factor.

The reliability/redundancy factor is applied in the load combination equations rather than in the base shear equation because stiffness and drift control requirements are not directly influenced.

◇   ◇   ◇   ◇   ◇   ◇   ◇

Example 6.2

The plan view of a one-story office building in San Francisco is shown. The building has a wood structural panel roof diaphragm and shear walls. The building is 14 ft (4.3 m) tall. The shear walls equally resist the seismic shear. (a) What is the value of the response modification factor? (b) What is the value of the seismic force amplification factor? (c) What is the value of the reliability/redundancy factor?

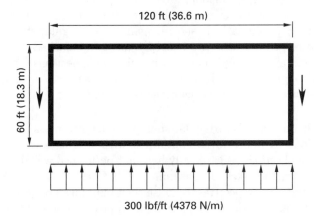

120 ft (36.6 m)

60 ft (18.3 m)

300 lbf/ft (4378 N/m)

Customary U.S. Solution

(a) This structure is a one-story, light, wood-framed bearing wall. Based on UBC-97 Table 16-N, for bearing wall systems with lateral-force-resisting elements that are wood structural panel walls and three stories or fewer in height, the value of the response modification factor (R) is 5.5.

(b) Overstrength as a characteristic of a structure occurs when the actual strength is greater than the design strength. The degree of overstrength greatly depends on structural material and system. According to UBC-97 Table 16-N, the value of the seismic force amplification (overstrength) factor, Ω_O, is 2.8.

(c) The UBC-97 requires that all structures be assigned a reliability/redundancy factor that accounts for the number of lateral-force-resisting elements, ground floor area of the building, and distribution of lateral forces to lateral-force-resisting elements. This factor can be calculated from Eq. 6.7 (equivalent to UBC-97 Formula 30-3).

$$\rho = 2 - \frac{20}{r_{max}\sqrt{A_{B,\text{ft}^2}}}$$

Based on the UBC-97 [Sec. 1630.1.1], the element-story shear ratio (r) is the ratio of the design story shear in the most heavily loaded single element divided by the total design story shear. For shear walls,

$$r_{max} = \frac{(\text{maximum wall shear})\left(\dfrac{10}{l_{w,\text{ft}}}\right)}{\text{total story shear}}$$

For this building, the shear walls equally resist the seismic shear forces. Thus,

$$r_{max} = (0.5)\left(\frac{10}{l_w}\right)$$
$$= (0.5)\left(\frac{10}{60\text{ ft}}\right)$$
$$= 0.0833\text{ ft}^{-1}$$

$$A_B = (60\text{ ft})(120\text{ ft})$$
$$= 7200\text{ ft}^2$$

$$\rho = 2 - \frac{20}{(0.0833\text{ ft}^{-1})\sqrt{7200\text{ ft}^2}}$$
$$= 2 - 2.83$$
$$= -0.83$$

However, ρ should not be less than 1.0 and need not be greater than 1.5 ($1.0 \leq \rho \leq 1.5$). Therefore, $\rho_{min} = 1.0$.

SI Solution

(a) This structure is a one-story, light, wood-framed bearing wall. Based on UBC-97 Table 16-N, for bearing wall systems with lateral-force-resisting elements which are wood structural panel walls and three stories or less in height, the value of the response modification factor (R) is 5.5.

(b) Overstrength as a characteristic of a structure occurs when the actual strength is greater than the design strength. The degree of overstrength greatly depends on structural material and system. According the UBC-97 Table 16-N, the value of the seismic force amplification (overstrength) factor, Ω_O, is 2.8.

(c) The UBC-97 requires that all structures be assigned a reliability/redundancy factor that accounts for the number of lateral-force-resisting elements, ground floor area of building, and distribution of lateral forces to lateral-force-resisting elements. This factor can be calculated from Eq. 6.7 (equivalent to UBC-97 Formula 30-3).

$$\rho = 2 - \frac{6.1}{r_{\max}\sqrt{A_{B,\text{m}^2}}}$$

Based on the UBC-97 [Sec. 1630.1.1], the element-story shear ratio (r) is the ratio of the design story shear in the most heavily loaded single element divided by the total design story shear. For shear walls,

$$r_{\max} = \frac{(\text{maximum wall shear})\left(\dfrac{3.05}{l_{w,\text{m}}}\right)}{\text{total story shear}}$$

For this building, the shear walls equally resist the seismic shear forces. Thus,

$$r_{\max} = (0.5)\left(\frac{3.05}{l_{w,\text{m}}}\right)$$
$$= (0.5)\left(\frac{3.05}{18.3\text{ m}}\right)$$
$$= 0.0833\text{ m}^{-1}$$

$$A_B = (18.3\text{ m})(36.6\text{ m})$$
$$= 669.78\text{ m}^2$$

$$\rho = 2 - \frac{6.1}{(0.0833\text{ m}^{-1})\sqrt{669.78\text{ m}^2}}$$
$$= 2 - 2.83$$
$$= -0.83$$

However, ρ should not be less than 1.0 and need not be greater than 1.5 ($1.0 \le \rho \le 1.5$). Therefore, $\rho_{\min} = 1.0$.

◇   ◇   ◇   ◇   ◇   ◇   ◇

6-23 REGULAR/IRREGULAR STRUCTURES

In designing a structure, selection of the structure's basic plan, shape, and configuration is a critical step. The decision will influence the ability of the structure to withstand earthquake ground shaking. While configuration cannot be presumed to be the sole reason for building inadequacy, it is usually a significant contributor. Previous earthquake performance of buildings clearly illustrates that all other parameters being identical, the simpler the building is, the better its seismic performance will be.

A structure is an assemblage of framing members designed to support vertical (gravity) loads and resist lateral forces. For building structures, there are many types and configurations, and a dominant ideal configuration for any particular type of building structure does not exist. Despite the relationship between configuration and seismic performance, it wasn't until the 1973 edition of the UBC that building *configuration* was addressed in a specific provision. In the UBC-97, Sec. 1629.5 provides guidelines for classifying building configuration for a structure. In Sec. 1629.5, structures are classified as either *structurally regular* or *structurally irregular*.

Regular structures in accordance with the UBC-97 [Sec. 1629.5.2] are described as structures having no significant physical discontinuities in plan or vertical configuration or in their force-resisting systems. Based on the UBC-97 [Sec. 1629.5.3], *irregular structures* are defined as structures having significant physical discontinuities in configuration or in their lateral-force-resisting systems.

Regular structures have a uniform and continuous distribution of mass, stiffness, strength, and ductility with no significant torsional forces or large height-width ratio or large changes in plan area from floor to floor. They have relatively shorter spans than irregular structures, simple structural subsystems, and balanced stiffness and strength between members, connections, and supports. The code permits static analysis for a regular structure with few exceptions (see Sec. 6-33).

Structures with irregular shapes, changes in mass from floor to floor, variable stiffness with height, and unusual setbacks historically have not performed well during earthquakes. This is unfortunate from aesthetic and creative perspectives, because such buildings are generally among the most pleasing in appearance and interesting to design. With few exceptions (see Sec. 6-33), the code specifies that a dynamic analysis is required for an irregular structure.[30]

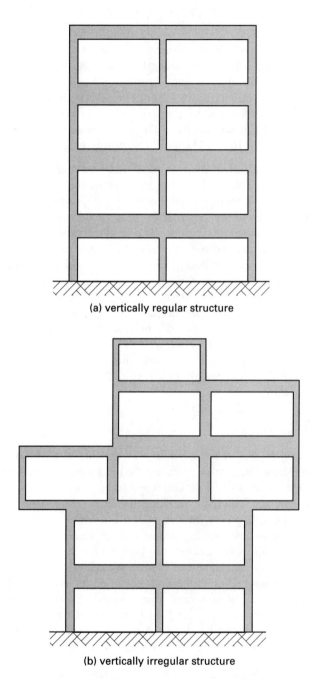

(a) vertically regular structure

(b) vertically irregular structure

Figure 6.9　Structural Configuration

To identify UBC-97 requirements affecting irregular structures, Table 6.11 (equivalent to UBC-97 Tables 16-L and 16-M, combined) lists major physical discontinuity types along with corresponding UBC-97 reference sections.

Table 6.11

Plan and Vertical Structural Irregularities
[UBC-97 Tables 16-L and 16-M]

structural irregularity	physical irregularity type	UBC-97 reference section
vertical	stiffness irregularity— soft story	1629.8.4, Item 2
	weight (mass) irregularity	1629.8.4, Item 2
	geometric irregularity	1629.8.4, Item 2
	in-plane discontinuity	1630.8.2
	discontinuity in capacity— weak story	1629.9.1
plan	torsional irregularity	1633.2.9, Item 6
	reentrant corners	1633.2.9, Items 6 and 7
	diaphragm discontinuity	1633.2.9, Item 6
	out-of-plane offsets	1630.8.2; 1633.2.9, Item 6; 2213.8
	nonparallel systems	1633.1

Reproduced from the 1997 edition of the *Uniform Building Code*, copyright © 1997, with the permission of the publisher, the International Conference of Building Officials.

6-24 VERTICAL STRUCTURAL IRREGULARITIES

UBC-97 Table 16-L lists and defines the following five types of *vertical structural irregularities*.

1. A *soft story* has a stiffness less than 70% of the story immediately above, or less that 80% of the average stiffness of the three stories above. (The requirement of UBC-97 [Sec. 1629.8.4, Item 2] should be considered.)

2. A story has a *mass (weight) irregularity* when its mass is more than 150% of the effective mass of a story above or below. (Roofs lighter than the floor immediately below are excluded.) (The requirement of UBC-97 [Sec. 1629.8.4, Item 2] should be considered.)

---

[30]Since the code is specific, some of the design criteria were set primarily by judgment, as neither historical nor empirical data were available. Thus, the criteria are somewhat "loose," and the intent is to penalize only the worst cases of structural irregularity and to allow less severe structural configurations.

3. A story has *vertical geometric irregularity* when the horizontal dimension of a story's lateral force-resisting system is more than 130% of that in an adjacent story. (One-story penthouses are excluded.) (The requirement of UBC-97 [Sec. 1629.8.4, Item 2] should be considered.)

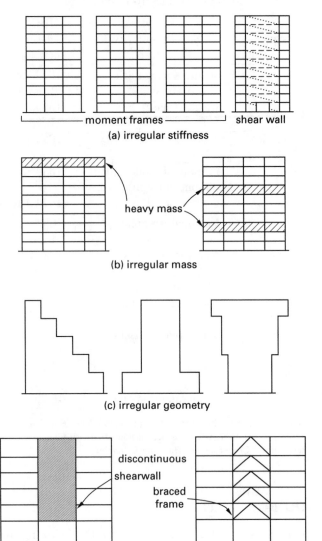

(a) irregular stiffness

(b) irregular mass

(c) irregular geometry

(d) in-plane discontinuity

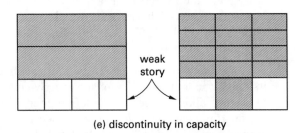

(e) discontinuity in capacity

Figure 6.10 Vertical Irregularities

4. An *in-plane discontinuity* exists at a story when there is an in-plane offset of the lateral load-resisting elements greater than the length of those elements. (The requirement of UBC-97 [Sec. 1630.8.2] should be considered.)

5. A *weak story* (discontinuity in capacity) exists when the story strength is less than 80% of that in the story above. The *story strength* is defined as the strength of all of the seismic resisting elements sharing the story shear in the direction of the earthquake. (The requirement of UBC-97 [Sec. 1629.9.1] should be considered.)

A structure that meets one of the five conditions of UBC-97 Table 16-L (corresponding to the five conditions listed above) would normally be considered irregular. However, a structure that would otherwise be considered irregular under the first two conditions listed can be considered regular if the story drift ratio (as calculated from the design lateral forces and neglecting torsional effects) for each floor is less than 1.3 times the story drift ratio for the floor above. (The story drift ratio is the ratio of the actual drift relative to the floor below and the floor-to-floor height.) Furthermore, this condition does not even have to be satisfied by the top two stories as long as all stories below the top two satisfy it [UBC-97 Sec. 1629.5.3, Item 2].

Example 6.3

Two 150-ft (45.7-m) building structures in California are shown in elevation. Structure I has a reinforced concrete shear wall for lateral loads in one direction and a special steel moment-resisting frame system in the orthogonal direction. Structure II has a combined special steel moment-resisting frame with reinforced concrete shear wall in each direction, but the shear wall is discontinuous at the second floor. (a) What is the applicable response modification factor, R, for structure I? (b) What is the applicable response modification factor, R, for structure II? (c) Which, if any, of the building structures would be classified as irregular by the UBC-97 and what are the implications?

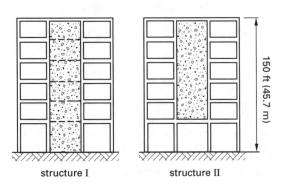

structure I   structure II

Solution

(a) Structure I is a building frame system with concrete shear walls in one direction and a special steel moment-resisting frame in the orthogonal direction. Where combinations along different axes exist, the UBC-97 [Sec. 1630.4.3] permits any combination of lateral systems for buildings less than 160 ft in height, and restricts R only for systems in combination with a bearing wall system in one direction. Based on UBC-97 Table 16-N,

> For a building frame system with concrete shear walls, $R = 5.5$.
> For special steel moment-resisting frame, $R = 8.5$.

Thus, R should be 5.5 for the elevation shown, and $R = 8.5$ should be used in the orthogonal direction.

(b) Structure II is a dual system in both directions. Where combinations along the same axes exist, the UBC-97 [Sec. 1630.4.4] requires that the value of seismic modification factor R for design in that direction not be greater than the least value for any of the structural systems utilized in that same direction. Based on UBC-97 Table 16-N,

> For reinforced concrete shear walls, $R = 4.5$
> For special steel moment-resisting frame, $R = 8.5$.

Thus, R should not be greater than 4.5 (least value of R for any of the structural systems), and the combined value of the seismic modification factor $R = 4.5$ should be used.

(c) Based on the UBC-97 [Sec. 1629.5.3], irregular structures bear significant physical discontinuities in configuration or in the lateral-force-resisting systems. Structure I is a regular structure based on given information and visual observation, whereas structure II has vertical irregularities. The observed irregularities are stiffness (soft story) and in-plane discontinuity in vertical lateral-force-resisting elements.

By inspection, the first story of structure II is considered a soft story. Based on the UBC-97 [Sec. 1629.8.4., Item 2], a soft story's lateral stiffness is less than 70% of that in the story above or less than 80% of the average stiffness of the three stories above. Also, it can be seen that the vertical lateral-force-resisting system is discontinuous. According to UBC-97 Table 16-L, this irregularity is categorized as an in-plane discontinuity in vertical lateral-resisting elements. The UBC-97 [Sec. 1630.8.2] requires concrete, masonry, steel, and wood elements (existing shown columns) supporting such discontinuous systems to have the design strength to resist the special seismic load combinations [UBC-97 Sec. 1612.4].

◇   ◇   ◇   ◇   ◇   ◇   ◇

6-25 PLAN STRUCTURAL IRREGULARITIES

UBC-97 Table 16-M lists and defines the following five types of *plan structural irregularities*.

1. *Torsional irregularity* exists when the maximum story drift (caused by the lateral load *and* the accidental torsion) at one end of the structure transverse to its axis is more than 1.2 times the average story drifts calculated from both ends. Only buildings with rigid diaphragms are affected by this type of irregularity.[31] (The requirement of UBC-97 [Sec. 1633.2.9, Item 6] applies here.)

$$\Delta_2 > 1.2 \left(\frac{\Delta_1 + \Delta_2}{2} \right)$$

The UBC-97 [Sec. 1633.1] specifies that earthquake forces should be considered from any direction other than the principal axes if torsional irregularity exists for both major axes.

2. A building has *reentrant corner irregularity* when one or more parts of the structure project beyond a reentrant corner a distance greater than 15% of the plan dimension in the given direction. (The requirement of UBC-97 [Sec. 1633.2.9, Items 6 and 7] applies here.)

> projecting wing $D_2 > 0.15D_1$
> projecting wing $L_2 > 0.15L_1$

3. *Diaphragm discontinuity* occurs with diaphragms having abrupt discontinuities or variations in stiffness, including when there are cutout, or open, areas greater than 50% of the gross diaphragm area, or when the stiffness of the diaphragm changes more than 50% from story to adjacent story. (The requirement of UBC-97 [Sec. 1633.2.9, Item 6] applies here.)

> opening area $> 0.50LD$

4. An *out-of-plane offset* is a discontinuity in the lateral force path—an out-of-plane offset of the vertical elements. (The requirements of UBC-97 [Secs. 1630.8.2; 1633.2.9, Item 6; and 2213.8] apply here.)

---

[31] As defined in UBC-97 Sec. 1630.6, a *flexible diaphragm* is one that has a maximum lateral deflection at a story more than two times the average story drift at that story.

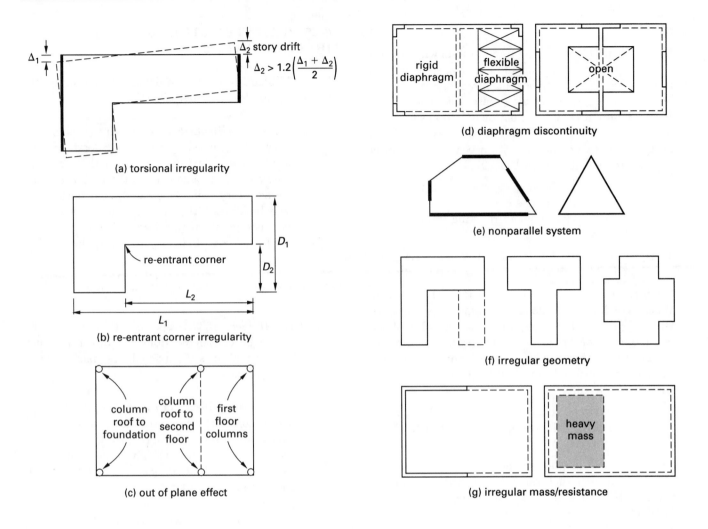

Figure 6.11　Plan Irregularities

5. A *nonparallel system* is one for which the vertical load-carrying elements are not parallel to or symmetrical about the major orthogonal axes of the lateral force-resisting system. This includes buildings in which a column is part of two or more intersecting lateral force-resisting systems (as would most likely occur in a corner column at the ends of two orthogonal frames), unless the axial column load due to seismic forces is less than 20% of the column allowable load. (The requirement of UBC-97 [Sec. 1633.1] applies here.)

6-26 BUILDING PERIOD: *T*

The UBC-97 gives two methods for determining the *building period*, *T*. The first is an approximate method

that implies that the natural period increases as the height of the structure increases (see Sec. 3-8). Equation 6.8 (corresponding to UBC-97 Formula 30-8) can be used for all buildings.[32] In Eq. 6.8 (known as *Method A*), h_n is the actual height (feet or meters) of the building above the base to the nth level. (The maximum allowable heights of the various types of structural systems are summarized in Table 6.10, Sec. 6-21 of this book, corresponding to UBC-97 Table 16-N.)[33]

---

[32]This first method will probably be used for almost all preliminary designs and many final designs.

[33]The 1988 Blue Book recommended that masonry bearing wall systems in zones 3 and 4 be limited to 120 ft (36.6 m). The 1988 UBC set the limit at 160 ft (49 m). Thus, it is apparent that height limits are somewhat subjective. The discrepancy was eliminated in the 1990 Blue Book.

C_t is 0.035 (0.0853) for steel moment-resisting frames, 0.030 (0.0731) for reinforced concrete moment-resisting frames and eccentrically braced frames (see Sec. 7-8), and 0.020 (0.0488) for all other buildings.[34]

$$T = C_t(h_n)^{3/4} \qquad [6.8]$$

The UBC-97 [Sec. 1630.2.2, Item 1] contains an alternate method to be used in finding C_t for structures with concrete or masonry shear walls.[35]

$$C_t = \frac{0.1}{\sqrt{A_{c,\text{ft}^2}}} \qquad \text{[U.S.]} \quad [6.9(a)]$$

$$C_t = \frac{0.0743}{\sqrt{A_{c,\text{m}^2}}} \qquad \text{[SI]} \quad [6.9(b)]$$

A_c is the combined effective area of the shear walls in the first story of the structure in square feet (square meters). The value of A_c can be obtained from Eq. 6.10 (equivalent to UBC-97 Formula 30-9).

$$A_c = \sum A_e \left[0.2 + \left(\frac{D_e}{h_n} \right)^2 \right] \qquad [6.10]$$

A_e is the horizontal cross-sectional area of a shear wall in the first story, and D_e is the length of a shear wall in the first story in the direction parallel to the applied forces. In Eq. 6.10, the value of D_e/h_n should be equal to or less than 0.9.

Another method for finding the period, T, known as *Method B*, is based on the deformation characteristics of the resisting elements and is a more rational determination. The UBC-97 Method B (UBC-97 Formula 30-10) is referred to by some engineers as the *Rayleigh method*. In addition to the UBC-97 method of determining the period, any other substantiated analysis method can be used. If a dynamic analysis is performed, the first modal period should be used for T.

If Method B is used to find the period, T, the UBC-97 [Sec. 1630.2.2, Item 2] requires that the value of T cannot be more than 30% greater than the value of T determined from the empirical period given by Eq. 6.8 (i.e., method A) in seismic zone 4 and 40% in seismic zones 1, 2, and 3.

---

[34]Most of the data supporting these values came from the 1971 San Fernando earthquake.

[35]When Eq. 6.8 is used with shear wall structures, the base shear formula may be overly conservative in the lower seismic zones.

Example 6.4

In designing a 72-ft (21.95-m) steel frame structure with cast-in-place concrete shear walls, the natural period is calculated to be 0.8 sec using the UBC-97's rational analysis. Utilizing the UBC-97's approximate method, what is the natural period for this building?

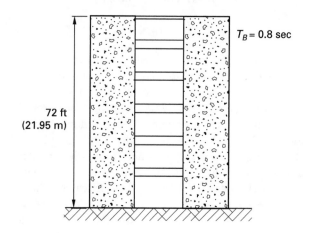

Customary U.S. Solution

The approximate formula (Method A) is given in the UBC-97 [Sec. 1630.2.2, Item 1]. UBC-97 Formula 30-8 is

$$T = C_t(h_n)^{3/4}$$

C_t is 0.020 for the building structure shown, and h_n is the building height.

$$T = (0.020)(72 \text{ ft})^{3/4}$$
$$= 0.49 \text{ sec}$$

SI Solution

The approximate formula (Method A) is given in the UBC-97 [Sec. 1630.2.2, Item 1]. The UBC-97 Formula 30-8 is

$$T = C_t(h_n)^{3/4}$$

C_t is 0.0488 for the building structure shown, and h_n is the building height.

$$T = (0.0488)(21.95 \text{ m})^{3/4}$$
$$= 0.49 \text{ s}$$

6-27 EARTHQUAKE LOADS: E_h AND E_v

In designing structures, the load effects of the horizontal and vertical components of the earthquake ground motion should be considered. The earthquake load is a function of horizontal and vertical seismic-induced forces. This load on an element of the structure is denoted as E and should be determined from Eq. 6.11 (equivalent to UBC-97 Formula 30-1).

$$E = \rho E_h + E_v \qquad [6.11]$$

E_h represents the forces associated with the horizontal component of the earthquake load. Horizontal earthquake forces are due to the base shear (V) or the design lateral force $(F_p$ or $F_x)$. E_v represents loads resulting from the vertical component of the earthquake ground motion. Vertical earthquake effect may not be considered $(E_v = 0)$ if the allowable stress design (ASD) method is used for proportioning structural elements. For strength design (LRFD), the vertical earthquake effect is equal to an increase of $0.5C_a I D$ over the dead load effect D, where C_a and I are the near-source and importance factors respectively, and ρ is the redundancy/reliability factor.

Based on the UBC-97 [Sec. 1630.1.1], the estimated maximum earthquake force (E_m) that can be developed in a structure should be determined from Eq. 6.12 (equivalent to UBC-97 Formula 30-2).

$$E_m = \Omega_O E_h \qquad [6.12]$$

Ω_O, as defined in the UBC-97 [Sec. 1630.3.1], represents the seismic force amplification factor that is required to account for structural overstrength. The seismic force amplification factor provides an upper bound approximation of actual seismic load acting on a structure with inelastic response capacity.

6-28 LOAD COMBINATIONS

For a structure, the basic contributing design loads are *floor live load* (L), *roof live load* (L_r), *dead load* (D), *earthquake load* (E), *estimated maximum earthquake force* (E_m), *snow load* (S), and *wind load* (W).[36] These loads are referred to as the *design loads*. The UBC-97 requires members to be designed for the most critical and unfavorable combination of loads. Thus, all combinations must be checked to find the controlling load combination.

The fundamental change from working stress (ASD) to strength design (LRFD) principles in the UBC-97 has resulted in a new load combination section of the UBC-97 [Sec. 1612]. The equations provided are consistent with 1997 NEHRP provisions. Basic factored load combinations and other loads are given in the UBC-97 for both strength design (load and resistance factor) and allowable stress design (working stress) [Secs. 1612.2.1 and 1612.3.1, respectively]. Both formats are given to accommodate the preferences of engineers to work with one or the other design method.

Beside providing basic load combinations, the UBC-97 permits that the alternate basic factored load combinations be used in designing structures and portions for the most critical effects resulting from the certain load combinations. In addition, for both allowable stress and strength design methods, the UBC-97 requires the use of special load combinations for seismic design of certain elements. The intent in specifying the special seismic load combinations is to cover conditions where ductility is lacking. Table 6.12 contains the basic, alternate, and special seismic load combinations using allowable stress and/or strength design methods as applicable and UBC-97-appropriate reference sections. The design loads are multiplied by appropriate load factors as given in Table 6.12. A load factor of 1.0 implies that the specified forces are at design levels and require no further amplification.

The parameter E_m in the special seismic load combinations represents the maximum earthquake force that can be developed in the structure based on UBC-97 Sec. 1630.1.1. It should be noted that for strength design, (1) for masonry and concrete elements, the factored basic loading combinations should be increased by a factor of 1.1 where load combinations include the seismic forces, and (2) for concrete elements, UBC-97 Sec. 1909.2 provides the appropriate load combinations for combinations with no seismic component.

---

[36]The other loads are: load due to fluids (F); load due to lateral pressure of soil and water in soil (H); ponding load (P); self-straining force and effects arising from contraction or expansion resulting from temperature change, shrinkage, or moisture change; creep in component materials; movement due to different settlement; or combinations thereof (T). The applicable load factors are given in the UBC-97 [Secs. 1612.2.2 and 1612.3.3].

Table 6.12
Combinations of Loads
[UBC-97 Sec. 1612.3.2]

design method	combination of loads	UBC-97 section	UBC-97 formula
basic load combinations			
LRFD	$1.4D$		(12-1)
	$1.2D + 1.6L + 0.5(L_r \text{ or } S)$		(12-2)
	$1.2D + 1.6(L_r \text{ or } S) + (f_1L \text{ or } 0.8W)$	1612.2.1	(12-3)
	$1.2D + 1.3W + f_1L + 0.5(L_r \text{ or } S)$		(12-4)
	$1.2D + 1.0E + (f_1L + f_2S)$		(12-5)
	$0.9D \pm (1.0E \text{ or } 1.3W)$		(12-6)
ASD (no 1.33 allowance for stress increase)	D		(12-7)
	$D + L + (L_r \text{ or } S)$		(12-8)
	$D + \left(W \text{ or } \dfrac{E}{1.4}\right)$	1612.3.1	(12-9)
	$0.9D \pm \dfrac{E}{14}$		(12-10)
	$D + 0.75\left[L + (L_r \text{ or } S) + \left(W \text{ or } \dfrac{E}{1.4}\right)\right]$		(12-11)
alternate basic load combinations for ASD			
ASD (1.33 allowance for stress increase permitted)	$D + L + (L_r \text{ or } S)$		(12-12)
	$D + L + \left(W \text{ or } \dfrac{E}{1.4}\right)$		(12-13)
	$D + L + W + \dfrac{S}{2}$	1612.3.2	(12-14)
	$D + L + S + \dfrac{W}{2}$		(12-15)
	$D + L + S + \dfrac{E}{1.4}$		(12-16)
	$0.9D \pm \dfrac{E}{1.4}$		(12-16-1)
special seismic load combinations			
LRFD & ASD	$1.2D + f_1L + 1.0E_m$	1612.4	(12-17)
	$0.9D \pm 1.0E_m$		(12-18)

Example 6.5

A special steel moment-resisting frame is located in seismic zone 3 of California. This site is not near active seismic sources, and its soil profile is S_B. Assume that the importance factor (I) and reliability/redundancy factor (ρ) are both equal to 1.0. (a) What is the earthquake load effect E_h on column A in the axial direction due to design base shear (i.e., resulting from the horizontal component of the earthquake ground motion)? (b) What is the load effect of the vertical component of the earthquake ground motion (i.e., vertical acceleration effect E_v) on the column? (c) What is the earthquake load effect, E, on one column in the axial direction resulting from the vertical and horizontal components of the earthquake ground motion? (d) What is the estimated maximum earthquake force that can be developed in the structure at the column?

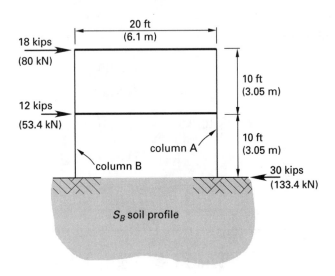

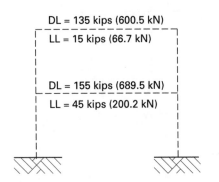

Customary U.S. Solution

(a) The earthquake load effect of the horizontal acceleration (design base shear) is found by summing moments about column B's footing.

$$E_h = \frac{\Sigma F_x h_x}{b}$$

$$= \frac{(18\text{ kips})(20\text{ ft}) + (12\text{ kips})(10\text{ ft})}{20\text{ ft}}$$

$$= 24\text{ kips}$$

In consideration of the structural redundancy in seismic zone 3, E_h will be

$$\rho E_h = (1.0)(24\text{ kips})$$

$$= 24\text{ kips}$$

(b) The earthquake load effect of the vertical acceleration is

$$E_v = 0.5 C_a I D$$

With soil-profile type S_B and seismic zone 3 ($Z = 0.3$), the seismic response coefficient C_a can be obtained from Table 6.9 (equivalent to UBC-97 Table 16-Q). Thus, $C_a = 0.30$.

$$D = \frac{135\text{ kips} + 155\text{ kips}}{2}$$

$$= 145\text{ kips}$$

For strength design, E_v is

$$E_v = (0.5)(0.30)(1.0)(145\text{ kips})$$

$$= 21.75\text{ kips}$$

However, for allowable stress design, $E_v = 0$ based on the UBC-97 [Sec. 1630.1.1].

(c) The earthquake load (E) is a function of horizontal and vertical seismic-induced forces and can be determined from Eq. 6.11 (equivalent to UBC-97 Formula 30-1).

$$E = \rho E_h + E_v$$

For strength design, E is

$$E = 24\text{ kips} + 21.75\text{ kips}$$

$$= 45.75\text{ kips}$$

For allowable stress design, E is

$$E = \frac{E_h}{1.4} = \frac{24\text{ kips}}{1.4}$$

$$= 17.14\text{ kips}$$

(d) The estimated maximum earthquake force can be computed from Eq. 6.12 (equivalent to UBC-97 Formula 30-2).

$$E_m = \Omega_O E_h$$

The earthquake force amplification factor Ω_O can be obtained from Table 6.10 (equivalent to UBC-97 Table 16-N). For special steel moment-resisting frame, Ω_O is equal to 2.8.

$$E_m = (2.8)(24 \text{ kips})$$
$$= 67.20 \text{ kips}$$

SI Solution

(a) The earthquake load effect of the horizontal acceleration (design base shear) is

$$E_h = \frac{\Sigma F_x h_x}{b}$$
$$= \frac{(80 \text{ kN})(6.1 \text{ m}) + (53.4 \text{ kN})(3.05 \text{ m})}{6.1 \text{ m}}$$
$$= 106.7 \text{ kN}$$

In consideration of the structural redundancy in seismic zone 3, ρE_h will be

$$\rho E_h = (1.0)(106.70 \text{ kN})$$
$$= 106.7 \text{ kN}$$

(b) The earthquake load effect of the vertical acceleration is

$$E_v = 0.5 C_a I D$$

With soil-profile type S_B in seismic zone 3 ($Z = 0.3$), the seismic response coefficient C_a can be obtained from Table 6.9 (equivalent to UBC-97 Table 16-Q). Thus, $C_a = 0.30$.

$$D = \frac{600.5 \text{ kN} + 689.5 \text{ kN}}{2}$$
$$= 645 \text{ kN}$$

For strength design, E_v is

$$E_v = (0.5)(0.30)(1.0)(645 \text{ kN})$$
$$= 96.75 \text{ kN}$$

However, for allowable stress design, $E_v = 0$ based on the UBC-97 [Sec. 1630.1.1].

(c) The earthquake load (E) is a function of horizontal and vertical seismic-induced forces and can be determined from Eq. 6.11 (equivalent to UBC-97 Formula 30-1).

$$E = \rho E_h + E_v$$

For strength design, E is

$$E = 106.7 \text{ kN} + 96.75 \text{ kN}$$
$$= 203.45 \text{ kN}$$

For allowable stress design, E is

$$E = \frac{E_h}{1.4} = \frac{106.7 \text{ kN}}{1.4}$$
$$= 76.21 \text{ kN}$$

It should be noted that simultaneous maximum response to horizontal and vertical accelerations should be regarded as improbable.

(d) The estimated maximum earthquake force can be computed from Eq. 6.12 (equivalent to UBC-97 Formula 30-2).

$$E_m = \Omega_O E_h$$

The earthquake force amplification factor Ω_O, which provides a reasonable approximation of actual forces acting in an inelastically responding structure, can be obtained from Table 6.10 (equivalent to UBC-97 Table 16-N). For special steel moment-resisting frame, Ω_O is equal to 2.8.

$$E_m = (2.8)(106.7 \text{ kN})$$
$$= 298.76 \text{ kN}$$

◇   ◇   ◇   ◇   ◇   ◇   ◇

6-29 TOTAL SEISMIC DEAD LOAD: *W*

The weight, W (in pounds or newtons), used to calculate base shears and building periods is normally the total seismic dead load of the structure.[37] This includes the weight of the ceiling, partitions, pipes, ducts, and equipment that are normally attached. W does not include full design roof and live loads. The objective in summing up seismic weight W is to include all contributions to mass likely to be present at the time of an earthquake. However, applicable portions of other loads should be included as follows [UBC-97 Sec. 1630.1.1].

◇ A minimum of 25% of the floor live load (i.e., storage) is added in warehouses and storage buildings [UBC-97 Sec. 1630.1.1, Item 1].

---

[37]As part of normal practice, all of the foundation weight and half of the first-story wall weight are commonly omitted in the analysis of seismic diaphragm loads. (See Sec. 6-34.) However, such a provision is not explicitly defined in the UBC-97 for calculating base shear, nor is such a provision mentioned in the Blue Book commentary.

⋄ No less than 10 lbf/ft² (0.48 kN/m²) must be added when partition loads are used in the design of the floor [UBC-97 Sec. 1630.1.1, Item 2].

⋄ Design snow loads exceeding 30 lbf/ft² (1.44 kN/m²) must be included, but may be reduced by up to 75% if approved by the local building official [UBC-97 Sec. 1630.1.1, Item 3].

⋄ The total weight of permanent equipment must be included [UBC-97 Sec. 1630.1.1, Item 4].

The units of seismic dead load, W, will determine the units of base shear, V. Thus, if weight is expressed in kips, the base shear will be in kips. Mass is seldom, if ever, used in practice, as all weights are given in pounds or kips. However, care must be taken in the unlikely event that the structure mass, as opposed to the structure weight, is specified. Equation 6.13 shows that the weight in pounds-force (lbf) is numerically the same as the mass in pounds-mass (lbm), although this is not true if other units of mass are used.

$$W_{\text{lbf}} = m_{\text{lbm}} \left(\frac{g}{g_c} \right) \qquad [6.13]$$

$$W_{\text{lbf}} = m_{\text{slug}} g \qquad [6.14]$$
$$g = 32.2 \text{ ft/sec}^2$$
$$g_c = 32.2 \text{ ft-lbm/lbf-sec}^2$$
$$W_{\text{newtons}} = m_{\text{kg}} g \qquad [6.15]$$
$$g = 9.81 \text{ m/s}^2$$

For design of a multilevel structure, the total seismic dead load W is the sum of the seismic dead loads of all levels.

$$W = \sum_{x=1}^{n} W_x \qquad [6.16]$$

6-30 SEISMIC BASE SHEAR COEFFICIENT

The formula $V = C_s W$, found in the 1997 NEHRP, determines the seismic base shear in a given direction. The seismic *base shear coefficient* (C_s) is a function of the natural period of structure (T) and the soil profile type and represents a point on the design response spectra, which is defined by the site-specific values of C_a and C_v. From UBC-97 design base shear Formula 30-4, the corresponding expression for C_s may be obtained.

$$C_s = \frac{C_v I}{RT} \qquad [6.17]$$

Equation 6.17 can be used for structures having a longer (velocity response) period. Base shear values will be greater for structures having natural periods of less than $T_s = C_v/2.5C_a$, whereas the natural period greater than T_s will result in a lower value. Coefficients C_a and C_v, which are site-dependent seismic response coefficients, are given in UBC-97 Tables 16-Q and 16-R.

For shorter (acceleration response) period structures, the seismic base shear coefficient is

$$C_s = \frac{2.5C_a I}{R} \qquad [6.18]$$

In addition to the natural period of the structure, T, the importance factor, I, and response modification factor, R, influence the value of the seismic base shear coefficient. An increase in the value of I increases the seismic base shear coefficient, whereas an increase in the value of R decreases it.

$$\frac{C_v I}{RT} \qquad\qquad \frac{2.5C_a I}{R}$$
velocity-controlled < acceleration-controlled
seismic   seismic
base shear coefficient   base shear coefficient

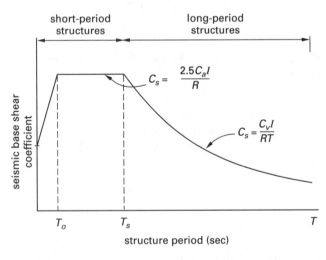

Figure 6.12 Minimum Design Base Shear Coefficient

From UBC-97 [Sec. 1630.2.1] and Formula 30-5, the maximum value of the seismic base shear coefficient is

$$C_{s,\text{max}} = \frac{2.5C_a I}{R} \qquad [6.19]$$

In Eq. 6.19, in all seismic zones for short period structures, C_a represents effective peak acceleration at grade with a maximum natural period of 1 sec.

For long-period structures based on the design base shear requirements of the UBC-97 [Sec. 1630.2.1] and Formula 30-6, the minimum seismic base shear coefficient in all seismic zones is specified as

$$C_{s,\min} = 0.11 C_a I \qquad [6.20]$$

In addition, for seismic zone 4, the UBC-97 requires that the minimum seismic base shear coefficient be further limited to account for near-source effects.[38] Based on UBC-97 Formula 30-7 [Sec. 1630.2.1], this minimum seismic base shear coefficient in seismic zone 4 is

$$C_{s,\min,\text{zone } 4} = \frac{0.8 Z N_v I}{R} \qquad [6.21]$$

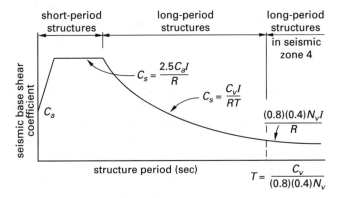

Figure 6.13 Minimum Design Base Shear Coefficient, Zone 4 Only

Example 6.6

In designing a ten-story steel special concentrically braced frame for a shelter in an emergency-preparedness center in seismic zone 4, a location about 5 mi (8 km) from an active seismic source with an undetermined soil profile is being considered. Determine (a) the seismic base shear coefficient, (b) the minimum seismic base shear coefficient, and (c) the maximum seismic base shear coefficient.

---

[38]In seismic zone 4, near-source effects will be significant at locations less than 9.3 mi (15 km) from a potential seismic source (fault).

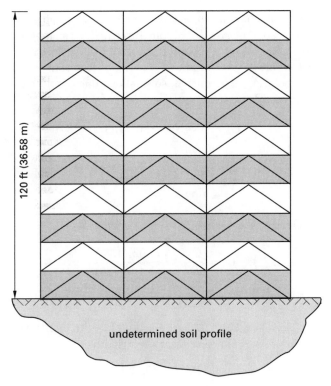

undetermined soil profile

Customary U.S. Solution

(a) The seismic base shear coefficient (C_s) can be determined from Eq. 6.17,

$$C_s = \frac{C_v I}{RT}$$

The natural period (T) can be determined from Method A of the UBC-97 [Sec. 1630.2.2, Item 1].

$$T = C_t (h_n)^{3/4}$$

For steel special concentrically braced frame, $C_t = 0.02$ and $h_n = 120$ ft. Thus,

$$T = (0.020)(120 \text{ ft})^{3/4}$$
$$= 0.725 \text{ sec}$$

From Table 6.10 for steel special concentrically braced frames, $R = 6.4$. From Table 6.5, the importance factor for structures and shelters in emergency-preparedness centers, $I = 1.25$. Based on the UBC-97 [Secs. 1629.3 and 1636.2], when the soil properties are not known in sufficient detail to determine the soil-profile type, type S_D should be assumed. (Type S_D has stiff soil-profile properties.) From Table 6.9, for soil-profile type S_D and for seismic zone factor $Z = 0.4$, the applicable acceleration and velocity-controlled seismic response coefficients are

$$C_a = 0.44 N_a$$
$$C_v = 0.64 N_v$$

From Table 6.7, for an active seismic source (fault), the seismic source type is A. From Table 6.8, for a site 5 mi from a known seismic source, the applicable near-source factors are $N_a = 1.2$ and $N_v = 1.6$. (Note that linear interpolation is permitted, so smaller values could have been used.)

$$C_a = 0.44N_a = (0.44)(1.2)$$
$$= 0.53$$

$$C_v = 0.64N_v = (0.64)(1.6)$$
$$= 1.02$$

Therefore, the seismic base shear coefficient is

$$C_s = \frac{C_v I}{RT} = \frac{(1.02)(1.25)}{(6.4)(0.725)}$$
$$= 0.2748$$

(b) Based on UBC-97 Formula 30-6 [Sec. 1630.2.1], the minimum seismic base shear coefficient is

$$C_{s,\text{min}} = 0.11C_a I = (0.11)(0.53)(1.25)$$
$$= 0.073$$

For seismic zone 4, the minimum value of the seismic base shear coefficient according to UBC-97 Formula 30-7 [Sec. 1630.2.1] is

$$C_{s,\text{min,zone 4}} = \frac{0.8ZN_v I}{R}$$
$$= \frac{(0.8)(0.4)(1.6)(1.25)}{6.4}$$
$$= 0.10$$

$C_{s,\text{min,zone 4}} = 0.10$ is greater than $C_{s,\text{min}} = 0.073$. Hence, the seismic base shear coefficient C_s cannot be less than 0.10.

(c) Based on UBC-97 Formula 30-5 [Sec. 1630.2.1], the maximum seismic base shear coefficient is

$$C_{s,\text{max}} = \frac{2.5C_a I}{R} = \frac{(2.5)(0.53)(1.25)}{6.4}$$
$$= 0.2588$$

$C_s = 0.2748$ is greater than $C_{s,\text{max}} = 0.2588$. Consequently, $C_s = 0.2588$ is the governing seismic base shear coefficient for this structure.

SI Solution

(a) The seismic base shear coefficient (C_s) can be determined from

$$C_s = \frac{C_v I}{RT}$$

The natural period (T) can be determined from Method A of the UBC-97 [Sec. 1630.2.2, Item 1].

$$T = C_t(h_n)^{3/4}$$

For steel special concentrically braced frame, $C_t = 0.0488$ and $h_n = 36.58$ m. Thus,

$$T = (0.0488)(36.58 \text{ m})^{3/4}$$
$$= 0.725 \text{ s}$$

From Table 6.10 for steel special concentrically braced frames, $R = 6.4$. From Table 6.5, the importance factor for structures and shelters in emergency-preparedness centers, $I = 1.25$. Based on the UBC-97 [Secs. 1629.3 and 1636.2], when the soil properties are not known in sufficient detail to determine soil-profile type, type S_D should be assumed. (Type S_D has stiff soil profile properties.) From Table 6.9, for soil-profile type S_D and for seismic zone factor $Z = 0.4$, the applicable acceleration and velocity-controlled seismic response coefficients are

$$C_a = 0.44N_a$$
$$C_v = 0.64N_v$$

From Table 6.7, for an active seismic source (fault), the seismic source type is A. From Table 6.8, for a site 8 km from a known seismic source, the applicable near-source factors are $N_a = 1.2$ and $N_v = 1.6$. (Note that linear interpolation is permitted, so smaller values could have been used.)

$$C_a = 0.44N_a = (0.44)(1.2)$$
$$= 0.53$$

$$C_v = 0.64N_v = (0.64)(1.6)$$
$$= 1.02$$

Therefore, the seismic base shear coefficient is

$$C_s = \frac{C_v I}{RT} = \frac{(1.02)(1.25)}{(6.4)(0.725)}$$
$$= 0.2748$$

(b) Based on UBC-97 Formula 30-6 [Sec. 1630.2.1], the minimum seismic base shear coefficient is

$$C_{s,\text{min}} = 0.11C_a I = (0.11)(0.53)(1.25)$$
$$= 0.073$$

For seismic zone 4, the minimum value of the seismic base shear coefficient according to UBC-97 Formula 30-7 [Sec. 1630.2.1] is

$$C_{s,\text{min,zone 4}} = \frac{0.8ZN_v I}{R}$$
$$= \frac{(0.8)(0.4)(1.6)(1.25)}{6.4}$$
$$= 0.10$$

$C_{s,\mathrm{min,zone\ 4}} = 0.10$ is greater than $C_{s,\mathrm{min}} = 0.073$. Hence, the seismic base shear coefficient C_s cannot be less than 0.10.

(c) Based on UBC-97 Formula 30-5 [Sec. 1630.2.1], the maximum seismic base shear coefficient is

$$C_{s,\mathrm{max}} = \frac{2.5C_aI}{R} = \frac{(2.5)(0.53)(1.25)}{6.4}$$
$$= 0.2588$$

$C_s = 0.2748$ is greater than $C_{s,\mathrm{max}} = 0.2588$. Consequently, $C_s = 0.2588$ is the governing seismic base shear coefficient for this structure.

6-31 LATERAL FORCE PROCEDURES

There are three alternate methods of determining the lateral forces in the UBC-97. The selection of the appropriate lateral-force procedure primarily depends on the type of structure (i.e., regular versus irregular), number of stories, and height, among other factors. These procedures are: (1) simplified static, (2) static (equivalent), and (3) dynamic lateral-force procedures.

Any structure may be designed using the dynamic lateral-force procedures of the UBC-97 [Sec. 1631]. Certain structures such as those defined in the UBC-97 [Sec. 1629.8.4] should be designed using only the dynamic lateral-force procedures. These structures include all structures, regular or irregular, with fundamental periods greater than 0.7 sec that are located on soil profile S_F after site-specific evaluation and consideration of soil-structure resonance based on the UBC-97 [Sec. 1631.2, Item 4]. The dynamic lateral-force procedure should also be used for those structures that have irregular features not characterized in the UBC-97 Tables 16-L or 16-M except as permitted by the UBC-97 [Sec. 1630.4.2], vertical combination.

6-32 STATIC LATERAL-FORCE PROCEDURE

The *static force procedure* is also referred to as the *equivalent static lateral-force procedure*. The structures considered for this procedure are mainly regular structures. UBC-97 Sec. 1630.2 provides the provisions for determining base shear by the static lateral-force procedure. The total design *base shear* in a given direction should be determined in accordance with Eq. 6.22 [UBC-97 Formula 30-4].

$$V = \left(\frac{C_vI}{RT}\right)W \qquad [6.22]$$

Equation 6.22 can be written in the form of base shear formula $V = C_sW$ described in UBC-97 Sec. 1630.2, where C_s is the seismic base shear coefficient. Examining UBC-97 Formula 30-4, the seismic base shear coefficient is

$$C_s = \frac{C_vI}{RT}$$

Based on the design response spectra identified in Sec. 6-30, the C_vI/RT portion of the spectrum is velocity controlled and is representative of longer period structures. The equation is capped for the acceleration-controlled portion of the spectrum, which controls shorter period structures. The code sets this limit by way of UBC-97 Formula 30-5.

$$V = \left(\frac{2.5C_aI}{R}\right)W \qquad [6.23]$$

From UBC-97 Formula 30-5, the equivalent of C_s (seismic base shear coefficient) is

$$C_s = \frac{2.5C_aI}{R}$$

In Eq. 6.23, in all seismic zones for short-period structures, C_a represents the effective peak acceleration at grade with a maximum natural period of 1 sec. For long-period structures based on the design base shear requirements of the UBC-97 [Sec. 1630.2.1] and Formula 30-6, the minimum total design base shear in all seismic zones can be determined from

$$V_{\mathrm{min}} = C_{s,\mathrm{min}}W = (0.11C_aI)W \qquad [6.24]$$

In addition, for seismic zone 4 only, the UBC-97 requires that the minimum total design base shear be more controlled to account for near-source effects. This minimum total design base shear in seismic zone 4 can be determined from UBC-97 Formula 30-7.

$$V_{\mathrm{min,zone\ 4}} = C_{s,\mathrm{min}}W = \left(\frac{0.8ZN_vI}{R}\right)W \qquad [6.25]$$

Example 6.7

A 100-ft (30.48-m), ten-story office building has a total weight of 15,000 kips (66 723 kN). The building is in seismic zone 4 at a location 3.11 mi (5 km) from a seismic source with a high rate of seismic activity. It is designed with a special moment-resisting steel frame system and is constructed on rock (soil type S_B). Use the UBC-97 static lateral-force procedure to calculate the total design base shear.

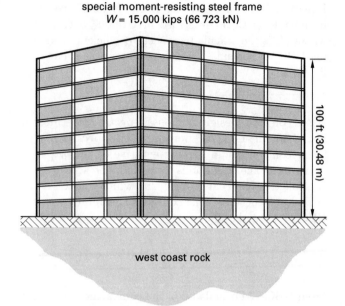

special moment-resisting steel frame
$W = 15{,}000$ kips (66 723 kN)

100 ft (30.48 m)

west coast rock

Customary U.S. Solution

The total design base shear (V) can be determined from Eq. 6.22.

$$V = \left(\frac{C_v I}{RT}\right) W$$

To compute V, the natural period (T), seismic response modification factor (R), importance factor (I), seismic response coefficient (C_v), and total seismic dead load (W) for this structure must be determined. The natural period (T) can be determined from Method A of the UBC-97 [Sec. 1630.2.2, Item 1].

$$T = C_t (h_n)^{3/4}$$

For special moment-resisting steel frame systems, $C_t = 0.035$ and $h_n = 100$ ft.

$$T = (0.035)(100 \text{ ft})^{3/4}$$
$$= 1.11 \text{ sec}$$

From Table 6.10, for special moment-resisting steel frame systems, $R = 8.5$. From Table 6.5, the importance factor for office buildings is $I = 1.00$. The total seismic dead load (W) is 15,000 kips. From Table 6.9, for soil-profile type S_B and for seismic zone factor $Z = 0.4$, the applicable acceleration and velocity-controlled seismic response coefficients are

$$C_a = 0.40 N_a$$
$$C_v = 0.40 N_v$$

From Table 6.7, for a seismic source with high rate of seismic activity, the seismic source type is classified as

type A. From Table 6.8, for a site 3.11 mi from a known seismic source, the applicable near-source factors are $N_a = 1.2$ and $N_v = 1.6$.

$$C_a = 0.40 N_a = (0.40)(1.2)$$
$$= 0.48$$

$$C_v = 0.40 N_v = (0.40)(1.6)$$
$$= 0.64$$

Therefore, the total design base shear is

$$V = \left(\frac{C_v I}{RT}\right) W$$
$$= \left[\frac{(0.64)(1.00)}{(8.5)(1.11 \text{ sec})}\right] (15{,}000 \text{ kips})$$
$$= 1017.49 \text{ kips}$$

Based on UBC-97 Formula 30-6 [Sec. 1630.2.1], the minimum total design base shear is

$$V_{\min} = 0.11 C_a I W$$
$$= (0.11)(0.48)(1.00)(15{,}000 \text{ kips})$$
$$= 792.0 \text{ kips}$$

In addition, for seismic zone 4, the above minimum value of total design base shear is further limited according to UBC-97 Formula 30-7 [Sec. 1630.2.1].

$$V_{\min, \text{zone 4}} = \left(\frac{0.8 Z N_v I}{R}\right) W$$
$$= \left[\frac{(0.8)(0.4)(1.6)(1.0)}{8.5}\right] (15{,}000 \text{ kips})$$
$$= 903.53 \text{ kips}$$

$V_{\min, \text{zone 4}} = 903.53$ kips is greater than $V_{\min} = 792.0$ kips. Therefore, the total design base shear V cannot be less than 903.53 kips. The calculated design base shear $V = 1017.49$ kips must be used in designing this structure because it is greater than $V_{\min, \text{zone 4}} = 903.53$ kips.

Based on UBC-97 Formula 30-5 [Sec. 1630.2.1], the total design base shear need not exceed the maximum total design base shear. The maximum total design base shear is

$$V_{\max} = \left(\frac{2.5 C_a I}{R}\right) W$$
$$= \left[\frac{(2.5)(0.48)(1.00)}{8.5}\right] (15{,}000 \text{ kips})$$
$$= 2117.65 \text{ kips}$$

The calculated design base shear $V = 1017.49$ kips is not greater than $V_{max} = 2117.65$ kips. Consequently, $V = 1017.49$ kips is the controlling total design base shear for this structure.

SI Solution

The total design base shear (V) can be determined from Eq. 6.22.

$$V = \left(\frac{C_v I}{RT}\right) W$$

To compute V, the natural period (T), seismic response modification factor (R), importance factor (I), seismic response coefficient (C_v), and total seismic dead load (W) for this structure must be determined. The natural period (T) can be determined from Method A of the UBC-97 [Sec. 1630.2.2, Item 1].

$$T = C_t(h_n)^{3/4}$$

For special moment-resisting steel frame systems, $C_t = 0.0853$ and $h_n = 30.48$ m.

$$T = (0.0853)(30.48 \text{ m})^{3/4}$$
$$= 1.11 \text{ s}$$

From Table 6.10, for special moment-resisting steel frame systems, $R = 8.5$. From Table 6.5, the importance factor for office buildings is $I = 1.00$. The total seismic dead load (W) is 66 723 kN. From Table 6.9, for soil-profile type S_B and for seismic zone factor $Z = 0.4$, the applicable acceleration and velocity-controlled seismic response coefficients are

$$C_a = 0.40 N_a$$
$$C_v = 0.40 N_v$$

From Table 6.7, for a seismic source with high rate of seismic activity, the seismic source type is classified as type A. From Table 6.8, for a site 5 km from a known seismic source, the applicable near-source factors are $N_a = 1.2$ and $N_v = 1.6$.

$$C_a = 0.40 N_a = (0.40)(1.2)$$
$$= 0.48$$

$$C_v = 0.40 N_v = (0.40)(1.6)$$
$$= 0.64$$

Therefore, the total design base shear is

$$V = \left(\frac{C_v I}{RT}\right) W$$
$$= \left[\frac{(0.64)(1.00)}{(8.5)(1.11 \text{ s})}\right] (66\,723 \text{ kN})$$
$$= 4526 \text{ kN}$$

Based on UBC-97 Formula 30-6 [Sec. 1630.2.1], the minimum total design base shear is

$$V_{min} = 0.11 C_a I W$$
$$= (0.11)(0.48)(1.00)(66\,723 \text{ kN})$$
$$= 3523 \text{ kN}$$

In addition, for seismic zone 4, the above minimum value of total design base shear is further limited according to UBC-97 Formula 30-7 [Sec. 1630.2.1].

$$V_{min,\text{zone } 4} = \left(\frac{0.8 Z N_v I}{R}\right) W$$
$$= \left[\frac{(0.8)(0.4)(1.6)(1.0)}{8.5}\right] (66\,723 \text{ kN})$$
$$= 4019 \text{ kN}$$

$V_{min,\text{zone } 4} = 4019$ kN is greater than $V_{min} = 3523$ kN. Therefore, the total design base shear V cannot be less than 4019 kN. The calculated design base shear $V = 4526$ kN must be used in designing this structure because it is greater than $V_{min,\text{zone } 4} = 4019$ kN.

Based on UBC-97 Formula 30-5 [Sec. 1630.2.1], the total design base shear need not exceed the maximum total design base shear. The maximum total design base shear is

$$V_{max} = \left(\frac{2.5 C_a I}{R}\right) W$$
$$= \left[\frac{(2.5)(0.48)(1.00)}{8.5}\right] (66\,723 \text{ kN})$$
$$= 9419.72 \text{ kN}$$

The calculated design base shear $V = 4526$ kN is not greater than $V_{max} = 9419.72$ kN. Consequently, $V = 4526$ kN is the controlling total design base shear for this structure.

◇　　◇　　◇　　◇　　◇　　◇　　◇

6-33 WHEN THE UBC PERMITS A STATIC ANALYSIS

The UBC-97 [Sec. 1629.8] permits two methods to be used in determining the seismic loading: static and dynamic. (See Sec. 4-1.) The UBC-97 is very specific about when the static method can be used. In general, any structure *may* be designed using the dynamic method at the option of the structural engineer, and some structures *must* use the dynamic method [Sec. 1629.8.4].

The static method may be used for buildings with the following characteristics:

◇ Structures in seismic zones 1 and 2 (see Sec. 3-2) with standard occupancy categories 4 and 5 (see UBC-97 Table 16-K), whether they are regular or irregular (see Sec. 6-23) [Sec. 1629.8.3, Item 1]

◇ Regular structures under 240 ft (73 m) in height using one of the lateral force-resisting systems listed in Table 6.10, Sec. 6-19 of this book (equivalent to UBC-97 Table 16-N) except regular structures located on soil profile S_F (see Table 6.6) which have natural periods greater than 0.7 sec [Sec. 1629.8.4, Item 4]

◇ Irregular structures less than or equal to five stories or 65 ft (20 m) in height [UBC-97 Sec. 1629.8.3, Item 3]

◇ Structures with flexible upper portions (e.g., towers) supported on a rigid lower portion if three conditions are met: (1) both portions, when considered individually, are regular, (2) the average story stiffness of the lower portion is at least ten times the average story stiffness of the upper portion, and (3) the period of the entire structure is no more than 1.1 times the period of the upper portion considered as a separate structure fixed at the base [UBC-97 Sec. 1629.8.3, Item 4]

All structures not meeting these requirements, including irregular buildings (see Sec. 6-23), must be designed using the dynamic method [UBC-97 Sec. 1629.8.4].

6-34 SIMPLIFIED LATERAL-FORCE PROCEDURE

As an alternative to using the UBC-97 static lateral-force procedure, the total design base shear may be determined by a simplified lateral-force procedure, which is given in UBC-97 Sec. 1630.2.3. The simplified lateral-force procedure is only applicable to structures with occupancy categories 4 and 5 (e.g., apartments, single-family residential dwellings, hotels, office buildings, garages, carports, and wholesale/retail commercial buildings) in any seismic zone of any construction material not exceeding two stories in height, or three stories in height of light-frame wood or steel construction materials. In general, this procedure is not intended for routine use by design engineers.

The total design base shear in a given direction should be determined in accordance with Eq. 6.26 (UBC-97 Formula 30-11).

$$V = \left(\frac{3.0C_a}{R}\right) W \qquad [6.26]$$

The simplified lateral-force procedure is a simple way to compute base shear and provides conservative results when compared to other UBC-97 lateral-force procedures. The following spectrum is provided to illustrate this comparison.

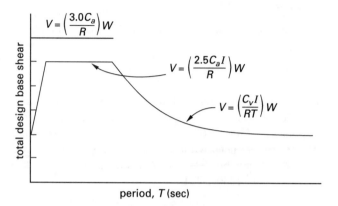

Figure 6.14   Simplified Design Base Shear

C_a in Eq. 6.26 is the site-dependent seismic response coefficient given in UBC-97 Table 16-Q. In earthquake-resistant design of a structure if soil profile is unknown, soil-profile type S_E in seismic zones 1 and 2 (A or B) or soil-profile type S_D in seismic zones 3 and 4 can be presumed.

From UBC-97 Table 16-S, the minimum value of the near-source factor, N_a, is 1.0. However, the maximum value of the near-source factor, N_a, in the same table is 1.5, which applies to structures located 1.24 mi (2 km) or closer to a type A seismic source. In determining the total design base shear for a structure in seismic zone 4 by the simplified lateral-force procedure, the value of the near-source factor, N_a, need not be greater than 1.3 if the structure does not contain structural irregularity types 1, 2, or 5 of UBC-97 Table 16-L, and/or types 1 or 4 of UBC-97 Table 16-M. Therefore, structures in seismic zone 4 must be regular in order to use the simplified lateral-force procedure.

Example 6.8

A single-family, light-framed wood residential dwelling is located on a site with very dense soil and soft rock soil profile. This house has a total weight of 45 kips (200.17 kN). Using the UBC-97 simplified lateral-force procedure, determine the total design base shear for this structure, assuming that (a) it is located in seismic zone 3, (b) it is located within 1.24 mi (2 km) of an active fault in seismic zone 4, and (c) it is underlain with an undetermined soil profile in seismic zone 2B.

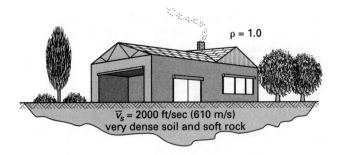

$\rho = 1.0$

$\overline{v}_s = 2000$ ft/sec (610 m/s)
very dense soil and soft rock

Customary U.S. Solution

The total design base shear (V) for any occupancy (including single-family dwelling) using the simplified lateral-force procedure can be determined from Eq. 6.26.

$$V = \left(\frac{3.0C_a}{R}\right)W$$

To compute V, the seismic response modification factor (R), seismic response coefficient (C_a), and total seismic dead load (W) for this structure must be determined.

(a) From Table 6.10, for light-framed wood structural bearing systems for structures three stories or fewer, $R = 5.5$. The total seismic dead load (W) is 45 kips. Based on the UBC-97 [Secs. 1629.3 and 1636], the S_C soil profile is defined as very dense soil and soft rock. From Table 6.9, for the soil-profile type S_C and for seismic zone factor $Z = 0.3$, the applicable acceleration-controlled seismic response coefficient is $C_a = 0.33$.

Therefore, the total design base shear is

$$V = \left(\frac{3.0C_a}{R}\right)W$$
$$= \left[\frac{(3.0)(0.33)}{5.5}\right](45 \text{ kips})$$
$$= 8.10 \text{ kips}$$

(b) $R = 5.5$ and $W = 45$ kips. From Table 6.9, for soil-profile type S_C and for seismic zone factor $Z = 0.4$, the applicable acceleration-controlled seismic response coefficient is
$$C_a = 0.40N_a$$

From Table 6.7, for a seismic source with potential seismic activity, the seismic source type is A. From Table 6.8, for a site 1.24 mi from a known seismic source, the applicable near-source factor $N_a = 1.5$. However, N_a need not exceed 1.3 for regular structures. Therefore, $N_a = 1.3$. Thus,

$$C_a = 0.40N_a = (0.40)(1.3)$$
$$= 0.52$$

Therefore, the total design base shear is

$$V = \left(\frac{3.0C_a}{R}\right)W$$
$$= \left[\frac{(3.0)(0.52)}{5.5}\right](45 \text{ kips})$$
$$= 12.8 \text{ kips}$$

(c) $R = 5.5$ and $W = 45$ kips. Based on the UBC-97 [Secs. 1629.3 and 1636], when the soil profile is unknown in seismic zone 2B, the soil-profile type S_D should be used. From Table 6.9, for soil-profile type S_D and for seismic zone 2B, the applicable acceleration-controlled seismic response coefficient is $C_a = 0.34$.

Therefore, the total design base shear is

$$V = \left(\frac{3.0C_a}{R}\right)W$$
$$= \left[\frac{(3.0)(0.34)}{5.5}\right](45 \text{ kips})$$
$$= 8.35 \text{ kips}$$

SI Solution

The total design base shear (V) for any occupancy (including single-family dwelling) using the simplified lateral-force procedure can be determined from Eq. 6.26.

$$V = \left(\frac{3.0C_a}{R}\right)W$$

To compute V, the seismic response modification factor (R), seismic response coefficient (C_a), and total seismic dead load (W) for this structure must be determined.

(a) From Table 6.10, for light-framed wood structural bearing systems for structures three stories or fewer, $R = 5.5$. The total seismic dead load (W) is 200.17 kN. Based on the UBC-97 [Secs. 1629.3 and 1636], the S_C soil profile is defined as very dense soil and soft rock. From Table 6.9, for the soil-profile type S_C and for seismic zone factor $Z = 0.3$, the applicable acceleration-controlled seismic response coefficient is $C_a = 0.33$.

Therefore, the total design base shear is

$$V = \left(\frac{3.0C_a}{R}\right)W$$
$$= \left[\frac{(3.0)(0.33)}{5.5}\right](200.17 \text{ kN})$$
$$= 36.0 \text{ kN}$$

(b) $R = 5.5$ and $W = 200.17$ kN. From Table 6.9, for soil-profile type S_C and for seismic zone factor $Z = 0.4$, the applicable acceleration-controlled seismic response coefficient is

$$C_a = 0.40 N_a$$

From Table 6.7, for a seismic source with potential seismic activity, the seismic source type is A. From Table 6.8, for a site 2 km from a known seismic source, the applicable near-source factor $N_a = 1.5$. However, N_a need not exceed 1.3 for regular structures. Therefore, $N_a = 1.3$. Thus,

$$C_a = 0.40 N_a = (0.40)(1.3)$$
$$= 0.52$$

Therefore, the total design base shear is

$$V = \left(\frac{3.0 C_a}{R}\right) W$$
$$= \left[\frac{(3.0)(0.52)}{5.5}\right] (200.17 \text{ kN})$$
$$= 56.8 \text{ kN}$$

(c) $R = 5.5$ and $W = 200.17$ kN. Based on the UBC-97 [Secs. 1629.3 and 1636], when the soil profile is unknown in seismic zone 2B, the soil-profile type S_D should be used. From Table 6.9, for soil-profile type S_D and for seismic zone 2B, the applicable acceleration-controlled seismic response coefficient is $C_a = 0.34$.

Therefore, the total design base shear is

$$V = \left(\frac{3.0 C_a}{R}\right) W$$
$$= \left[\frac{(3.0)(0.34)}{5.5}\right] (200.17 \text{ kN})$$
$$= 37.12 \text{ kN}$$

$\diamond \quad \diamond \quad \diamond \quad \diamond \quad \diamond \quad \diamond \quad \diamond$

6-35 VERTICAL DISTRIBUTION OF BASE SHEAR TO STORIES

In the absence of a justifiable more rigorous approach, the base shear, V, is distributed to the n stories in accordance with Eq. 6.29 (corresponding to UBC-97 Formula 30-15). The F_x forces increase linearly with height above the base, as Fig. 6.15 illustrates. F_t is an additional force that is applied to the top level (i.e., the roof) in addition to the F_x force at that level. The F_t

accounts for higher-mode effects. It is zero for $T \leq 0.7$ sec. If $T > 0.7$ sec, F_t can be obtained from Eq. 6.27 (corresponding to UBC-97 Formula 30-14). The story shear, V_x, is the sum of the top force, F_t, and the forces F_x above that story. [UBC-97 1630.6]

$$F_t = 0.07TV \quad [F_t \leq 0.25V] \qquad [6.27]$$

$$F_t = 0 \quad [T \leq 0.7 \text{ sec}] \qquad [6.28]$$

$$F_x = \frac{(V - F_t) w_x h_x}{\sum\limits_{i=1}^{n} w_i h_i} \qquad [6.29]$$

$$V_x = F_t + \sum\limits_{i=x}^{n} F_x \quad [\text{story shear}] \qquad [6.30]$$

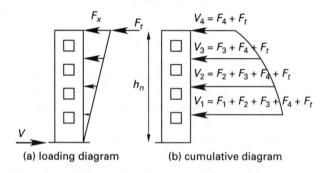

Figure 6.15 Vertical Distribution of Story Shears

As Fig. 6.15 shows, the distribution of the F_x is linear.[39] This distribution depends on the following assumptions. Structures that do not meet these requirements must receive a dynamic analysis.

1. The building is regular and nearly symmetrical.

2. The lateral stiffnesses of each floor are approximately the same.

3. The lateral stiffnesses of each floor are uniformly distributed in plan.

4. The weight of each floor is approximately the same and is uniformly distributed in plan.

5. There is a continuous load path between members.

6. Torsional components of drift are small compared to translational components.

7. The first deflection mode (see Sec. 4-16) can be approximately represented as a straight line.

---

[39]There have been some attempts, such as in the ATC and NEHRF documents, to introduce an exponent that varies from linear to quadratic as a function of the building period. There is no consensus that this improves the results, although it certainly increases the complexity.

Example 6.9

A five-story building is constructed with 12-ft (3.7-m) story heights. The base shear has been calculated as 160 kips (0.71 MN). Each story floor has a weight of 800 kips (3.6 MN), and the roof has a weight of 700 kips (3.1 MN). The natural period of oscillation is 0.5 sec. What are the story forces?

Customary U.S. Solution

The top force, F_t, is zero since $T < 0.7$ sec.

A table is the easiest way to set up the data for calculating and recording the story shears.

level x	h_x (ft)	w_x (kips)	$h_x w_x$ (ft-kips)	$\dfrac{h_x w_x}{\sum h_x w_x}$	F_x (kips)
5 (roof)	60	700	42,000	0.304	48.6
4	48	800	38,400	0.278	44.5
3	36	800	28,800	0.209	33.4
2	24	800	19,200	0.139	22.2
1	12	800	9,600	0.070	11.2
TOTALS		3900	138,000	1.000	160.0

The value of F_5 is given by Eq. 6.12.

$$F_5 = (V - F_t)\left(\frac{w_5 h_5}{\Sigma w_i h_i}\right)$$
$$= (160 \text{ kips} - 0 \text{ kips})(0.304)$$
$$= 48.6 \text{ kips}$$

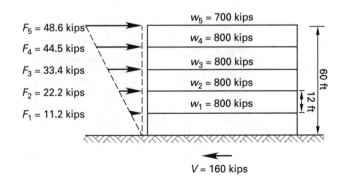

SI Solution

The top force, F_t, is zero since $T < 0.7$ s.

A table is the easiest way to set up the data for calculating and recording the story shears.

level x	h_x (m)	w_x (MN)	$h_x w_x$ (m·MN)	$\dfrac{h_x w_x}{\sum h_x w_x}$	F_x (MN)
5 (roof)	18.3	3.1	56.73	0.304	0.22
4	14.6	3.6	52.56	0.278	0.20
3	11.0	3.6	39.60	0.209	0.15
2	7.3	3.6	26.28	0.139	0.09
1	3.7	3.6	13.32	0.070	0.05
TOTALS	17.5		188.49	1.000	0.71

The value of F_5 is given by Eq. 6.12.

$$F_5 = (V - F_t)\left(\frac{w_5 h_5}{\Sigma w_i h_i}\right)$$
$$= (0.71 \text{ MN} - 0 \text{ MN})(0.304)$$
$$= 0.216 \text{ MN}$$

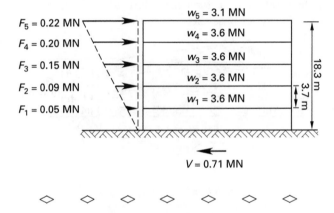

6-36 SIMPLIFIED VERTICAL DISTRIBUTION OF BASE SHEAR TO STORIES

Once the total design base shear (V) for simple structures (see Sec. 6-34) is determined by the simplified lateral-force procedure, the lateral forces at each level may be determined from Eq. 6.31 (corresponding to UBC-97 Formula 30-12).

$$F_x = \left(\frac{3.0 C_a}{R}\right) w_i \qquad [6.31]$$

w_i is the portion of W (total seismic dead load) at level i. The seismic response coefficient, C_a, can be found from UBC-97 Table 16-Q for a soil-profile type based on the UBC-97 [Sec. 1630.2.3.2]. From UBC-97 Formula 30-11, the $3.0 C_a/R$ coefficient simply equates to V/W. Therefore, Eq. 6.31 (corresponding to UBC-97 Formula 30-12) can be rewritten as Eq. 6.32.

$$F_x = \left(\frac{V}{W}\right) w_i \qquad [6.32]$$

In the simplified static procedure, F_x forces do not increase linearly with height above the base as they do in the standard static procedure. For the simplified vertical base shear distribution, natural period (T), P-Δ effects, and story drift consideration may not be required.

◇   ◇   ◇   ◇   ◇   ◇   ◇

Example 6.10

A two-story office building in Los Angeles, CA, with a masonry bearing shear wall system is located on a site with a stiff soil profile. The nearest distance between this site to a known seismic source is 6.21 mi (10 km). This particular seismic source is capable of producing earthquakes with moment magnitude greater than 7.0. Determine (a) the total design base shear using the simplified static lateral-force procedure, and (b) the vertical distribution of base shear to stories.

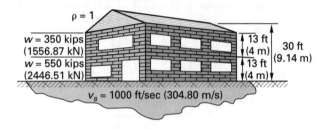

Customary U.S. Solution

(a) The total design base shear (V) for any occupancy using the simplified lateral-force procedure can be determined from Eq. 6.26.

$$V = \left(\frac{3.0 C_a}{R}\right) W$$

To compute V, the seismic response modification factor (R), seismic response coefficient (C_a), and total seismic dead load (W) for this structure must be determined.

From Table 6.10, for a masonry bearing shear wall system, $R = 4.5$. Los Angeles is in seismic zone 4. Based on the UBC-97 [Secs. 1629.3 and 1636], the S_D soil profile is defined as stiff soil. The total seismic dead load (W) for this building is

$$W = 550 \text{ kips} + 350 \text{ kips}$$
$$= 900 \text{ kips}$$

From Table 6.9, for soil-profile type S_D and for seismic zone factor $Z = 0.4$, the applicable acceleration-controlled seismic response coefficient is

$$C_a = 0.44 N_a$$

From Table 6.7, for a seismic source with capability of producing earthquakes with moment magnitude greater than 7.0, the seismic source type is classified as A. From Table 6.8, for a site 6.21 mi from the seismic source A, the applicable near-source factor is $N_a = 1.0$. Thus,

$$C_a = 0.44 N_a = (0.44)(1.0)$$
$$= 0.44$$

Therefore, the total design base shear is

$$V = \left(\frac{3.0 C_a}{R}\right) W$$
$$= \left[\frac{(3.0)(0.44)}{4.5}\right] (900 \text{ kips})$$
$$= 264 \text{ kips}$$

(b) Because the simplified static lateral-force procedure is used to determine the design base shear, the vertical distribution of base shear to stories is also based on the simplified vertical force distribution procedure of the UBC-97 [Sec. 1630.2.3.3]. The forces at each level can be obtained from Eq. 6.31.

$$F_x = \left(\frac{3.0 C_a}{R}\right) w_i$$

$$F_1 = \left[\frac{(3.0)(0.44)}{4.5}\right] (550 \text{ kips})$$
$$= 161.33 \text{ kips}$$

$$F_2 = \left[\frac{(3.0)(0.44)}{4.5}\right] (350 \text{ kips})$$
$$= 102.67 \text{ kips}$$

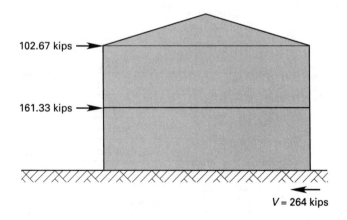

SI Solution

(a) The total design base shear (V) for any occupancy using the simplified lateral-force procedure can be determined from Eq. 6.26.

$$V = \left(\frac{3.0 C_a}{R}\right) W$$

To compute V, the seismic response modification factor (R), seismic response coefficient (C_a), and total seismic dead load (W) for this structure must be determined.

From Table 6.10, for a masonry bearing shear wall system, $R = 4.5$. Los Angeles is in seismic zone 4. Based on the UBC-97 [Secs. 1629.3 and 1636], the S_D soil profile is defined as stiff soil. The total seismic dead load (W) for this building is

$$W = 2446.51 \text{ kN} + 1556.87 \text{ kN}$$
$$= 4003.38 \text{ kN}$$

From Table 6.9, for soil-profile type S_D and for seismic zone factor $Z = 0.4$, the applicable acceleration-controlled seismic response coefficient is

$$C_a = 0.44N_a$$

From Table 6.7, for a seismic source with capability of producing earthquakes with moment magnitude greater than 7.0, the seismic source type is classified as type A. From Table 6.8, for a site 10 km from the seismic source A, the applicable near-source factor is $N_a = 1.0$. Thus,

$$C_a = 0.44N_a = (0.44)(1.0)$$
$$= 0.44$$

Therefore, the total design base shear is

$$V = \left(\frac{3.0C_a}{R}\right)W$$
$$= \left[\frac{(3.0)(0.44)}{4.5}\right](4003.38 \text{ kN})$$
$$= 1174.32 \text{ kN}$$

(b) Because the simplified static lateral-force procedure is used to determine the design base shear, the vertical distribution of base shear to stories will also be based on the simplified vertical force distribution procedure of the UBC-97 [Sec. 1630.2.3.3]. The forces at each level can be obtained from Eq. 6.31.

$$F_x = \left(\frac{3.0C_a}{R}\right)w_i$$

$$F_1 = \left[\frac{(3.0)(0.44)}{4.5}\right](2446.51 \text{ kN})$$
$$= 717.64 \text{ kN}$$

$$F_2 = \left[\frac{(3.0)(0.44)}{4.5}\right](1556.87 \text{ kN})$$
$$= 456.68 \text{ kN}$$

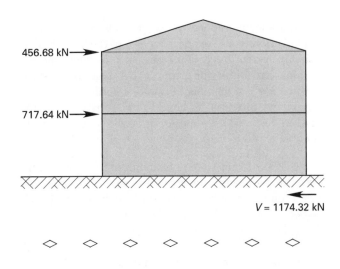

456.68 kN →

717.64 kN →

$V = 1174.32$ kN

6-37 UBC PROVISIONS FOR VERTICAL GROUND ACCELERATION

For distant earthquakes, the vertical component of ground motion may be taken two-thirds of the corresponding horizontal acceleration. When the near-source factor N_a exceeds 1.0, the UBC-97 [Sec. 1631.2, Item 5] requires that the site-specific vertical response spectra be used in lieu of the factor of two-thirds. Alternative factors may be used if they can be rationally substantiated.

The UBC-97 [Sec. 1630.11] provides for determining the effects of vertical acceleration on cantilever components and long-span prestressed components only in seismic zones 3 and 4. Specifically, horizontal cantilevers must be designed for a net upward force of $0.7C_aIW_p$. For gravity load, alone or in combination with the lateral force effects, only 50% of the dead load of horizontal prestressed components can be used or considered in their design (i.e., the dead load may be removed, driving the upper surface of horizontal beams into the tension region).

6-38 UBC PROVISIONS FOR DIRECTION OF THE EARTHQUAKE

In Sec. 1630.1.1, the UBC-97 requires that the ground motion producing structural response and seismic forces in any horizontal direction (i.e., x- and y-direction) be evaluated for structures. Usually directions parallel and perpendicular to the major structure faces, regardless of the structure's orientation with respect to a major fault, will be used. It is assumed that a structure will have enough lateral strength to resist an *oblique earthquake* when requirements for these two orthogonal directions have been met.

A simultaneous application of earthquake forces from two directions could overstress parts of the lateral force-resisting system in certain cases, such as when there is torsional irregularity or a nonparallel structural system, or when a given member—usually a corner column—is part of two intersecting lateral force-resisting systems. UBC-97 Sec. 1633.1 requires an analysis of such directional effects.

The UBC-97 seems to require an infinite number of analyses by requiring the structure to be designed for forces coming from "... any horizontal direction" [UBC-97 Sec. 1630.1.1]. However, later in the same section, the code clarifies the concept by specifying that, except for certain types of irregular buildings, the "... seismic forces may be assumed to act nonconcurrently in the direction of each principal axis..." The UBC-97 does not generally require the structure to be designed for earthquakes coming from more than these two orthogonal directions.

6-39 UBC PROVISIONS FOR HORIZONTAL TORSIONAL MOMENT

The UBC-97 provisions for torsion apply to rigid diaphragms only (see Sec. 7-4), where increased shears result from horizontal torsion. When the center of mass (C_M) does not coincide with the center of rigidity (C_R) of the vertical resisting elements in a story, a torsional moment is induced. For nonflexible diaphragms, the UBC-97 [Sec. 1630.7] requires that a certain amount of accidental torsion (see Sec. 5-14) be planned for, even in regular buildings. Specifically, the center of mass at each level is assumed to be displaced from the calculated center of mass in each direction a distance of 5% of the building dimension at that level perpendicular to the direction of the seismic force. Thus, the accidental eccentricity will be different for the two orthogonal directions.

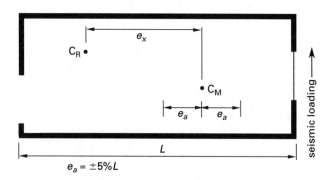

Figure 6.16 Torsional Moments

The accidental torsion is not a minimum value as it was in earlier UBCs. It is a value to be either added to or subtracted from the actual calculated eccentricity. The term "design eccentricity" is used to represent the algebraic sum of the actual and accidental (5%) eccentricities ($e_x + e_a$).

Based on the UBC-97 [Sec. 1630.7], the accidental torsion may have to be increased above the 5% level for torsionally irregular structures by use of an amplification factor (A_x), determined from UBC-97 Formula 30-16.

$$A_x = \left(\frac{\delta_{\max}}{1.2\delta_{\text{avg}}} \right)^2 \qquad [6.33]$$

$$A_x \leq 3.0$$

$\delta_{\max}$ is the maximum displacement at level x, and δ_{avg} is the average of the displacements at the extreme points of the structure at level x.

6-40 UBC STORY DRIFT DETERMINATION

Story drift is the lateral displacement of one level of a structure relative to the level above or below. In the UBC-97, drift requirements are based on the strength design method to conform with newly developed seismic base shear forces. In that regard, complete inelastic response drifts rather than force level drifts are used. Based on the UBC-97 [Sec. 1630.10.1], story drifts should be determined using the maximum inelastic response displacement, Δ_M, which is defined as the maximum total drift or total story drift caused by the design-level earthquake. Displacement includes both elastic and inelastic contributions to the total deformation. The maximum inelastic response displacement, Δ_M, should be computed from Eq. 6.34 (corresponding to UBC-97 Formula 30-17).

$$\Delta_M = 0.7R\Delta_S \qquad [6.34]$$

Δ_S is a design level elastic response displacement found from the elastic static analsysis of UBC-97 Sec. 1630.2.1 or the elastic dynamic analysis of UBC-97 Sec. 1631. The resulting deformations (Δ_S) should be determined at all critical locations in the structure under consideration. In calculation of Δ_S, translational and torsional deflections should be included. Figure 6.17 illustrates the concept behind determination of the elastic and inelastic response deformation Δ_M from elastic deformation Δ_S.

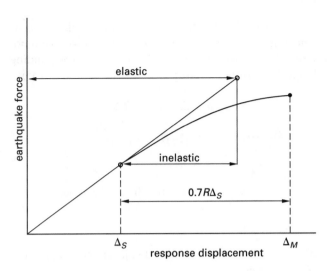

Figure 6.17   Elastic and Inelastic Response Deformation to Design Earthquakes

Table 6.13 provides UBC-97 limitations imposed on the calculated story drift using Δ_M. These limitations are intended to control inelastic deformations and potential instabilities in both structural and nonstructural elements that could affect life safety.

Table 6.13

UBC-97 Story Drift Limitations

structure's natural period	calculated story drift using Δ_M	UBC-97 provisions
$T < 0.7$ sec (short period structures)	$\Delta_M \leq 0.025h$ (2.5% of story height)	
		Sec. 1630.10.2
$T \geq 0.7$ sec (long period structures)	$\Delta_M \leq 0.020h$ (2.0% of story height)	

If both structural and nonstructural elements can tolerate greater drift without threat to life safety, the drift limitations presented in Table 6.13 may be exceeded. There is no story drift limitation for single-story steel frame structures with primary use as storage facilities, factories, workshops, and warehouses.[40]

---

[40]For example, many prefabricated, single-story steel buildings (commonly referred to as "Butler buildings") do not meet the code drift limit. Such deviations from the code may be permitted, particularly in industrial or warehouse space.

Furthermore, based on UBC-97 Sec. 1630.2.3.4, when the simplified static lateral-force procedure is used to determine the total design base shear, the UBC-97's drift and story drift limitation provisions need not be applied, except at the request of the building official, for structures having relatively flexible structural systems. When the simplified analysis is used, the maximum inelastic response displacement (Δ_M) can be obtained from Eq. 6.35.

$$\Delta_M = 0.01h \qquad [6.35]$$

Example 6.11

A 4-story special moment-resisting steel structure is located on a site with rock-like soil in seismic zone 4. This structure has equal story heights of 12 ft (3.66 m) at each story and a natural period (T) of 0.70 sec. The displacements at each story are indicated in the following figure.

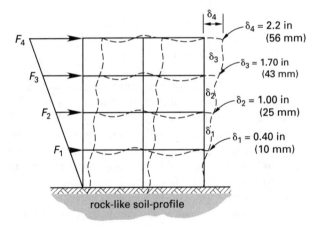

Determine (a) the design level response displacement in the third story, (b) the maximum inelastic response displacement in the first story, (c) the design level response story drift ratio in the top story, (d) the maximum allowable design level response story drift ratio, and (e) the design level response story drift that can be tolerated if architectural cladding panels are to be installed at the second story, which can accommodate only 0.71 in (18 mm) of movement.

Customary U.S. Solution

(a) The design level response displacement (Δ_S) in the third story is

$$\begin{aligned}\Delta_{S,\text{third story}} &= \delta_3 - \delta_2 \\ &= 1.7 \text{ in} - 1.0 \text{ in} \\ &= 0.7 \text{ in}\end{aligned}$$

(b) The maximum inelastic response displacement (Δ_M) in the first story can be obtained from Eq. 6.34.

$$\Delta_M = 0.7R\Delta_{S,\text{first story}}$$

From Table 6.10, for special moment-resisting steel structures, $R = 8.5$. The design level response displacement (Δ_S) in the first story is

$$\Delta_{S,\text{first story}} = \delta_1$$
$$= 0.4 \text{ in}$$
$$\Delta_M = (0.7)(8.5)(0.4 \text{ in})$$
$$= 2.38 \text{ in}$$

Based on the UBC-97 [Sec. 1630.10.2], for structures with natural period of 0.7 sec or greater, the calculated story drift using Δ_M should not be greater than 0.020 times the story height. Thus,

$$\Delta_M \leq 0.020 h_s$$
$$= (0.020)(12 \text{ ft})\left(\frac{12 \text{ in}}{1 \text{ ft}}\right)$$
$$= 2.88 \text{ in}$$

Because $\Delta_M = 2.38$ in does not exceed $\Delta_M = 2.88$ in, it is considered permissible. Of course, it should be noted that the drift limits may be exceeded when it can be shown that greater drift can be tolerated by both structural elements and nonstructural elements with due consideration of P-Δ effects. Any instability in structure can affect life safety.

(c) The design level response story drift in the top story is

$$\Delta_{S,\text{top story}} = \delta_4 - \delta_3$$
$$= 2.2 \text{ in} - 1.7 \text{ in}$$
$$= 0.5 \text{ in}$$

According to the UBC-97 [Sec. 1627], the story drift ratio is the story drift divided by the story height. Hence,

$$\Delta_{SR} = \frac{0.5 \text{ in}}{(12 \text{ ft})\left(\dfrac{12 \text{ in}}{1 \text{ ft}}\right)}$$
$$= 0.0035$$

(d) The maximum allowable design level response story drift ratio for each story is

$$\Delta_{MR} = \frac{0.020 h_s}{h_s}$$
$$= 0.020$$

(e) The external architectural cladding panels and their connections for this structure are designed to accommodate 0.71 in movement ($\delta = 0.71$ in). The corresponding design level response story drift in the second story is

$$\delta = \Delta_M = 0.7R\Delta_{S,\text{second story}}$$
$$\Delta_{S,\text{second story}} = \frac{\delta}{0.7R} = \frac{0.71 \text{ in}}{(0.7)(8.5)}$$
$$= 0.12 \text{ in}$$
$$\delta_2 - \delta_1 = 0.6 \text{ in} > 0.12 \quad \text{[no good]}$$

SI Solution

(a) The design level response displacement (Δ_S) in the third story is

$$\Delta_{S,\text{third story}} = \delta_3 - \delta_2$$
$$= 43 \text{ mm} - 25 \text{ mm}$$
$$= 18 \text{ mm}$$

(b) The maximum inelastic response displacement (Δ_M) in the first story can be obtained from Eq. 6.34.

$$\Delta_M = 0.7R\Delta_{S,\text{first story}}$$

From Table 6.10, for special moment-resisting steel structures, $R = 8.5$. The design level response displacement (Δ_S) in the first story is

$$\Delta_{S,\text{first story}} = \delta_1$$
$$= 10 \text{ mm}$$
$$\Delta_M = (0.7)(8.5)(10 \text{ mm})$$
$$= 60 \text{ mm}$$

Based on the UBC-97 [Sec. 1630.10.2], for structures with natural period of 0.7 sec or greater, the calculated story drift using Δ_M should not be greater than 0.020 times the story height. Thus,

$$\Delta_M \leq 0.020 h_s$$
$$= (0.020)(3.66 \text{ m})\left(\frac{1000 \text{ mm}}{1 \text{ m}}\right)$$
$$= 73 \text{ mm}$$

Because $\Delta_M = 60$ mm does not exceed $\Delta_M = 73$ mm, it is considered permissible. Of course, it should be noted that the drift limits may be exceeded when it can be shown that greater drift can be tolerated by both structural elements and nonstructural elements with due consideration of P-Δ effects. Any instability in structure can affect life safety.

(c) The design level response story drift in the top story is

$$\Delta_{S,\text{top story}} = \delta_4 - \delta_3$$
$$= 56 \text{ mm} - 43 \text{ mm}$$
$$= 13 \text{ mm}$$

According to the UBC-97 [Sec. 1627], the story drift ratio is the story drift divided by the story height. Hence,

$$\Delta_{SR} = \frac{13 \text{ mm}}{(3.66 \text{ m}) \left(\dfrac{1000 \text{ mm}}{1 \text{ m}}\right)}$$
$$= 0.0036$$

(d) The maximum allowable design level response story drift ratio for each story is

$$\Delta_{MR} = \frac{0.020 h_s}{h_s}$$
$$= 0.020$$

(e) The external architectural cladding panels and their connections for this structure are designed to accommodate 18 mm movement ($\delta = 18$ mm). The corresponding design level response story drift in the second story is

$$\delta = \Delta_M = 0.7 R \Delta_{S,\text{second story}}$$
$$\Delta_{S,\text{second story}} = \frac{\delta}{0.7R} = \frac{18 \text{ mm}}{(0.7)(8.5)}$$
$$= 3 \text{ mm}$$
$$\delta_2 - \delta_1 = 15 \text{ mm} > 3 \text{ mm} \quad \text{[no good]}$$

◇   ◇   ◇   ◇   ◇   ◇   ◇

6-41 UBC PROVISIONS FOR BUILDING SEPARATIONS

The UBC-97 requires all structures to be separated from adjoining structures. *Separations* should permit structures to react to seismic forces independently, so that damaging impact between adjacent structures' elements will be avoided. According to the UBC-97 [Sec. 1633.2.11], the separation should allow for the maximum inelastic response displacement (Δ_M). The minimum code requirement for separation of two adjacent buildings on the same property can be computed from Eq. 6.36 (corresponding to UBC-97 Formula 33-2).

$$\Delta_{MT} = \sqrt{(\Delta_{M1})^2 + (\Delta_{M2})^2} \qquad [6.36]$$

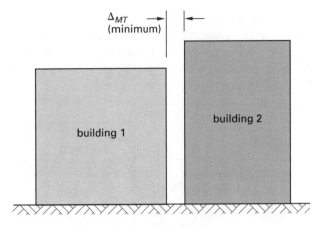

Figure 6.18   Building Separation

Δ_{M1} and Δ_{M2} are the maximum inelastic response displacements of the adjacent buildings given by Eq. 6.34 (UBC-97 Formula 30-17).

When computing building separations, the UBC-97 requires the following.

◇ Structures that adjoin a property line not common to a public way must have a minimum *setback* from the property line equal to the displacement Δ_M of that structure.

◇ Rational analyses based on maximum expected ground motions can justify smaller separations or property line setbacks.

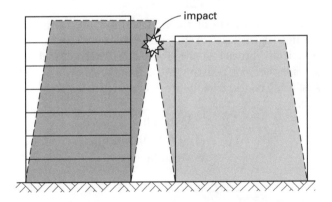

Figure 6.19   Inadequate Building Separation

Example 6.12

Two office buildings will be built adjacent to each other on the same property in Northern California. A special concentrically braced dual system consisting of steel and special moment-resisting steel frames is proposed for structure I. It is decided that structure II will be a special concentrically steel braced frame. The elastic displacements for the two buildings are known. What should be the sufficient distance between all parts of these two buildings to avoid damaging impact while the buildings are responding to earthquake motion, independently?

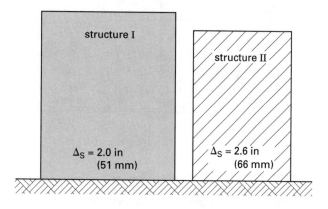

Customary U.S. Solution

Based on the UBC-97 [Sec. 1633.2.11], all structures are required to be separated from adjoining structures on the same property, and the separation should allow for the maximum inelastic response displacement (Δ_M). The minimum required separation can be computed from Eq. 6.36.

$$\Delta_{MT} = \sqrt{(\Delta_{M1})^2 + (\Delta_{M2})^2}$$
$$\Delta_{M1} = 0.7R\Delta_{S1}$$
$$\Delta_{M2} = 0.7R\Delta_{S2}$$

The displacements of these structures are 2.0 in and 2.6 in, respectively. From Table 6.10, for a special concentrically braced dual system consisting of steel and special moment-resisting steel frames, $R = 7.5$. For a special concentrically steel braced frame, $R = 6.4$. Thus,

$$\Delta_{M1} = 0.7R\Delta_{S1}$$
$$= (0.7)(7.5)(2.0 \text{ in})$$
$$= 10.50 \text{ in}$$

$$\Delta_{M2} = 0.7R\Delta_{S2}$$
$$= (0.7)(6.4)(2.6 \text{ in})$$
$$= 11.65 \text{ in}$$

The minimum required separation is

$$\Delta_{MT} = \sqrt{(\Delta_{M1})^2 + (\Delta_{M2})^2}$$
$$= \sqrt{(10.50 \text{ in})^2 + (11.64 \text{ in})^2}$$
$$= 15.7 \text{ in}$$

SI Solution

Based on the UBC-97 [Sec. 1633.2.11], all structures are required to be separated from adjoining structures

on the same property, and the separation should allow for the maximum inelastic response displacement (Δ_M). The minimum required separation can be computed from Eq. 6.36.

$$\Delta_{MT} = \sqrt{(\Delta_{M1})^2 + (\Delta_{M2})^2}$$
$$\Delta_{M1} = 0.7R\Delta_{S1}$$
$$\Delta_{M2} = 0.7R\Delta_{S2}$$

The displacements of these structures are 51 mm and 66 mm, respectively. From Table 6.10, for a special concentrically braced dual system consisting of steel and special moment-resisting steel frames, $R = 7.5$. For a special concentrically steel braced frame, $R = 6.4$. Thus,

$$\Delta_{M1} = 0.7R\Delta_{S1}$$
$$= (0.7)(7.5)(51 \text{ mm})$$
$$= 267.75 \text{ mm}$$

$$\Delta_{M2} = 0.7R\Delta_{S2}$$
$$= (0.7)(6.4)(66 \text{ mm})$$
$$= 295.68 \text{ mm}$$

The minimum required separation is

$$\Delta_{MT} = \sqrt{(\Delta_{M1})^2 + (\Delta_{M2})^2}$$
$$= \sqrt{(267.75 \text{ mm})^2 + (295.68 \text{ mm})^2}$$
$$= 399 \text{ mm}$$

◇   ◇   ◇   ◇   ◇   ◇   ◇

6-42 UBC PROVISIONS FOR OVERTURNING MOMENT

The UBC-97 requires that every designed structure be capable of resisting overturning effects induced by earthquake forces. At any level, the overturning moment must be determined using the seismic forces (F_t and F_x) that act on all of the levels above the level under consideration [UBC-97 Sec. 1630.8.1]. These forces are F_t and F_x as defined in Eq. 6.27 and 6.29. The overturning effects on every lateral-force-resisting element must be carried down to the foundation.

The incremental increases of the design overturning moment at each higher level should be distributed to the various lateral-force-resisting elements. The distribution should be in the same proportion as the distribution of the horizontal shears to those resisting elements. The effects of uplift caused by seismic loads must also be analyzed. Any net tension must be resisted by interaction with the soil (e.g., by use of friction piles that resist uplift).

In regular structures, the top force, F_t, can be omitted in calculating the overturning effects at the soil-foundation interface, including the calculation of soil pressure under typical footings and the soil-pile frictional forces during uplift [UBC-97 Sec. 1809.4].[41] This omission is permitted because the F_t force represents higher mode forces, and the moments at the base associated with the higher modes are unlikely to occur simultaneously with mode 1 response.

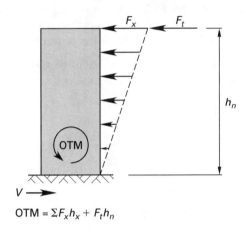

$$OTM = \Sigma F_x h_x + F_t h_n$$

Figure 6.20 Overturning Moment at the Base

6-43 UBC PROVISIONS FOR P-Δ EFFECTS

The P-Δ effect is defined as the secondary effect on shears, axial forces, and moments of frame members induced by the gravity loads acting on the laterally displaced structure frame. The UBC-97 [Sec. 1630.1.3] requires that the resulting member forces and moments, and the story drifts, be used to evaluate the overall structural frame stability caused by the P-Δ effects. For the overall structural frame stability evaluation, it is required that the forces producing the displacements of Δ_S, design level response displacement, be used.

The 1997 NEHRP presents the concept of a *stability coefficient*, θ, which is the ratio of secondary moments to primary moments.[42] If this stability coefficient is equal to or less than 0.10, the P-Δ effects can be disregarded.

---

[41]However, the omission of F_t is not permitted in the design of structural or foundation elements, including the design of the footings and piles that resist uplift and of their connections with the building.

[42]The ratio may be evaluated for any story as the product of the total dead load, floor live load, and snow load above the story and the seismic drift in that story, divided by the product of the seismic shear in that story and the height of that story.

Otherwise, a rational analysis must be used to evaluate the P-Δ effects.

$$\theta = \frac{P_x \Delta}{V_x h_{sx} C_d} \quad [6.37]$$

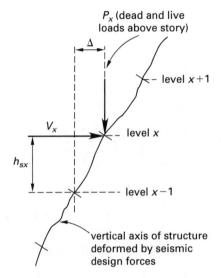

Figure 6.21 Stability Coefficient Variables

C_d is the *deflection amplification factor*. Based on the UBC-97 [Sec. 1630.1.3], P-Δ effects need not be considered when the following are true.

⋄ The ratio of secondary moment (M_s) to primary moment (M_p) does not exceed 0.10.

⋄ In seismic zones 3 and 4, the ratio of story drift to story height (story drift ratio) does not exceed $0.02/R$.

Based on the UBC-97 [Sec. 1630.2.3.4], when the simplified lateral-force procedure is used to determine the total design base shear, the UBC-97's P-Δ effects provisions need not be considered for structures having relatively flexible structural systems, except at the request of the building official.

Example 6.13

The columns of an ordinary moment-resisting steel frame (OMRF) and a special moment-resisting steel frame (SMRF), shown, are in seismic zones 2A and 4, respectively. Should the P-Δ effects be considered for (a) the detailed ordinary moment-resisting steel frame? (b) the detailed special moment-resisting steel frame?

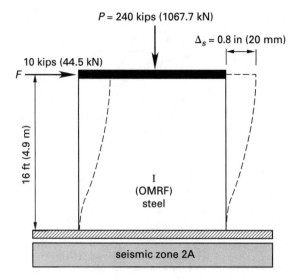

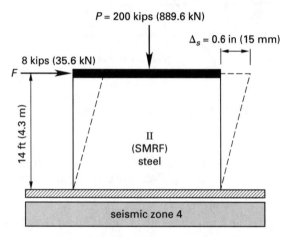

The ratio of the secondary moment to the primary moment is

$$\frac{M_s}{M_p} = \frac{8 \text{ ft-kips}}{160 \text{ ft-kips}}$$

$$= 0.05$$

Thus, for structure I in seismic zone 2A, the P-Δ effects need not be considered.

(b) Based on the UBC-97 [Sec. 1630.1.3], in seismic zones 3 and 4, the P-Δ effects need to be considered in the evaluation of overall structural stability if the story drift ratio exceeds $0.02/R$ for all stories.

From Table 6.10, for special moment-resisting steel frames, $R = 8.5$. The design level response displacement, Δ_S, is 0.60 in, and $h_s = 14$ ft. The story drift ratio (Δ_S/h_s) is

$$\Delta_{SR} = \frac{\Delta_S}{h_s} = \frac{0.60 \text{ in}}{(14 \text{ ft})\left(\frac{12 \text{ in}}{1 \text{ ft}}\right)}$$

$$= 0.0036$$

$$\frac{0.02}{R} = \frac{0.02}{8.5}$$

$$= 0.0024$$

$$\Delta_{SR} \le 0.02/R \text{ to neglect } P\text{-}\Delta$$

$$0.0036 > 0.0024$$

Thus, for structure II in seismic zone 4, the P-Δ effects should be considered.

Customary U.S. Solution

(a) Structure I is located in seismic zone 2A. Based on the UBC-97 [Sec. 1630.1.3], the P-Δ effects (i.e., the effects of the secondary moments) should be considered in the evaluation of overall structural stability when the ratio of secondary moment to primary moment exceeds 0.1 for any story.

For structure I, from Table 6.10, for ordinary moment-resisting steel frames, $R = 4.5$. The design level response displacement, Δ_S, is 0.80 in, and $h_s = 16$ ft. The lateral seismic force, F, is 10 kips, and the axial load, P, is 240 kips. The primary moment (M_p) and secondary moment (M_s) are

$$M_p = Fh_s$$

$$= (10 \text{ kips})(16 \text{ ft})$$

$$= 160 \text{ ft-kips}$$

$$M_s = P\Delta_S$$

$$= (120 \text{ kips})(0.80 \text{ in})\left(\frac{1 \text{ ft}}{12 \text{ in}}\right)$$

$$= 8 \text{ ft-kips}$$

SI Solution

(a) Structure I is located in seismic zone 2A. Based on the UBC-97 [Sec. 1630.1.3], the P-Δ effects (i.e., the effects of the secondary moments) should be considered in the evaluation of overall structural stability when the ratio of secondary moment to primary moment exceeds 0.1 for any story.

For structure I, from Table 6.10, for ordinary moment-resisting steel frames, $R = 4.5$. The design level response displacement, Δ_S, is 20 mm, and $h_s = 4.9$ m. The lateral seismic force, F, is 44.5 kN, and the axial load, P, is 1067.7 kN. The primary moment (M_p) and secondary moment (M_s) are

$$M_p = Fh_s$$

$$= (44.5 \text{ kN})(4.9 \text{ m})$$

$$= 218 \text{ kN·m}$$

$$M_s = P\Delta_S$$

$$= \left(\frac{1067.7 \text{ kN}}{2}\right)(20 \text{ mm})\left(\frac{1 \text{ m}}{1000 \text{ mm}}\right)$$

$$= 10.7 \text{ kN·m}$$

The ratio of the secondary moment to the primary moment is

$$\frac{M_s}{M_p} = \frac{10.7 \text{ kN·m}}{218 \text{ kN·m}}$$
$$= 0.05$$

Thus, for structure I in seismic zone 2A, the P-Δ effects need not be considered.

(b) Based on the UBC-97 [Sec. 1630.1.3], in seismic zones 3 and 4, the P-Δ effects need to be considered in the evaluation of overall structural stability if the story drift ratio exceeds $0.02/R$ for all stories.

From Table 6.10, for special moment-resisting steel frames, $R = 8.5$. The design level response displacement, Δ_S, is 15 mm, and $h_S = 4.3$ m. The story drift ratio (Δ_S/h_s) is

$$\Delta_{SR} = \frac{\Delta_S}{h_s} = \frac{15 \text{ mm}}{(4.3 \text{ m}) \left(\dfrac{1000 \text{ mm}}{1 \text{ m}} \right)}$$
$$= 0.0035$$
$$\frac{0.02}{R} = \frac{0.02}{8.5}$$
$$= 0.0024$$
$$\Delta_{SR} \le 0.02/R \text{ to neglect } P\text{-}\Delta$$
$$0.0036 > 0.0024$$

Thus, for structure II in seismic zone 4, the P-Δ effects should be considered.

◇   ◇   ◇   ◇   ◇   ◇   ◇

6-44 SEISMIC FORCE ON ELEMENTS OF STRUCTURES, NONSTRUCTURAL COMPONENTS, AND EQUIPMENT SUPPORTED BY STRUCTURES

Examples of nonstructural elements are unbraced parapet walls, interior bearing and nonbearing walls, partitions, and penthouses. Examples of nonstructural components are signs, billboards, and exterior and interior ornamentations, chimneys, stacks, storage racks over 6 ft (183 cm), anchorage, and lateral bracing for suspended ceilings and light fixtures. Equipment includes tanks and vessels with support systems; electrical, mechanical, and plumbing equipment; and emergency communication equipment.[43]   Table 6.14

---

[43]Design of attachments for floor- or roof-mounted equipment weighing less than 400 lbm (181 kg) is not required.

(corresponding to UBC-97 Table 16-O) provides a comprehensive list of these categories.

To ensure life safety and preserve the functionality of essential facilities, these components should be designed to resist the total design seismic forces of the UBC-97 [Sec. 1632.2]. The seismic loading, F_p, on these elements and their attachments can be calculated from Eqs. 6.38 or 6.39 (corresponding to UBC-97 Formulas 32-1 and 32-2, respectively).

$$F_p = 4.0 C_a I_p W_p \qquad [6.38]$$

Alternatively,

$$F_p = \left(\frac{a_p C_a I_p}{R_p} \right) \left[1 + (3) \left(\frac{h_x}{h_r} \right) \right] W_p \qquad [6.39]$$

C_a is the seismic response coefficient, I_p is the importance factor, a_p is the amplification factor associated with the dynamic characteristics of the component, R_p is the ductility factor associated with the component, and W_p is the weight of the component.

The minimum and maximum values of F_p are given by Eq. 6.40 (corresponding to UBC-97 Formula 32-3).

$$0.7 C_a I_p W_p \le F_p \le 4.0 C_a I_p W_p \qquad [6.40]$$

The value of the seismic response coefficient (acceleration controlled), C_a, is the same as that used for the building, and depends on the seismic zone, proximity of the site to active seismic sources, and site soil-profile characteristics. The value of the seismic importance factor, I_p, is either 1.0 or 1.5, depending on the occupancy or function of the structure. Appendix K of this book (corresponding to UBC-97 Table 16-K) gives the values of I_p for different types of facilities. A value of $I_p = 1.5$ is used for anchoring machinery and equipment required for life-safety systems, as well as for tanks containing toxic and explosive substances. W_p is the weight of the elements, components, or equipment. The natural period of the elements, components, and equipment is not a factor in calculating the force on the elements, components, and equipment, except indirectly in the determination of a_p.

The values of the *in-structure component amplification factor*, a_p, are given in Table 6.14 (corresponding to UBC-97 Table 16-O). The values of a_p vary from 1.0 to 2.5. Rigid components experience little amplification and are assigned an a_p of 1.0. Alternatively, dynamic properties or empirical data of the component and the structure that supports it may be used to determine this factor; however, its value may not be less than 1.0.

Table 6.14
Horizontal Force Factors, a_p and R_p
(UBC-97 Table 16-O)

elements of structures and nonstructural components and equipment[1]	a_p	R_p	footnote
1. Elements of structures			
A. Walls including the following:			
(1) Unbraced (cantilevered) parapets	2.5	3.0	
(2) Exterior walls at or above the ground floor and parapets braced above their centers of gravity	1.0	3.0	2
(3) All interior-bearing and nonbearing walls	1.0	3.0	2
B. Penthouse (except when framed by an extension of the structural frame)	2.5	4.0	
C. Connections for prefabricated structural elements other than walls (see also Sec. 1632.2)	1.0	3.0	3
2. Nonstructural components			
A. Exterior and interior ornamentations and appendages	2.5	3.0	
B. Chimneys, stacks, and trussed towers supported on or projecting above the roof:			
(1) Laterally braced or anchored to the structural frame at a point below their centers of mass	2.5	3.0	
(2) Laterally braced or anchored to the structural frame at or above their centers of mass	1.0	3.0	
C. Signs and billboards	2.5	3.0	
D. Storage racks (including contents) over 6 ft (1829 mm) tall	2.5	4.0	4
E. Permanent floor-supported cabinets and book stacks (including contents) more than 6 ft (1829 mm) in height	1.0	3.0	5
F. Anchorage and lateral bracing for suspended ceilings and light fixtures	1.0	3.0	3, 6, 7, 8
G. Access floor systems	1.0	3.0	4, 5, 9
H. Masonry or concrete fences over 6 ft (1829 mm) high	1.0	3.0	
I. Partitions	1.0	3.0	
3. Equipment			
A. Tanks and vessels (including contents), including support systems	1.0	3.0	
B. Electrical, mechanical, and plumbing equipment and associated conduit and ductwork and piping	1.0	3.0	5, 10, 11, 12, 13, 14, 15, 16
C. Any flexible equipment laterally braced or anchored to the structural frame at a point below its center of mass	2.5	3.0	5, 10, 14, 15, 16
D. Anchorage of emergency power supply systems and essential communications equipment. Anchorage and support systems for battery racks and fuel tanks necessary for operation of emergency equipment. See also Sec. 1632.2.	1.0	3.0	17, 18
E. Temporary containers with flammable or hazardous materials	1.0	3.0	19
4. Other components			
A. Rigid components with ductile material and attachments	1.0	3.0	1
B. Rigid components with nonductile material or attachments	1.0	1.5	1
C. Flexible components with ductile material and attachments	2.5	3.0	1
D. Flexible components with nonductile material and attachments	2.5	1.5	1

[1]See Sec. 1627 for definitions of flexible components and rigid components.

[2]See Secs. 1633.2.4 and 1633.2.8 for concrete and masonry walls and Sec. 1632.2 for connections for panel connectors for panels.

[3]Applies to seismic zones 2, 3, and 4 only.

[4]Ground-supported steel storage racks may be designed using the provisions of Sec. 1634. Chapter 22, Div. X, may be used for design, provided seismic design forces are equal to or greater than those specified in Sec. 1632.2 or 1634.2, as appropriate.

[5]Only anchorage or restraints need be designed.

[6]Ceiling weight shall include all light fixtures and other equipment or partitions that are laterally supported by the ceiling. For purposes of determining the seismic force, a ceiling weight of less than 4 psf (0.19 kN/m^2) shall be used.

[7]Ceilings constructed of lath and plaster or gypsum board screw or nail attached to suspended members that support a ceiling at one level extending from wall to wall need not be analyzed, provided the walls are not over 50 ft (15 240 mm) apart.

[8]Light fixtures and mechanical services installed in metal suspension systems for acoustical tile and lay-in panel ceilings shall be independently supported from the structure above as specified in UBC-97 Standard 25-2, Part III.

[9]W_p for access floor systems shall be the dead load of the access floor system plus 25% of the floor live load plus a 10-psf (0.48-kN/m^2) partition load allowance.

[10]Equipment includes, but is not limited to, boilers, chillers, heat exchangers, pumps, air-handling units, cooling towers, control panels, motors, switchgear, transformers, and life-safety equipment. It shall include major conduit, ducting, and piping, which services such machinery and equipment and fire sprinkler systems. See Sec. 1632.2 for additional requirements for determining a_p for nonrigid or flexibly mounted equipment.

(continued)

Table 6.14

Horizontal Force Factors, a_p and R_p

(continued)

[11]Seismic restraints may be omitted from piping and duct supports if all the following conditions are satisfied.

    [11.1]Lateral motion of the piping or duct will not cause damaging impact with other systems.

    [11.2]The piping or duct is made of ductile material with ductile connections.

    [11.3]Lateral motion of the piping or duct does not cause impact of fragile appurtenances (e.g., sprinkler heads) with any other equipment, piping, or structural member.

    [11.4]Lateral motion of the piping or duct does not cause loss of system vertical support.

    [11.5]Rod-hung supports of less than 12 in (305 mm) in length have top connections that cannot develop moments.

    [11.6]Support members cantilevered up from the floor are checked for stability.

[12]Seismic restraints may be omitted from electrical raceways, such as cable trays, conduit and bus ducts, if all the following conditions are satisfied:

    [12.1]Lateral motion of the raceway will not cause damaging impact with other systems.

    [12.2]Lateral motion of the raceway does not cause loss of system vertical support.

    [12.3]Rod-hung supports of less than 12 inches (305 mm) in length have top connections that cannot develop moments.

    [12.4]Support members cantilevered up from the floor are checked for stability.

[13]Piping, ducts and electrical raceways, which must be functional following an earthquake, spanning between different buildings or structural systems shall be sufficiently flexible to withstand relative motion of support points assuming out-of-phase motions.

[14]Vibration isolators supporting equipment shall be designed for lateral loads or restrained from displacing laterally by other means. Restraint shall also be provided, which limits vertical displacement, such that lateral restraints do not become disengaged. a_p and R_p for equipment supported on vibration isolators shall be taken as 2.5 and 1.5, respectively, except that if the isolation mounting frame is supported by shallow or expansion anchors, the design forces for the anchors calculated by Formula (32-1),(32-2) or (32-3) shall be additionally multiplied by a factor of 2.0.

[15]Equipment anchorage shall not be designated such that lateral loads are resisted by gravity friction (e.g., friction clips).

[16]Expansion anchors, which are required to resist seismic loads in tension, shall not be used where operational vibrating loads are present.

[17]Movement of component within electrical cabinets, rack- and skid-mounted equipment and portions of skid-mounted electromechanical equipment that may cause damage to other components by displacing, shall be restricted by attachment to anchored equipment or support frames.

[18]Batteries on racks shall be restrained against movement in all directions due to earthquake forces.

[19]Seismic restraints may include straps, chains, bolts, barriers or other mechanisms that prevent sliding, falling and breach of containment of flammable and toxic materials. Friction forces may not be used to resist lateral loads in these restraints unless positive uplift restraint is provided which ensures that the friction forces act continuously.

Values of the component response modification factor, R_p, are furnished in Table 6.14 (corresponding to UBC-97 Table 16-O). The maximum value of R_p, 4.0, is reserved for steel storage racks. Most components and equipment are allowed an R_p of 3.0. A value of $R_p = 1.5$ is used for equipment with shallow anchors, including expansion, chemical, and embedded anchor bolts with an embedment length-to-diameter ratio less than 8. For nonductile anchors and anchors used with adhesive, $R_p = 1.0$.

Equation 6.39 includes an amplification factor based on building height. h_x and h_r signify elevations with respect to grade. h_x denotes the height of element or component attachment above base, and h_r represents the height of the structure roof above base. The resulting factor is 1.0 at the base and 4.0 at the roof.

The design seismic forces obtained from Eqs. 6.38 and 6.39 (corresponding to UBC-97 Formulas 32-1 and 32-2, respectively) should be distributed in proportion to the mass distribution of the element or component. The computed seismic forces, F_p, are to be used to design the members and connections that transfer these forces into the structure seismic-resisting systems.

Exterior cladding panels must resist a seismic force F_p calculated from Eq. 6.38 or 6.39, and must accommodate movements of the structure based on Δ_M. Panel connectors (e.g., bolts, inserts, welds, and dowels) and connection bodies (e.g., angles, bars, and plates) are designed for the seismic force F_p determined from Eq. 6.39, where for connectors, the values of R_p and a_p are 1.0 for both [UBC-97 Sec. 1633.2.4.2, Item 5], and for connection bodies, the values of R_p and a_p are 3.0 and 1.0, respectively [UBC-97 Sec. 1633.2.4.2, Item 4]. Drift, movement, and other standards also apply [UBC-97 Sec. 1633.2.4.2].

The seismic force on a diaphragm is specified elsewhere, in UBC-97 Sec. 1633.2.9. Equations 6.38 and 6.39 are not used. UBC-97 Formula 33-1 specifies the method to calculate the force on the diaphragm at level x, for diaphragm design. (w_{px} is the weight of the diaphragm plus any elements tributary to it [Sec. 1633.2.9, Item 2].)

$$F_{px} = \left(\frac{F_t + \sum\limits_{i=x}^{n} F_i}{\sum\limits_{i=x}^{n} w_i} \right) w_{px} \qquad [6.41]$$

The diaphragm should be designed to resist force F_{px}. However, F_{px} may not be less than $0.5C_aIw_{px}$, nor need the force exceed $1.0C_aIw_{px}$.

Example 6.14

A parapet wall extends 4 ft (1.22 m) above the roof line of a one-story concrete-walled commercial building, as shown. The walls are 8 in (203 mm) thick. Consider all connections between the roof, walls, and footings to be pinned (free to rotate). The building is located in seismic zone 4 on a site with a stiff soil profile. The proximity of this site to the San Andreas Fault is 6.2 mi (10 km). Consider earthquake forces perpendicular to the wall face. Determine (a) the force that the wall-to-roof connection must be designed for and (b) the moment at the base of the parapet wall.

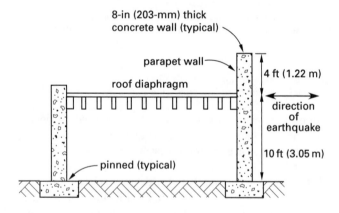

Customary U.S. Solution

(a) The wall length is not given, so work with a 1-ft strip of wall. Concrete has a mass of approximately 150 lbf/ft³, so the distributed weight per foot of width is

$$w_p = \gamma(\text{volume})$$
$$= \gamma(\text{width})(\text{height})(\text{thickness})$$
$$= \left(150 \, \frac{\text{lbf}}{\text{ft}^3} \right) (1 \, \text{ft})(14 \, \text{ft})(8 \, \text{in}) \left(\frac{1 \, \text{ft}}{12 \, \text{in}} \right)$$
$$= 1400 \, \text{lbf} \quad [\text{per foot of wall}]$$

Although only the force on the parapet is wanted, the entire wall must first be analyzed. From App. K of

this book (corresponding to UBC-97 Table 16-K), the seismic importance factor, I_p, is 1.0. From Table 6.3, the seismic zone coefficient, Z, is 0.40. From Table 6.6, for stiff soil-profile, the soil-profile type is S_D. From Table 6.9, for $Z = 0.40$ and soil-profile type S_D, the seismic response coefficient, C_a, is $0.44N_a$. The San Andreas Fault is capable of producing large-magnitude events and has a high rate of seismic activity. Thus, from Table 6.7, the seismic source type is A. From Table 6.8, for seismic source type A and a distance of 6.2 mi to a known seismic source, the nearest-source factor, N_a, is 1.0. Therefore, the seismic response coefficient C_a is

$$C_a = (0.44)(1.0)$$
$$= 0.44$$

From Eq. 6.38, the distributed seismic force per foot of wall width is

$$F_p = 4.0C_aI_pw_p$$
$$= (4.0)(0.44)(1.0)(1400 \, \text{lbf})$$
$$= 2464 \, \text{lbf} \quad [\text{per foot of wall}]$$

For the purpose of finding reactions, this force can be assumed to act at the mid-height of the wall, at

$$\left(\tfrac{1}{2} \right)(14 \, \text{ft}) = 7 \, \text{ft}$$

Summing moments about the wall base,

$$\sum M = (7 \, \text{ft}) \left(2464 \, \frac{\text{lbf}}{\text{ft}} \right) - F_{\text{anchor}}(10 \, \text{ft}) = 0$$
$$F_{\text{anchor}} = 1724.8 \, \text{lbf} \quad [\text{per foot of wall}]$$

For seismic zones 1, 2, and 3, the UBC-97 [Secs. 1605.2.3, 1611.4, and 1633.2.8.1] require the roof-wall diaphragm connection to withstand a minimum force of 280 lbf/ft. However, for seismic zone 4, the UBC-97 [Sec. 1633.2.8.1] sets a minimum anchorage force of 420 lbf/ft. The calculated value of 1724.8 lbf/ft controls. (See Sec. 12-9.)

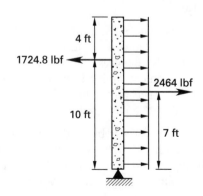

(b) The weight of the parapet alone is

$$w_p = \gamma(\text{volume})$$
$$= \gamma(\text{width})(\text{height})(\text{thickness})$$
$$= \left(150 \ \frac{\text{lbf}}{\text{ft}^3}\right)(1 \ \text{ft})(4 \ \text{ft})(8 \ \text{in})\left(\frac{1 \ \text{ft}}{12 \ \text{in}}\right)$$
$$= 400 \ \text{lbf} \quad \text{[per foot of wall]}$$

From Eq. 6.38, the distributed seismic force per meter of parapet wall width is

$$F_p = 4.0 C_a I_p w_p$$
$$= (4.0)(0.44)(1.0)(400 \ \text{lbf})$$
$$= 704 \ \text{lbf} \quad \text{[per foot of wall]}$$

For the purpose of determining the moment at the base of the parapet, this force can be assumed to act at the mid-height, 2 ft up from the base. In effect, the parapet acts as a vertical cantilever wall. The net moment at the parapet base (i.e., where it joins the roof) is

$$M = (2 \ \text{ft})(704 \ \text{lbf})$$
$$= 1408 \ \text{ft-lbf} \quad \text{[per foot of wall]}$$

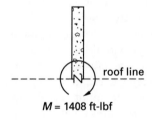

roof line

M = 1408 ft-lbf

SI Solution

(a) The wall length is not given, so work with a 1-m strip of wall. Concrete has a mass of approximately 2400 kg/m³, so the distributed weight per meter of width is

$$w_p = \gamma(\text{volume})$$
$$= \gamma(\text{width})(\text{height})(\text{thickness})$$
$$= \left(2400 \ \frac{\text{kg}}{\text{m}^3}\right)\left(9.81 \ \frac{\text{m}}{\text{s}^2}\right)(1 \ \text{m})(4.27 \ \text{m})$$
$$\times (203 \ \text{mm})\left(\frac{1 \ \text{m}}{1000 \ \text{mm}}\right)$$
$$= 20.4 \ \text{kN} \quad \text{[per meter of wall]}$$

Although only the force on the parapet is wanted, the entire wall must first be analyzed. From App. K of this book (corresponding to UBC-97 Table 16-K), the seismic importance factor, I_p, is 1.0. From Table 6.3,

the seismic zone coefficient, Z, is 0.40. From Table 6.6, for stiff soil-profile, the soil-profile type is S_D. From Table 6.9, for $Z = 0.40$ and soil-profile type S_D, the seismic response coefficient, C_a, is $0.44N_a$. The San Andreas Fault is capable of producing large-magnitude events and has a high rate of seismic activity. Thus, from Table 6.7, the seismic source type is A. From Table 6.8, for seismic source type A and a distance of 10 km to a known seismic source, the nearest-source factor, N_a, is 1.0. Therefore, the seismic response coefficient C_a is

$$C_a = (0.44)(1.0)$$
$$= 0.44$$

From Eq. 6.38, the distributed seismic force per foot of wall width is

$$F_p = 4.0 C_a T_p w_p$$
$$= (4.0)(0.44)(1.0)(20.4 \ \text{kN})$$
$$= 35.9 \ \text{kN} \quad \text{[per meter of wall]}$$

For the purpose of finding reactions, this force can be assumed to act at the mid-height of the wall, at

$$\left(\tfrac{1}{2}\right)(4.27 \ \text{m}) = 2.14 \ \text{m}$$

Summing moments about the wall base,

$$\sum M = (2.14 \ \text{m})\left(35.9 \ \frac{\text{kN}}{\text{m}}\right) - F_{\text{anchor}}(3.05 \ \text{m}) = 0$$
$$F_{\text{anchor}} = 25.2 \ \text{kN} \quad \text{[per meter of wall]}$$

For seismic zones 1, 2, and 3, the UBC-97 [Secs. 1605.2.3, 1611.4, and 1633.2.8.1] require the roof-wall diaphragm connection to withstand a minimum force of 4.09 kN/m. However for seismic zone 4, the UBC-97 [Sec. 1633.2.8.1] sets a minimum anchorage force of 6.1 kN/m. The calculated value of 25.2 kN/m controls. (See Sec. 12-9.)

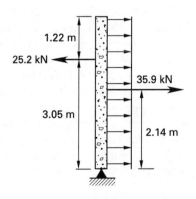

1.22 m

25.2 kN

35.9 kN

3.05 m

2.14 m

(b) The weight of the parapet alone is

$$w_p = \gamma(\text{volume})$$
$$= \gamma(\text{width})(\text{height})(\text{thickness})$$
$$= \left(2400 \ \frac{\text{kg}}{\text{m}^3}\right)\left(9.81 \ \frac{\text{m}}{\text{s}^2}\right)\left(\frac{1 \ \text{kN}}{1000 \ \text{N}}\right)$$
$$\times (1 \ \text{m})(1.22 \ \text{m})(203 \ \text{mm})\left(\frac{1 \ \text{m}}{1000 \ \text{mm}}\right)$$
$$= 5.83 \ \text{kN} \quad [\text{per meter of wall}]$$

From Eq. 6.38, the distributed seismic force per meter of parapet wall width is

$$F_p = 4.0 C_a I_p w_p$$
$$= (4.0)(0.44)(1.0)(5.83 \ \text{kN})$$
$$= 10.3 \ \text{kN} \quad [\text{per meter of wall}]$$

For the purpose of determining the moment at the base of the parapet, this force can be assumed to act at the mid-height, 0.61 m up from the base. In effect, the parapet acts as a vertical cantilever wall. The net moment at the parapet base (i.e., where it joins the roof) is

$$M = (0.61 \ \text{m})(10.3 \ \text{kN})$$
$$= 6.28 \ \text{kN·m} \quad [\text{per meter of wall}]$$

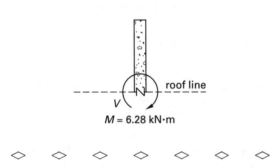

◇ ◇ ◇ ◇ ◇ ◇ ◇

6-45 UBC PROVISIONS FOR TIES AND CONTINUITY

UBC-97 Sec. 1633.2.5, Ties and Continuity, is basically concerned with ensuring continuity in designed structures. The provisions are intended to provide a minimum design force for connections tying portions of a structure together, for example, ties at projecting wings, or beam splices. Often, such tie forces are difficult to calculate by rational method without separating the structure or system into two parts at the tie. The UBC-97 specifies a minimum continuity force to use in designing such connections. For example, as a minimum, connections tying any smaller portion of the structure to the remainder of the structure should have a minimum strength to resist F_p, as defined in Eq. 6.42.

$$F_p \geq 0.5 C_a I W_{\text{smaller portion}} \qquad [6.42]$$

For beams, girders, or trusses, the UBC-97 [Sec. 1633.2.5] requires that positive connections be provided to resist horizontal forces acting parallel to the members. This horizontal force can be obtained from Eq. 6.43.

$$F_p \geq 0.3 C_a I W_{\text{total dead load+live load on the member}}$$
$$[6.43]$$

6-46 NONBUILDING STRUCTURES

Nonbuilding structures [UBC-97 Sec. 1634] are self-supporting structures other than buildings that, nevertheless, come under the jurisdiction of the local building official.[44] Covered by the UBC-97 are non-building structures that (1) have building-like structural systems such as those described in UBC-97 Sec. 1629.6, (2) are rigid structures (i.e., those with natural period, T, less than 0.06 sec), or (3) are specifically mentioned in UBC-97 Table 16-P, for example, tanks, silos, chimneys, signs, and towers. (To qualify as having a building-like structural system, the structure must have one or more levels—floors and roof—at which the mass is concentrated, and the framing system must extend between the levels.)

Most nonbuilding structures, even though they are not designed to accommodate people, are supported by structural systems traditionally found in occupied building structures. Nonbuilding structures carry gravity loads and resist the effects of earthquakes. Consequently, the strength required to resist the displacements caused by the minimum design seismic forces on nonbuilding structures is calculated in the same manner as those for building structures. For nonbuilding structures, as opposed to building structures, the standard building drift limitations need not be met. However, P-Δ effects should be evaluated [UBC-97 Sec. 1634.1.5].

There are some differences, however, in the total design base shear equations for building and nonbuilding structures. The natural period, T, must be determined by rational methods. For example, Method B of the UBC-97 [Sec. 1630.2.2, Item 2] with UBC-97 Formula 30-10, or dynamic analysis, can be used. However, the 80% of the Method A lower limit does not apply [UBC-97 Sec. 1634.1.4].

The total design base shear given in Eq. 6.44 is used for nonbuilding structures with natural period greater

---

[44]Other items specifically not included in the UBC-97 are offshore platforms, electrical transmission towers, dams, and highway and railroad bridges. These structures are not normally within the jurisdiction of the local building official.

than 0.06 sec. For reduction in these forces, the values of the response modification factor (R) and seismic force amplification factor (Ω_O) are given in Table 6.15 (corresponding to UBC-97 Table 16-P). Generally speaking, the values of R assigned to nonbuilding structures are less than for building structures. This is considered justified because nonbuilding structures do not have structural redundancy of multiple bays and nonstructural panels that effectively give buildings greater strength and damping than is considered in the design process. The weight, W, includes the weight of the full contents (if any) of the structure [UBC-97 Sec. 1634.1.3].

$$V = \left(\frac{C_v I}{RT}\right) W \qquad [6.44]$$

The reliability/redundancy factor (ρ) is calculated from UBC-97 Formula 30-3. For total design base shear, the code sets the limit not to exceed UBC-97 Formula 30-5 and not to be less than UBC-97 Formula 30-6.

$$V_{max} = \left(\frac{2.5 C_a I}{R}\right) W \qquad [6.45]$$

$$V_{min} = (0.11 C_a I) W \qquad [6.46]$$

Rigid nonbuilding structures, however, are handled differently. Rigid nonbuilding structures are structures with a natural period (T) of less than 0.06 sec. The natural period is the determining factor of whether the structure is rigid. An example would be a concrete pedestal structure at grade level. Rigid structures are covered in the UBC-97 [Sec. 1634.3], which specifies that $\rho = 1$. The lateral force, V, on rigid structures and their anchorages is given by UBC-97 Formula 34-1.

$$V = 0.7 C_a I W \qquad [6.47]$$

The force, V, is distributed over the height according to the distribution of the mass. It is assumed to act in any horizontal direction.

The size of the supporting structural system for some short (i.e., less than 50 ft (15.24 m) in height) nonbuilding structures is determined by the footprint of the structure, vibration limitation, or other operational considerations rather than traditional lateral loadings. In such cases, the support can be much stronger than required for seismic resistance, and it can be expected to remain in the elastic range during a maximum earthquake. Therefore, ductility is not a consideration. The UBC-97 [Sec. 1634.2] permits these applications (with some restrictions) to be evaluated as intermediate moment-resisting frames (IMRF). The value $R = 2.8$ (corresponding to low ductility demand during ground motion) must be used where this allowance is made [UBC-97 Sec. 1634.2, Exception].

Table 6.15

Values of R and Ω_O for Nonbuilding Structures [UBC-97 Table 16-P]

structure type	R	Ω_O
1. Vessels, including tanks and pressurized spheres, on braced or unbraced legs	2.2	2.0
2. Cast-in-place concrete silos and chimneys having walls continuous to the foundation	3.6	2.0
3. Distributed mass cantilever structures such as stacks, chimneys, silos, and skirt-supported vertical vessels	2.9	2.0
4. Trussed towers (freestanding or guyed), guyed stacks, and chimneys	2.9	2.0
5. Cantilevered column-type structures	2.2	2.0
6. Cooling towers	3.6	2.0
7. Bins and hoppers on braced or unbraced legs	2.9	2.0
8. Storage racks	3.6	2.0
9. Signs and billboards	3.6	2.0
10. Amusement structures and monuments	2.2	2.0
11. All other self-supporting structures not otherwise covered	2.9	2.0

Reproduced from the 1997 edition of the *Uniform Building Code*™, copyright © 1997, with the permission of the publisher, the International Conference of Building Officials.

Certain concrete pedestal-type structures can also be expected to remain in the elastic range during a maximum earthquake. However, the code recommends that some ductility be included in the design. Such ductility is obtained (during the design stage) by providing sufficient transverse reinforcement to avoid brittle shear and/or development failures and by providing continuity and development of longitudinal reinforcement.

Tanks with ground support are also provided with detailed specific provisions [UBC-97 Sec. 1634.4]. Flat-bottom tanks and other tanks with supported bottoms at or below grade (as opposed to tanks on legs) are considered to be nonbuilding structures. Due primarily to the fact that liquid contents slosh around and add their own dynamic forces, the seismic performance of tanks is more complex than their simple appearance would suggest.[45]

The UBC-97 [Sec. 1634.4] allows three different types of analysis for tanks. First, the contents can be assumed to be rigid (i.e., have a period of less than or equal to 0.06 sec) and the design base shear calculated as

---

[45]Sloshing has very little damping (i.e., 0.1% or less).

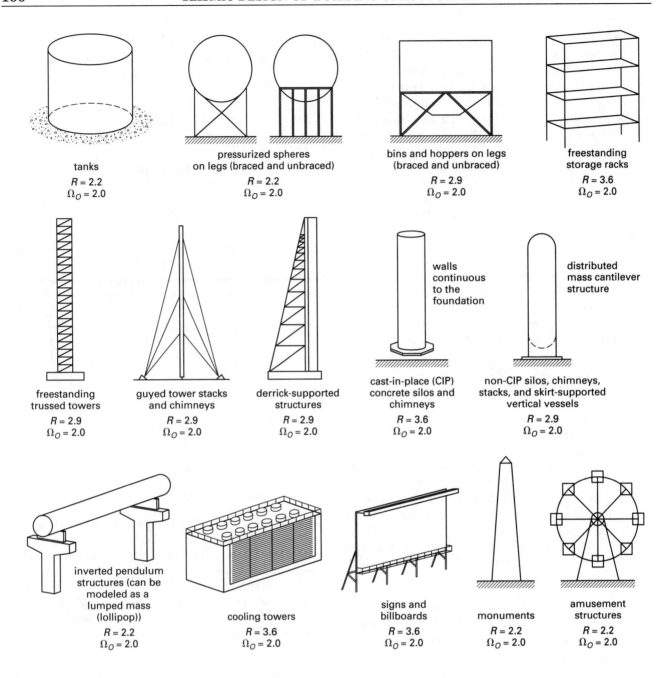

Figure 6.22   Typical Nonbuilding Structures

$V = 4.0C_a I_p W_p$ [UBC-97 Sec. 1632.2, Formula 32-1]. Second, a response spectrum analysis that includes the inertial effects of the tank contents can be performed. Third, an approved national standard for design of tank supports can be used.[46]

Particular attention must be paid to preventing uplift of tanks from their cradles or supports. Specifically,

tanks must be anchored to their foundations. Furthermore, sloshing and freeboard should be considered in the design of the tank. Sloshing can be reduced by including baffles in the tank. Sufficient freeboard should be included in open tanks to prevent the contents from spilling out over the top.

Many nonbuilding structures are bolted to their foundations. In order to ensure ductile response, these bolts must stretch inelastically without failure and without being pulled from their concrete embedment. The bolt

---

[46]For example, American Petroleum Institute (API) Publication 650, App. E, is an approved standard.

size (diameters and lengths) and placement patterns should be chosen so that the bolts achieve their full strength in a maximum earthquake without failure, using the loads specified in UBC-97 Sec. 1634.

Nonbuilding structures other than rigid structures and tanks should be designed to resist design seismic forces no less than those that are determined in accordance with the provisions of UBC-97 Sec. 1630, with the additional considerations of Eqs. 6.48 and 6.49.

$$V_{\min} = 0.56 C_a I W \qquad [6.48]$$

Equation 6.48 corresponds to UBC-97 Formula 34-2. Additionally, for seismic zone 4, the minimum total design base shear is given by Eq. 6.49 (corresponding to UBC-97 Formula 34-3).

$$V_{\min} = \frac{1.6 Z N_v I W}{R} \qquad [6.49]$$

For other nonbuilding structures, when applicable, the seismic zone, occupancy category, soil-profile type, seismic source classification, near-source factor, seismic response coefficient, and response modification factor are the same as for building structures. The vertical distribution of design seismic forces may be determined either by Eq. 6.31 (corresponding to UBC-97 Formula 30-12) or from a dynamic analysis [UBC-97 Sec. 1631].

◇   ◇   ◇   ◇   ◇   ◇   ◇

Example 6.15

A simple billboard-type sign is constructed on S_C soil-profile in seismic zone 4. It has a total weight of 3000 lbf (13.34 kN) distributed evenly across its width and 18 ft (5.5 m) of height. The total effective cross-sectional moment of inertia at the base of its supports is 0.05 ft^4 (4.3×10^{-4} m^4). The modulus of elasticity of the three support posts is 2×10^6 psi (1.38×10^4 MPa). What is the UBC-97 total design base shear for an earthquake acting perpendicular to the sign face?

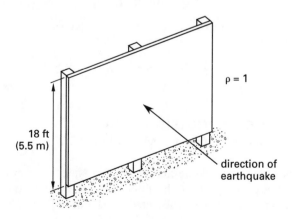

Customary U.S. Solution

One rational method of determining the sign's natural period is to consider the sign as an SDOF system. The supporting elements (i.e., the three posts) are essentially uniformly loaded cantilevers. From standard beam deflection tables, the tip deflection, x, for such a configuration with distributed load w per unit length is

$$x = \frac{wL^4}{8EI}$$

The stiffness, k, at the tip is

$$k = \frac{\text{total load}}{x} = \frac{wL}{x} = \frac{8EI}{L^3}$$
$$= \frac{(8) \left(2 \times 10^6 \, \dfrac{\text{lbf}}{\text{in}^2} \right) \left(144 \, \dfrac{\text{in}^2}{\text{ft}^2} \right) (0.05 \, \text{ft}^4)}{(18 \, \text{ft})^3}$$
$$= 19{,}750 \, \text{lbf/ft}$$

From Eqs. 4.11 and 4.12, the natural period of oscillation is

$$T = \frac{2\pi}{\omega} = 2\pi \sqrt{\frac{W}{kg}}$$
$$= 2\pi \sqrt{\frac{3000 \, \text{lbf}}{\left(19{,}750 \, \dfrac{\text{lbf}}{\text{ft}} \right) \left(32.2 \, \dfrac{\text{ft}}{\text{sec}^2} \right)}}$$
$$= 0.431 \, \text{sec}$$

This is greater than 0.06 sec, so the billboard is not rigid as defined in UBC-97 Sec. 1634.3.

From Table 6.15, the response modification factor $R = 3.6$. From Table 6.9, for a soil profile S_C and $Z = 0.4$, the seismic response coefficient is $C_v = 0.56 N_v$. From Table 6.7, for a seismic source with a high rate of seismic activity, the seismic source type is classified as type A. From Table 6.8, for this billboard 3.1 mi from a known seismic source, the applicable near-source factor $N_v = 1.6$. Thus,

$$C_v = 0.56 N_v = (0.56)(1.6)$$
$$= 0.90$$

The design base shear is determined from Eq. 6.44.

$$V = \left(\frac{C_v I}{RT} \right) W$$
$$= \left[\frac{(0.90)(1.0)}{(3.6)(0.431)} \right] (3000 \, \text{lbf})$$
$$= 1740.14 \, \text{lbf}$$

However, based on the UBC-97 requirements,

$$V_{max} = \left(\frac{2.5 C_a I}{R} \right) W$$

$$V_{min} = 0.11 C_a I W$$

From Table 6.9, for a soil profile S_C and $Z = 0.4$, the seismic response coefficient is $C_a = 0.40 N_a$. From Table 6.8, the applicable near-source factor is $N_a = 1.2$.

$$C_a = 0.40 N_a = (0.40)(1.2)$$
$$= 0.48$$

$$V_{max} = \left[\frac{(2.5)(0.48)(1.0)}{3.6} \right] (3000 \text{ lbf})$$
$$= 1000 \text{ lbf}$$

$$V_{min} = (0.11)(0.48)(1.0)(3000 \text{ lbf})$$
$$= 158.40 \text{ lbf}$$

For this billboard, V_{max} controls. Hence, $V_{design} = 1000$ lbf.

SI Solution

One rational method of determining the sign's natural period is to consider the sign as an SDOF system. The supporting elements (i.e., the three posts) are essentially uniformly loaded cantilevers. From standard beam deflection tables, the tip deflection, x, for such a configuration with distributed load w per unit length is

$$x = \frac{wL^4}{8EI}$$

The stiffness, k, at the tip is

$$k = \frac{\text{total load}}{x} = \frac{wL}{x} = \frac{8EI}{L^3}$$

$$= \frac{(8)(1.38 \times 10^4 \text{ MPa})\left(1000 \frac{\text{kPa}}{\text{MPa}} \right)(4.3 \times 10^{-4} \text{ m}^4)}{(5.5 \text{ m})^3}$$

$$= 288.5 \text{ kN/m}$$

From Eqs. 4.11 and 4.12, the natural period of oscillation is

$$T = \frac{2\pi}{\omega} = 2\pi \sqrt{\frac{W}{kg}}$$

$$= 2\pi \sqrt{\frac{13.34 \text{ kN}}{\left(288.5 \frac{\text{kN}}{\text{m}} \right)\left(9.81 \frac{\text{m}}{\text{s}^2} \right)}}$$

$$= 0.431 \text{ s}$$

This is greater than 0.06 s, so the billboard is not rigid as defined in UBC-97 Sec. 1634.3.

From Table 6.15, the response modification factor $R = 3.6$. From Table 6.9, for a soil profile S_C and $Z = 0.4$, the seismic response coefficient is $C_v = 0.56 N_v$. From Table 6.7, for a seismic source with a high rate of seismic activity, the seismic source type is classified as type A. From Table 6.8, for this billboard 5 km from a known seismic source, the applicable near-source factor $N_v = 1.6$. Thus,

$$C_v = 0.56 N_v = (0.56)(1.6)$$
$$= 0.90$$

The design base shear is determined from Eq. 6.44.

$$V = \left(\frac{C_v I}{RT} \right) W$$

$$= \left[\frac{(0.90)(1.0)}{(3.6)(0.431)} \right] (13.34 \text{ kN})$$

$$= 7.73 \text{ kN}$$

However, based on the UBC-97 requirements,

$$V_{max} = \left(\frac{2.5 C_a I}{R} \right) W$$

$$V_{min} = 0.11 C_a I W$$

From Table 6.9, for a soil profile S_C and $Z = 0.4$, the seismic response coefficient is $C_a = 0.40 N_a$. From Table 6.8, the applicable near-source factor is $N_a = 1.2$.

$$C_a = 0.40 N_a = (0.40)(1.2)$$
$$= 0.48$$
$$V_{max} = \left[\frac{(2.5)(0.48)(1.0)}{3.6} \right] (13.34 \text{ kN})$$
$$= 4.45 \text{ kN}$$

$$V_{min} = (0.11)(0.48)(1.0)(13.34 \text{ kN})$$
$$= 0.7 \text{ kN}$$

For this billboard, V_{max} controls. Hence, $V_{design} = 4.45$ kN.

◇　　◇　　◇　　◇　　◇　　◇　　◇

6-47 UBC PROVISIONS FOR DEFORMATION COMPATIBILITY

For structural framing elements and their connections that are not part of the lateral-force-resisting system but are nonetheless subjected to the deformations resulting from seismic forces, the UBC-97 requires design and detailing to be adequate to maintain support of design gravity (dead plus live) loads under expected seismic deformations. The UBC-97 provisions [Sec. 1633.2.4] are more rigorous than previous UBC editions. This is because the importance of *deformation compatibility* has been demonstrated in recent earthquakes in population regions. For example, many vertical load-carrying structural framing elements and their connections performed poorly in the 1994 Northridge earthquake (see App. L).

When computing expected deformations for the above-mentioned elements, the UBC-97 requires the following.

◇ P-Δ effects on such elements should be considered.

◇ Expected deformations should be the greater of the maximum inelastic response displacement (Δ_M) considering P-Δ effects or the deformation caused by story drift $= 0.0025h$ (story height).

◇ The stiffening effect of such elements should not be neglected.

◇ The forces induced by the expected deformation may be considered as ultimate or factored forces.

◇ For elements constructed with concrete or masonry, the presumed flexural and shear stiffness properties should not exceed one-half of the gross section properties unless a rational cracked-section analysis is performed.

◇ Additional deformations that may result from foundation flexibility and diaphragm deflections should be considered.

6-48 PENALTIES FOR STRUCTURAL IRREGULARITY

The so-called "penalty" for some irregular structures is the requirement of a dynamic lateral-force analysis. For example, buildings greater than five stories in height in seismic zones 3 and 4 with vertical stiffness, mass, or geometry irregularities (types 1, 2, and 3 in UBC-97 Table 16-L) must receive a dynamic treatment.

However, adjustments for other types of irregularity are not adequately accomplished by a dynamic analysis. The UBC-97 penalizes irregular structures both by imposing additional requirements and by eliminating special allowances given to regular structures. For example, the total base shear can be reduced by 10% for regular structures when a dynamic analysis is performed according to the method specified by the UBC-97 [Sec. 1631.5.4, Items 1 and 2]. When a structure is irregular, this bonus is not permitted.

Irregularity is also discouraged, for example, where weak stories are prohibited [UBC-97 Sec. 1629.9.1] in buildings over two stories in height when the strength ratio based on the story above is below 65%. Heights are more limited for irregular structures [UBC-97 Sec. 1630.1.3]. Discontinuous lateral-force-resisting systems (i.e., discontinuous shear walls) must receive special checking and detailing to ensure ductile behavior [UBC-97 Sec. 1630.8].

The omission of the higher mode F_t force in overturning moment calculations is not permitted for irregular structures in calculation of soil pressure and foundation design [UBC-97 Sec. 1809.4].

Torsional irregularity is penalized by requiring the use of an increased accidental eccentricity [UBC-97 Sec. 1630.7]. Reduced stress limits are required for connections of diaphragms to vertical elements and to drag members and between drag members themselves with plan irregularities [UBC-97 Sec. 1633.2.9, Item 6]. Seismic forces in two orthogonal directions must be evaluated when torsional irregularities exist in both principal horizontal directions [UBC-97 Sec. 1633.1].

6-49 UBC PROVISIONS FOR SCALING OF RESULTS

The static lateral-force analysis is influenced by structural period, ductility, and energy dissipation for various structural systems. To reflect these influences on a consistent basis for dynamic analysis design is difficult. For this reason, the UBC-97 [Sec. 1631.5.4] provides an appropriate scaling factor for a dynamic response. This scaling depends on a response spectrum constructed either in accordance with design response spectra of UBC-97 Fig. 16-3 or a site-specific elastic design response spectrum. For regular and irregular structures, Table 6.16 provides scaling factors for corresponding design base shears.

Table 6.16
Scaling Factors

structure	response spectrum	
	UBC-97 Figure 16-3	site-specific
regular	$V_{\text{dynamic}} \geq 0.9 V_{\text{static}}$ scale factor $= \dfrac{0.9 V_{\text{static}}}{V_{\text{dynamic}}}$	$V_{\text{dynamic}} \geq 0.8 V_{\text{static}}$ scale factor $= \dfrac{0.8 V_{\text{static}}}{V_{\text{dynamic}}}$
irregular	$V_{\text{dynamic}} \geq V_{\text{static}}$	$V_{\text{dynamic}} \geq V_{\text{static}}$

A site-specific elastic design response spectrum is associated with the specific site considering the geologic, tectonic, seismologic, and soil characteristics. The design response spectrum should be developed for a damping ratio of 0.05. Based on the UBC-97 [Sec. 1631.2, Item 2], a different value of damping ratio may be shown to be consistent with the anticipated structural behavior at the intensity of shaking established for the site. If so, the design response spectrum should be developed for that damping ratio.

6-50 UBC DYNAMIC ANALYSIS PROCEDURE

Although the details of how to perform an elastic dynamic analysis are beyond the scope of this book, the basic dynamic analysis procedure required by the UBC-97 consists of three steps: (1) the static base shear is calculated; (2) a dynamic analysis using an elastic response spectrum is performed to determine the building period, base shear, story shears, and drifts; and (3) the results (with the exception of the period) are scaled upward in the ratio of the static to dynamic base shears [UBC-97 Sec. 1631.5.4].[47]

The upward scaling provision of the UBC-97 [Sec. 1631.5.4] is criticized by some structural engineers as being nonconservative. Depending on the site and geology, the scaling can double the forces for which the building must be designed.

Since the results are scaled, the magnitude of the design response spectrum is not as important as its shape (i.e., frequency content) and duration. Three different forcing inputs are permitted. (1) UBC-97 Fig. 16-3 contains an elastic design (representing 5% damping) that can be used. (2) If available, site-specific design spectrum from the actual building location can be used. (3) A spectrum ground motion time history analysis using accelerometer data from one or more actual earthquakes can be performed.[48] Enough modes must be included in the analysis to achieve a minimum 90% participation factor [UBC-97 Secs. 1631.5.1 and 1631.5.2]. (See Sec. 4-20.)

When performing a dynamic analysis, the minimum ground input must have a 10% (or greater) probability of occurring in a 50-year period [UBC-97 Sec. 1631.2]. This corresponds approximately to a Loma Prieta-sized

---

[47]Downward scaling is permitted but not required. However, the base shear cannot be scaled to less than 90% (for regular buildings) or 100% (for irregular buildings) of the static design value [UBC-97 Sec. 1631.5.4].

[48]The incremental response of the structure should be digitized with a time step that is 3 to 10 times smaller than the shortest effective modal period.

earthquake. More significant events near 8.0 on the Richter scale (e.g., the 1906 San Francisco and the 1985 Mexico City earthquakes) are considered to be exceptional situations.

Dynamic analysis is usually performed on a computer. However, the following steps can be used to carry out a manual dynamic analysis on a simple multistory structure when desired. (It is not practical to perform a dynamic analysis on structures with irregularities by hand.)

step 1: Construct a lumped-mass, two-dimensional model of the structure. (i represents the mode index; x represents the floor index.)

step 2: Calculate the mode shape factors, $\phi_{i,m}$. (See Sec. 4-18.) Normalize the mode shape factors so that $\phi = 1$ at the highest level.

step 3: Calculate the period, T_m, for each mode.

step 4: For each mode shape, calculate

$$L_m = \sum_{i=1}^{n} \left(\frac{W_i}{g} \right) \phi_{i,m} \qquad [6.50]$$

$$M_m = \sum_{i=1}^{n} \left(\frac{W_i}{g} \right) \phi_{i,m}^2 \qquad [6.51]$$

step 5: Calculate the spectral acceleration ($S_{a,m}$) and seismic design coefficient for each mode from the UBC-97 normalized response spectra [UBC-97 Fig. 16-3].

$$C_m = \frac{S_{a,m}I}{R} = \frac{C_v I}{RT} \qquad [6.52]$$

step 6: Calculate the base shear for each mode.

$$W_m = \frac{L_m^2 g}{M_m} \quad \text{[effective weight]} \qquad [6.53]$$

$$V_m = C_m \left(\frac{W_m}{g} \right) \qquad [6.54]$$

step 7: Calculate the participating mass fraction for each mode.

$$PM_m = \frac{L_m^2 g}{M_m W_t} \qquad [6.55]$$

$$W_t = \sum W_x \qquad [6.56]$$

step 8: Combine the base shears into the design dynamic lateral force, V_{dynamic}, using the SRSS (i.e., square root of the sum of the squares) method, with as many modes as are necessary to include at least 90% of the participating mass of the structural (i.e., until $\sum (PM) \geq 0.90$).

step 9: Calculate the lateral force, V_{static}, according to the static provisions of the UBC-97. Use Method A to determine the building period for the first mode.

step 10: Determine the scale factor.

$$\text{scale factor} = (0.9)\left(\frac{V_{static}}{V_{dynamic}}\right) \geq 1$$

If site-specific spectrum is used or the building is irregular, $V_{dynamic} \geq V_{static}$.

step 11: Scale $V_{dynamic}$ upward as is required by the UBC-97 in Sec. 1631.5.4.

$$V_{dynamic} = \sqrt{(V_1^2 + V_2^2 + \cdots + V_n^2)^2} \quad [6.57]$$

step 12: Distribute the scaled-up base shear to each level.

$$F_{x,m} = V_{dynamic}\left[\frac{W_x \phi_{x,m}}{\sum(W_i \phi_{i,m})}\right] \quad [6.58]$$

step 13: Determine the raw deflections, moments, and shears for each mode [UBC-97 Sec. 1631.4.1].

step 14: Use SRSS to combine the raw deflections, moments, and shears into effective values [UBC-97 Sec. 1631.5.3].

6-51 ALLOWABLE STRESS LEVELS

The allowable stress (working stress) method is a method of proportioning structural elements. The primary requirement of this design method is that calculated stresses produced in the elements by the allowable stress design load combinations (service level loads) do not exceed specified allowable stress limits. The calculated allowable stress is based on the yield stress and a reasonable factor of safety. When the loading combinations include seismic or wind loads, the UBC-97 permits an increase (usually one-third) in the allowable stress[49] because these loads are transient loads of short duration in nature that occur occasionally.

There are two one-third increases in stress when working in timber design. One is related to wind and seismic forces, and the other is specific to wood members by way of the load duration factor. The UBC-97 does not

permit an increase in allowable stresses to be used with the basic ASD load combinations [Sec. 1612.3.1], except as specifically allowed by Sec. 1809.2 (Foundation Construction). Specifically, UBC-97 Table 18-I-A (allowable foundation and lateral pressure), Ftn. 2, refers to the permitted increase.

When using the alternate basic load combinations for all materials other than wood, the UBC-97 [Secs. 1612.3.2 and 1612.3.3] permits a one-third increase in allowable stresses for all combinations including W (wind) or E (earthquake). Per UBC-97 Sec. 2316.2, Amendment 5, the one-third increase should not be used concurrently with the duration of load increase.

The UBC-97 sections governing the use of a one-third increase in allowable stresses for load combinations including W or E are as given in Table 6.17.

Table 6.17
UBC-97 Allowable Stress Sections

structural material	UBC-97 section
lightweight metals	2001.1
steel	2209, Amendments, Note 3
masonry	2106.1.3 (indirectly)
concrete	not applicable (strength design)
timber	2301.2.1 (indirectly) and 2305.4, which refers to Chap. 23, Div. III, in which Part I is 1991 NDS[1]

[1] The National Design Specification (NDS) for Wood Construction offers a preferable description of the various factors applicable to wood design. The UBC-97 [Sec. 2316.2] contains UBC amendments to the NDS. The NDS has 1.6 load duration factor for wind and seismic, coupled with 1.1 factor for diaphragms, but wind and seismic factors of 1.33 are recognized by the UBC-97 [Sec. 2316.2, Amendment 5 and Table 2.3.2]. For this, the corresponding diaphragm factor is 1.3, which gives about the same result ($1.6 \times 1.1 \sim 1.33 \times 1.3$).

It is important to recognize that in some cases, this one-third stress allowance has already been "built in" to tables provided by the UBC-97 and vendors. For example, the wood structural panel diaphragm nailing requirements (as in Table 12.1, UBC-97 Table 23-II-H) have already considered this increase, as have the connector/connection/strap/tie recommendations published by certain vendors. The table footnotes should be read to determine if this increase has already been included in published data.

For elements supporting discontinuous systems, the UBC-97 [Sec. 1630.8.2] specifies that for allowable stress design (ASD), the strength design (SD) may be determined using an increase factor of 1.7 for allowable stress

---

[49]Unfortunately, the sections of the UBC-97 that used to clearly state that a one-third increase was permitted have been obscured in the transition to move to strength design methods in the UBC-97. With strength design, stresses are not computed.

with a resistance factor (ϕ) of unity.[50] The one-third allowable stress increase of the UBC-97 [Sec. 1612.3.2] does not apply here.

There are exceptions in considering the duration of load increase and the one-third stress increase permitted in allowable stresses for elements resisting earthquake forces, such as connections of diaphragms to the vertical elements in structures having a plan irregularity in seismic zones 3 and 4 (see Sec. 6-25) according to the UBC-97 [Sec. 1633.2.9, Item 6]. For collector elements, UBC-97 Sec. 1633.2.6 specifies the same conditions as for elements supporting discontinuous systems.

---

[50]It is not appropriate to apply this reduction in LRFD load combinations because most likely it is taken into account by the various load combinations.

7

DIAPHRAGM THEORY

7-1 DIAPHRAGM ACTION

The story shears calculated by Eq. 6.29 are assumed to be applied to a lumped mass representing the floor/ceiling layer in a building. The ceiling does not actually resist the story shear, but it does distribute the force among the resisting elements (e.g., shear walls, columns, moment-resisting frames, and other structural systems).

Ceilings and floors that transmit lateral forces to the resisting elements are known as *horizontal diaphragms*. The diaphragm's function of distributing the story shears is known as *diaphragm action*. It is common to refer to the story shear as the *diaphragm force*. However, it should be recognized that the diaphragm force may include some of the story shears (including the top force, F_t) for the level *and above* [UBC-97 Sec. 1633.2.9, Item 2].

Depending on the displacement of the diaphragm relative to the supporting shear walls, diaphragms can be considered either flexible or rigid.

Wood structural panel diaphragms are almost always considered flexible. Generally, concrete slab floors and diaphragms are considered to be rigid. Steel deck diaphragms and poured-in-place gypsum floors can be either rigid or flexible, depending on their design.

7-2 SEISMIC WALL AND DIAPHRAGM FORCES

Figure 7.1 shows a simple (regular) one-story box building with a flexible diaphragm roof. (While this discussion is applicable to larger buildings, Secs. 7-2 through 7-19 are primarily concerned with simple one-story buildings with masonry or reinforced concrete walls.) Both walls are identical. An earthquake acceleration occurs in the direction shown by the ground motion arrow.

When discussing seismic forces in structures with diaphragms (e.g., one- or two-story masonry buildings with wood structural panel diaphragm floors and ceilings), it is important to distinguish between forces in the parallel and perpendicular walls. (See Fig. 7.1.) The forces in the parallel walls are shear forces, while the forces in the perpendicular walls are normal forces (i.e., perpendicular to the face of the wall). This section is primarily concerned with the shear force in the parallel walls.

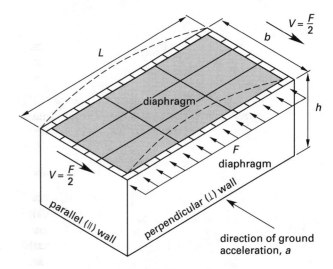

Figure 7.1 Simplified Building and Roof Diaphragm

The seismic shear force acting on the parallel walls depends on the mass being accelerated, which consists of the diaphragm weight and some portion, usually assumed to be half, of the total wall weight. (It is assumed that the seismic shear force from the remaining half of the total wall weight passes directly to the foundation without stressing the wall.) The story weight includes the weight of any equipment mounted on the roof, anything suspended inside the building from the roof, and anything mounted on the upper half of the walls. The wall weight includes all weight of any parapet that projects above the roof line.

Forces are calculated from the UBC-97 equations (see Eqs. 6.38 and 6.41).[1] Both depend on the weight, W, of the structure. (See Sec. 6-29.) Openings, such as windows and doors, in the walls that reduce the wall weight are usually disregarded when determining wall weight.

The total seismic force resisted by the two parallel walls near the ground level is the sum of seismic forces resulting from the diaphragm and wall weights. In the simple illustration of Fig. 7.1, the force on one wall is half of the total force for both rigid and flexible diaphragms.

The portion of seismic load originating from the acceleration of the perpendicular walls is given by Eq. 7.1. The symbols $\perp$ and $\parallel$ refer to "perpendicular" and "parallel," respectively.

$$F_{\perp\text{walls}} = \tfrac{1}{2}(4.0)C_a I_p W_{\perp\text{walls}} \qquad [7.1(a)]$$

$$F_{\perp\text{walls}} = \tfrac{1}{2}\left(\frac{W_{\perp\text{walls}}}{g}\right)a \qquad [7.1(b)]$$

The portion of seismic load originating from the acceleration of the parallel walls is

$$F_{\parallel\text{walls}} = \tfrac{1}{2}(4.0)C_a I_p W_{\parallel\text{walls}} \qquad [7.2(a)]$$

$$F_{\parallel\text{walls}} = \tfrac{1}{2}\left(\frac{W_{\parallel\text{walls}}}{g}\right)a \qquad [7.2(b)]$$

All of the inertial load from the accelerating walls and roof masses must be carried by the wall-roof connections. In calculating the total shear force on the parallel walls, the total walls and diaphragm weights should be used in design seismic base shear formulas.

$$W_{\text{total}} = W_{\text{diaphragm}} + W_{\perp\text{walls}} + W_{\parallel\text{walls}} \qquad [7.3]$$

$$F_{\text{total}} = \tfrac{1}{2}(4.0)C_a I_p W_{\text{total}} \qquad [7.4(a)]$$

$$F_{\text{total}} = \tfrac{1}{2}\left(\frac{W_{\text{total}}}{g}\right)a \qquad [7.4(b)]$$

There are two reasons for calculating the forces from the diaphragm and parallel walls separately. The first reason is to distinguish between the two for the purpose of subsequent calculations; that is, the parallel wall force does not contribute to chord loads and diaphragm shear where the diaphragm is flexible. The second reason is

---

[1]Equation 6.22 cannot be used to calculate the seismic force on walls, which are *elements* of the structure. Equation 6.38 (for elements of structures) must be used.

to emphasize the timing difference that occurs in a real earthquake.

The perpendicular and parallel walls experience an almost immediate force due to ground acceleration. However, the parallel walls receive the diaphragm force only after some delay. Unfortunately, an accurate analysis of this aspect of seismic behavior is almost impossible. For simple structures with three or fewer floors, the simple method of adding all forces together is used for convenience.

As with any seismic analysis, the diaphragm force must be evaluated in both orthogonal directions.

Based on the UBC-97 [Sec. 1633.2.9], floor and roof diaphragms in multi-story buildings should be designed to resist the forces determined from Eq. 7.5 (equivalent to UBC-97 Formula 33-1). (See Sec. 6-44.)

$$F_{px} = \left(\frac{F_t + \sum\limits_{i=x}^{n} F_i}{\sum\limits_{i=x}^{n} w_i}\right) w_{px} \qquad [7.5]$$

The determined force F_{px} should be equal to or greater than $0.5 C_a I w_{px}$, and need not exceed $C_a I w_{px}$. The seismic base shear should be determined in accordance with the UBC-97 [Sec. 1630.2]. For simple structures, Eq. 6.26 (equivalent to UBC-97 Formula 30-11) can be used; however, for other structures, Eq. 6.22 (equivalent to UBC-97 Formula 30-4) applies.

7-3 WALL SHEAR STRESS

In the simple building shown in Fig. 7.1, the rigidities (for a rigid diaphragm) and tributary areas (for a flexible diaphragm) are identical for the two walls. Therefore, half of the total seismic force is carried by each parallel wall. The shear stress, v, in a parallel wall of thickness t is

$$v_{\text{total}} = \frac{F_{\text{total}}}{2b} \qquad \text{[per unit length]} \qquad [7.6(a)]$$

$$v_{\text{total}} = \frac{F_{\text{total}}}{2bt} \qquad \text{[per unit area]} \qquad [7.6(b)]$$

Shear walls located on adjoining levels should be structurally continuous and should not be offset. There should be a complete transmission path from a shear wall on one level to another shear wall below.

Horizontally and vertically stacked openings in shear walls need special attention. Vertical shears need to be transferred to adjacent piers or boundary columns.

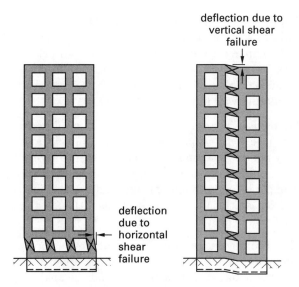

Figure 7.2   Stacked Openings

The UBC-97 [Sec. 2107.1.7] requires masonry shear walls in seismic zones 3 and 4 to be designed to resist 150% of the calculated seismic force.

7-4 RIGID DIAPHRAGM ACTION

A *rigid diaphragm* does not change its plan shape when subjected to lateral loads. It remains the same size, and square corners remain square. There is no flexure. Rigid diaphragms are capable of transmitting torsion to the major resisting elements (usually the outermost elements). The lateral story shear is distributed to the resisting elements in proportion to the rigidities of those elements.

Figure 7.3 illustrates a simple arrangement of a rigid diaphragm distributing the seismic load to two shear walls.

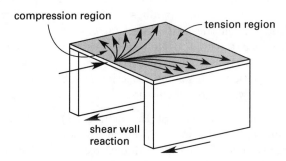

Figure 7.3   Rigid Diaphragm Action

Since the diaphragm force is distributed to the resisting elements in proportion to the rigidities of those elements, the rigidities must be determined. In practice, a few guidelines are needed to do so.

1. The relative rigidities of masonry or concrete structures can be calculated using Eqs. 4.8 and 4.9. Alternatively, Apps. D and E can be used. There is no need to use actual values of E and G, since only relative values are needed.

2. If a wall extends above roof level (i.e., has a *parapet*), the distance above the roof (i.e., the parapet height) should be disregarded when calculating the rigidity.

3. Shear walls with openings such as doors and windows require special attention. As a first approximation, such a wall can be treated as a solid shear wall. However, other methods (see Sec. 7-5) exist for evaluating the overall wall rigidity.

4. The rigidities of *transverse walls* (i.e., walls running perpendicular to the direction of the lateral force) are usually disregarded for calculating direct loads. This is called "omitting the *weak walls*." However, rigidities of all walls must be known in order to calculate torsional loads.

Example 7.1

Two walls—wall A with rigidity of 3.278 and wall B with rigidity of 7.895—support a rigid diaphragm roof in a one-story reinforced masonry building. A total seismic force of 120,000 lbf (533.8 kN) is applied parallel to the walls in such a way that there is no rotation. Determine the shear carried by each wall.

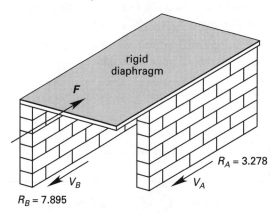

Customary U.S. Solution

The total seismic force is distributed to the walls in proportion to the rigidities.

$$V_A = \left(\frac{R_A}{R_A + R_B} \right) F$$
$$= \left(\frac{3.278}{3.278 + 7.895} \right) (120,000 \text{ lbf})$$
$$= 35,210 \text{ lbf}$$

$$V_B = \left(\frac{R_B}{R_A + R_B}\right) F$$
$$= \left(\frac{7.895}{3.278 + 7.895}\right) (120{,}000 \text{ lbf})$$
$$= 84{,}790 \text{ lbf}$$

SI Solution

The total seismic force is distributed to the walls in proportion to the rigidities.

$$V_A = \left(\frac{R_A}{R_A + R_B}\right) F$$
$$= \left(\frac{3.278}{3.278 + 7.895}\right) (533.8 \text{ kN})$$
$$= 156.6 \text{ kN}$$

$$V_B = \left(\frac{R_B}{R_A + R_B}\right) F$$
$$= \left(\frac{7.895}{3.278 + 7.895}\right) (533.8 \text{ kN})$$
$$= 377.2 \text{ kN}$$

◇   ◇   ◇   ◇   ◇   ◇   ◇

7-5 CALCULATING WALL RIGIDITY

In order to determine the rigidity of a wall with openings, it is necessary to divide the wall into piers and beams. A *pier* is a vertical portion of the wall whose height is taken as the smaller of the heights of the openings on either side of it. A *beam* is a horizontal portion left after the piers have been located.

Figure 7.4 illustrates a wall with two windows. P_1, P_2, and P_3 are piers. B_1 and B_2 are beams.

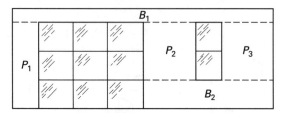

Figure 7.4 Wall with Openings

There are several methods of calculating wall rigidity from the characteristics of the wall's piers and beams, all of which yield slightly different answers.[2] The actual

---

[2]These other methods are approximate and often do not agree. Also, it is not uncommon for the methods to determine the rigidity of a wall—particularly a wall with fixed-pier assumptions—with openings to be greater than a solid wall (an obvious impossibility). The rigidity of a wall with openings should be compared to the rigidity of a solid wall of the same dimensions.

rigidity for any particular wall in a structure is unimportant. What is important is the *relative rigidity* of the wall compared to all other resisting walls in the structure. Thus, it is essential that the same method be used to calculate the rigidities of all walls. Two methods for calculating wall rigidity are discussed below.

Although it is generally assumed that cantilever conditions (see Sec. 4-4) prevail for the walls in one- and two-story buildings taken in their entireties, piers within the wall can be considered either fixed or cantilevered. For example, piers between openings may be considered to be fixed at their tops and bottoms although the wall taken as a whole is cantilevered.

The accuracy in wall rigidity calculations is not great. There are many assumptions made about material properties and wall performance, and the analysis procedure, though formalized, is less than rigorous. Therefore, values with more than three or four significant digits are unwarranted.

Method A

With this fast and simple method (used only for preliminary analyses), the rigidity of a wall is calculated as the sum of the rigidities of the individual piers framed between openings in a wall. All piers are assumed to be fixed. The pier height is the height of the shortest adjacent opening.[3] Beams and wall portions above and below the openings are not considered.

Method B

By far the most commonly used method of evaluating wall rigidities calculates "deflections" from standardized values of force, thickness, and modulus of elasticity. These deflections are recognized as being the reciprocals of rigidity.

To start, the gross deflection of the solid wall is calculated, ignoring all openings and assuming cantilever action. Then the strip deflection of an interior strip having length equal to the wall length and height equal to the tallest opening is calculated, again assuming cantilever action. This strip deflection is subtracted from the solid wall's gross deflection.

Next, the rigidities (not the deflections) of all piers (assuming fixed ends) within the removed strip are summed, and the pier deflection correction is calculated as the reciprocal of the sum. The pier deflection correction is added to the difference of the gross and strip deflections to give the net deflection. The wall rigidity is the reciprocal of this net deflection.

---

[3]A common error is to use the ground-to-ceiling distance.

The rigidity of the wall must be calculated by this method one opening at a time, considering the fixed pier adjacent to the opening and the wall section below the opening. The calculation of the pier deflection correction becomes recursive when openings in the wall are of different heights. For this reason, Method B can take a long time if the wall is relatively complex.

◇   ◇   ◇   ◇   ◇   ◇   ◇

Example 7.2

The masonry wall shown in Fig. 7.4 and dimensioned below has a uniform thickness and is part of a one-story building. Determine the rigidity using the two methods described in Sec. 7-5.

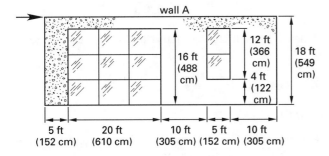

wall A

16 ft (488 cm)   12 ft (366 cm)   4 ft (122 cm)   18 ft (549 cm)

5 ft (152 cm)   20 ft (610 cm)   10 ft (305 cm)   5 ft (152 cm)   10 ft (305 cm)

Customary U.S. Solution

Method A

The total rigidity of the wall is the sum of all the pier rigidities. Beams are disregarded.

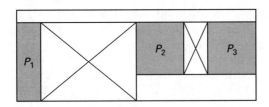

P_1 P_2 P_3

The rigidities, R, of all piers (assumed fixed) are obtained from Eq. 4.8 or App. D after calculating the height/depth ratio, h/d, of each.

element	h	d	h/d	R
P_1	16	5	3.2	0.236
P_2	12	10	1.2	1.877
P_3	12	10	1.2	1.877

In this method, the rigidity of the wall is the sum of the individual fixed-pier rigidities.

$$R = 0.236 + 1.877 + 1.877 = 3.99$$

Method B

First determine the deflection of a solid wall. The height/depth ratio of the entire wall is

$$\frac{h}{d} = \frac{18 \text{ ft}}{50 \text{ ft}} = 0.36$$

Since this is a one-story building, the wall taken as a whole is assumed to be cantilevered. From App. E, the rigidity is 7.895. The "deflection" of the solid wall is

$$\Delta_{\text{solid}} = \frac{1}{R} = \frac{1}{7.895} = 0.1267$$

The highest opening has a height of 16 ft, so a "mid-strip" 16 ft high and 50 ft long is removed. Assuming a solid wall, the height/depth ratio is

$$\frac{h}{d} = \frac{16 \text{ ft}}{50 \text{ ft}} = 0.32$$

The entire wall was assumed to be a cantilever pier, and this mid-strip represents the majority of the wall. Therefore, it also is a cantilever member. Appendix E gives the rigidity as 9.165. The "deflection" of an assumed solid mid-strip is

$$\Delta_{\text{mid-strip}} = \frac{1}{9.165} = 0.109$$

The mid-strip, however, is not solid. It consists of two functional parts: a 5 ft by 16 ft solid pier (P_1) at the left and a larger section with a small window at the right. The large window section running floor-to-ceiling is assumed to contribute no rigidity to the wall. The rigidity of this mid-strip is the sum of the rigidities of pier P_1 and the larger section.

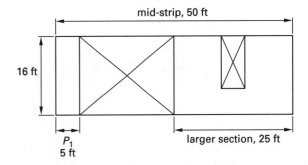

mid-strip, 50 ft

16 ft

P_1 5 ft

larger section, 25 ft

For the solid pier, P_1, at the left,

$$\frac{h}{d} = \frac{16 \text{ ft}}{5 \text{ ft}} = 3.2$$

For the same reason for considering the mid-strip to be a cantilever, this pier is assumed to act as a cantilever. From App. E, $R = 0.071$. (The "deflection" of pier P_1 is not needed.)

The larger section at the right of the wall has dimensions of 16 ft by 25 ft. Taken as a solid cantilever pier, the gross rigidity and deflection are

$$\frac{h}{d} = \frac{16 \text{ ft}}{25 \text{ ft}} = 0.64$$

$$R_{\text{larger section,gross}} = 3.37$$

$$\Delta_{\text{larger section,gross}} = \frac{1}{3.37} = 0.297$$

However, the larger section is not itself solid. There is a 12-ft-high window. For a 12 ft by 25 ft solid window strip assumed to act as a cantilever,

$$\frac{h}{d} = \frac{12 \text{ ft}}{25 \text{ ft}} = 0.48$$

$$R_{\text{window strip}} = 5.31$$

$$\Delta_{\text{window strip}} = \frac{1}{5.31} = 0.188$$

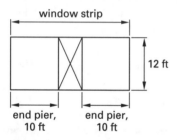

The window strip contributes stiffness, as there are two 12 ft by 10 ft end piers. Although cantilever performance could be argued as well, these two end piers are assumed to be fixed. Their combined rigidity and deflection are

$$\frac{h}{d} = \frac{12 \text{ ft}}{10 \text{ ft}} = 1.2$$

$$R_{\text{end piers}} = (2)(1.877) = 3.754$$

$$\Delta_{\text{end piers}} = \frac{1}{3.754} = 0.266$$

Now that all the pieces have been evaluated, the rigidity of the entire wall can be built up.

The net deflection and net rigidity of the larger section is

$$\Delta_{\text{larger section,net}} = 0.297 - 0.188 + 0.266$$
$$= 0.375$$

$$R_{\text{larger section,net}} = \frac{1}{0.375} = 2.67$$

(Check that 2.67 is less than the gross value of 3.37.)

Since the highest opening in the mid-strip extends the full height, the rigidity is merely the sum of the rigidities of pier P_1 and the larger section.

$$R_{\text{mid-strip,net}} = 0.071 + 2.67 = 2.74$$

$$\Delta_{\text{mid-strip,net}} = \frac{1}{2.74} = 0.365$$

(Check that 2.74 is less than the gross value of 9.165.)

The deflection and relative rigidity of the entire wall is

$$\Delta_{\text{entire wall}} = 0.127 - 0.109 + 0.365 = 0.383$$

$$R_{\text{entire wall}} = \frac{1}{0.383} = 2.61$$

(Check that 2.61 is less than the gross value of 7.895.)

SI Solution

Method A

The total rigidity of the wall is the sum of all the pier rigidities. Beams are disregarded.

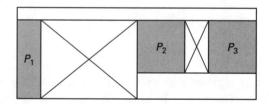

The rigidities, R, of all piers (assumed fixed) are obtained from Eq. 4.8 or App. D after calculating the height/depth ratio, h/d, of each.

element	h	d	h/d	R
P_1	488	152	3.2	0.236
P_2	366	305	1.2	1.877
P_3	366	305	1.2	1.877

In this method, the rigidity of the wall is the sum of the individual fixed-pier rigidities.

$$R = 0.236 + 1.877 + 1.877 = 3.99$$

Method B

First determine the deflection of a solid wall. The height/depth ratio of the entire wall is

$$\frac{h}{d} = \frac{549 \text{ cm}}{1524 \text{ cm}} = 0.36$$

Since this is a one-story building, the wall taken as a whole is assumed to be cantilevered. From App. E, the rigidity is 7.895. The "deflection" of the solid wall is

$$\Delta_{\text{solid}} = \frac{1}{R} = \frac{1}{7.895} = 0.1267$$

The highest opening has a height of 488 cm, so a "mid-strip" 488 cm high and 1524 cm long is removed. Assuming a solid wall, the height/depth ratio is

$$\frac{h}{d} = \frac{488 \text{ cm}}{1524 \text{ cm}} = 0.32$$

The entire wall was assumed to be a cantilever pier, and this mid-strip represents the majority of the wall. Therefore, it also is a cantilever member. Appendix E gives the rigidity as 9.165. The "deflection" of an assumed solid mid-strip is

$$\Delta_{\text{mid-strip}} = \frac{1}{9.165} = 0.109$$

The mid-strip, however, is not solid. It consists of two functional parts: a 152 cm by 488 cm solid pier (P_1) at the left and a larger section with a small window at the right. The large window section running floor-to-ceiling is assumed to contribute no rigidity to the wall. The rigidity of this mid-strip is the sum of the rigidities of pier P_1 and the larger section.

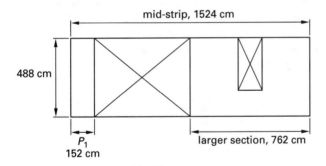

For the solid pier, P_1, at the left,

$$\frac{h}{d} = \frac{488 \text{ cm}}{152 \text{ cm}} = 3.2$$

For the same reason for considering the mid-strip to be a cantilever, this pier is assumed to act as a cantilever. From App. E, $R = 0.071$. (The "deflection" of pier P_1 is not needed.)

The larger section at the right of the wall has dimensions of 488 cm by 762 cm. Taken as a solid cantilever pier, the gross rigidity and deflection are

$$\frac{h}{d} = \frac{488 \text{ cm}}{762 \text{ cm}} = 0.64$$

$$R_{\text{larger section,gross}} = 3.37$$

$$\Delta_{\text{larger section,gross}} = \frac{1}{3.37} = 0.297$$

However, the larger section is not itself solid. There is a 366-cm-high window. For a 366 cm by 762 cm solid window strip assumed to act as a cantilever,

$$\frac{h}{d} = \frac{366 \text{ cm}}{762 \text{ cm}} = 0.48$$

$$R_{\text{window strip}} = 5.31$$

$$\Delta_{\text{window strip}} = \frac{1}{5.31} = 0.188$$

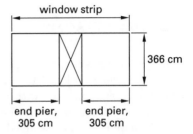

The window strip contributes stiffness, as there are two 366 cm by 305 cm end piers. Although cantilever performance could be argued as well, these two end piers are assumed to be fixed. Their combined rigidity and deflection are

$$\frac{h}{d} = \frac{366 \text{ cm}}{305 \text{ cm}} = 1.2$$

$$R_{\text{end piers}} = (2)(1.877) = 3.754$$

$$\Delta_{\text{end piers}} = \frac{1}{3.754} = 0.266$$

Now that all the pieces have been evaluated, the rigidity of the entire wall can be built up.

The net deflection and net rigidity of the larger section is

$$\Delta_{\text{larger section,net}} = 0.297 - 0.188 + 0.266$$
$$= 0.375$$

$$R_{\text{larger section,net}} = \frac{1}{0.375} = 2.67$$

(Check that 2.67 is less than the gross value of 3.37.)

Since the highest opening in the mid-strip extends the full height, the rigidity is merely the sum of the rigidities of pier P_1 and the larger section.

$$R_{\text{mid-strip,net}} = 0.071 + 2.67 = 2.74$$

$$\Delta_{\text{mid-strip,net}} = \frac{1}{2.74} = 0.365$$

(Check that 2.74 is less than the gross value of 9.165.)

The deflection and relative rigidity of the entire wall is

$$\Delta_{\text{entire wall}} = 0.127 - 0.109 + 0.365 = 0.383$$

$$R_{\text{entire wall}} = \frac{1}{0.383} = 2.61$$

(Check that 2.61 is less than the gross value of 7.895.)

7-6 FLEXIBLE DIAPHRAGMS

A flexible diaphragm changes shape when subjected to lateral loads. Its tension chord bends outward, and its compression chord bends inward, with a deflection shape similar to that of a simply supported beam loaded uniformly. Flexible diaphragms are assumed to be incapable of transmitting torsion to the resisting elements. (Also, see Ch. 5, Ftn. 9.)

A flexible diaphragm distributes the diaphragm force in proportion to the tributary areas of the diaphragm, as opposed to distributing it in proportion to the rigidities of the vertical resisting elements, as does a rigid diaphragm.

As defined by the UBC-97, a *flexible diaphragm* is one that has a maximum lateral deflection more than two times the average story drift [UBC-97 Sec. 1630.6]. To determine if a diaphragm is flexible, compare the in-plane deflection at the midpoint of the diaphragm to the story drift of the adjoining vertical resisting elements under equivalent tributary load.

7-7 FRAMING TERMINOLOGY

Figure 7.5 illustrates such common wood framing terms as *sheathing, girder, beam, purlin,* and *joist* (or *sub-purlin*), as well as the *bridging* and *blocking* that are used to prevent lateral buckling. Usually blocking (e.g., often cut from the same material as the joists, although other blocking techniques are used) frames into joists or sub-purlins (two-by-sixes, two-by-eights, etc.); joists frame into purlins (e.g., four-by-eights); purlins frame into beams (e.g., four-by-fourteens); beams frame into girders (e.g., glulams); and girders frame into the walls.[4]

---

[4]These terms are not so rigidly defined that they preclude incorrect usage.

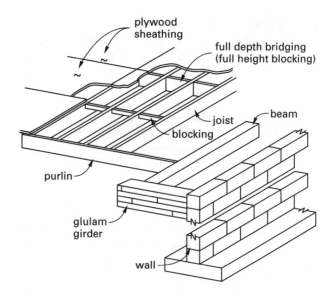

Figure 7.5   Framing Members

7-8 COLLECTORS

It is required that *collectors* (also known as *drag struts, braces,* or merely *struts* or *ties*) be used to transmit diaphragm reactions to shear walls at points of discontinuity (irregularity) in the plan [UBC-97 Sec. 1633.2.9, Item 4]. These collectors effectively separate the diaphragm into subdiaphragms that are analyzed independently. Diaphragm sheathing alone may not be relied on to transfer loads between chords [UBC-97 Sec. 2315.5.2].

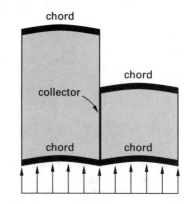

Figure 7.6   Use of a Collector

7-9 COLLECTOR FORCE

The collector force is the product of the diaphragm load (per unit area) and the areas tributary to the collector. Tributary areas are usually taken as some fraction of the subdiaphragm areas located on either side of the collector. The force is not the same along the length of the collector but increases to a maximum at the point where the collector frames into a shear wall.

7-10 COLUMNS SUPPORTING PARTS OF FLEXIBLE DIAPHRAGMS

An interior or exterior column can be used to support the vertical roof or floor load, but such a column usually provides no lateral support when the diaphragm is flexible. (The seismic response coefficient, R, is 2.2 for systems where the column is presumed to be part of the lateral system of the building.) A girder that frames into such a column at one girder end and a shear wall at the other girder end almost always acts as a collector for seismic forces parallel to the girder direction.

Example 7.3

The plan view of an irregular building is shown. (a) Determine the tributary areas for an earthquake in the north-south direction. (b) Determine where the force in the collector is maximum.

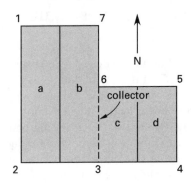

Solution

(a) The walls along sides 1-2, 6-7, and 4-5 will resist the seismic force. (Walls 1-7, 5-6, and 2-3-4 are *perpendicular walls*. See Sec. 7-2.) The collector between areas b and c splits the diaphragm into two rectangular diaphragms: a-b and c-d.

Area a is tributary to wall 1-2. Area b is tributary to wall 6-7. The forces transmitted to the collector are carried by the collector back to wall 6-7.

Area d is tributary to wall 4-5. Area c is tributary to the collector, which transmits all of the area c diaphragm force into wall 6-7.

(b) The collector carries half of area b's diaphragm force and all of area c's diaphragm force. The maximum value of this force occurs at point 6, where the collector frames into the shear wall 6-7.

7-11 FLEXIBLE DIAPHRAGM CONSTRUCTION

A flexible diaphragm is a relatively thin structural element such as a roof or floor attached to relatively rigid walls. It can be constructed as a braced frame with non-structural covering, or as joists sheathed with wood structural panel, boards, or gypsum sheets.

Horizontal diaphragms are most common and are the simplest to analyze. However, diaphragms do not need to be horizontal in order to resist shear. A plywood roof can be inclined, peaked (i.e., folded-plate), or curved. In such cases, the roof trusses act as web stiffeners. A non-horizontal diaphragm is analyzed according to its footprint (plan) dimensions.

There are three structural requirements imposed on flexible diaphragms.

1. The diaphragm must be strong enough to remain intact under the action of wind and seismic loads.

2. The diaphragm must be securely attached to a wall in order to resist forces parallel to the wall.

3. The diaphragm must be securely attached to a wall in order to react to forces perpendicular to the wall.

Based on the UBC-97 [Sec. 1633.2.9], flexible diaphragms providing lateral supports for walls or frames of either concrete or masonry should be designed using a value of seismic response coefficient R not exceeding 4.0.

7-12 FLEXIBLE DIAPHRAGM TORSION

There is no torsional shear stress (see Sec. 5-14) from eccentric mass placement in either the walls or diaphragm because flexible diaphragms are not considered capable of distributing torsional shear stresses.

7-13 DIAPHRAGM SHEAR STRESS

In most cases, the criterion by which diaphragm construction and connections is evaluated[5] is the *diaphragm shear stress*, $v_{\text{diaphragm}}$.[6] The stress is assumed

---

[5] The larger of the seismic and wind shear stresses will be used in the design, but not both.

[6] The maximum allowable shear stress on wood structural panel diaphragms depends on the nail size and spacing, panel thickness, and width of framing members, and on whether or not the panel

to exist uniformly across the length b, known as the *diaphragm depth*. (The total force is shared by the two parallel walls, each of depth b. The perpendicular walls do not resist the applied seismic force.) The parallel wall force is not included in the diaphragm shear loading.

$$v_{\text{diaphragm}} = \frac{F_{\text{diaphragm}}}{2b} \quad \text{[per unit length]} \quad [7.7]$$

7-14 DIAPHRAGM FLEXURE

A flexible diaphragm is designed to withstand shear in its plane. It has no bending strength of its own. Rather, the diaphragm relies on the stiffness of its chords to limit overall diaphragm deflection. A common analogy is to assume the diaphragm acts like a girder, where the flanges (i.e., the chord members) resist the bending moment, and the web (i.e., the diaphragm) resists the shear. This is illustrated in Fig. 7.7.

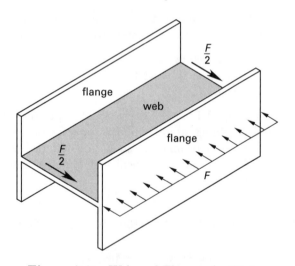

Figure 7.7   Web and Flange of a Girder

The diaphragm shear stress is assumed to be linearly distributed from zero at the midpoint (i.e., at $L/2$) to $v_{\text{diaphragm}}$ at the parallel walls. (Distance L in Fig. 7.8 is known as the *diaphragm span*). At any particular point, the shear stress is uniform across the diaphragm between perpendicular walls. (The girder analogy fails here since the shear stress distribution between perpendicular walls is not parabolic.) Edge nailing of the wood

edges are blocked, among other factors. Typical values range from 100 to 800 lbf/ft (1460 to 11 675 N/m), with the lower values (i.e., 100 to 300 lbf/ft (1460 to 4380 N/m)) applying to unblocked diaphragms. (See Table 12.1.) Refer to the code for exact shear stress limits and required nail spacing [UBC-97 Table 23-II-H].

structural panel to the framing keeps the shear resistance continuous across the diaphragm.

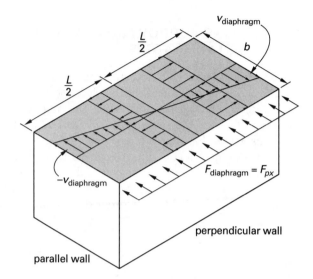

Figure 7.8   Shear Stress Distribution on a Diaphragm

7-15 DEFLECTION OF FLEXIBLE DIAPHRAGMS

A flexible diaphragm will deflect. The beam analogy described in Sec. 7-14 is also valid here, as the diaphragm assumes the deflected shape of a simply supported beam loaded by a uniform load. The deflection is resisted by the perpendicular walls. Since the perpendicular wall deflection is the same as the diaphragm deflection, this deflection may be the factor that limits the force that can be safely applied to the diaphragm.

Actual determination of the deflection in a wood structural panel diaphragm is complex, as it is for any wood/timber structural member, but procedures are available. The following equation [UBC-97 Standards 23-2 and 23-3] calculates the plywood diaphragm deflection as a sum of *flexural distortion* (the first term), *shear distortion* (the second term), and *nail distortion* (the third term). In some cases, such as in wood-framed perpendicular walls where a wood double-plate serves as the chord, the fourth term may be added to account for *chord-splice slip* values. Such slippage is neglected with masonry walls.

$$\text{deflection (in)} = \frac{5vL^3}{8EAb} + \frac{vL}{4Gt} + 0.188Le_n + \frac{\sum(\Delta_c X)}{2b}$$
$$\text{[U.S.]} \quad [7.8(a)]$$

$$\text{deflection} = \frac{52vL^3}{EAb} + \frac{vL}{4Gt} + 0.614Le_n + \frac{\sum(\Delta_c X)}{2b}$$
$$\text{[SI]} \quad [7.8(b)]$$

v = maximum shear in pounds per foot
(lbf/ft or N/m)

L = diaphragm length (ft or m)

E = modulus of elasticity of chord
(psi or kPa)

b = diaphragm width or depth (ft or m)

A = section area of chord (in^2 or cm^2)

t = wood structural panel thickness
(in or cm)

e_n = nail slip, at load per nail (in or cm)

$\sum(\Delta_c X)$ = sum of individual chord-splice slip values
on both sides of the diaphragm, each
multiplied by its distance to the nearest
support

G = wood structural panel modulus of rigidity
(psi or kPa, typically taken as 90,000 psi
(620 530 kPa) for wood structural panel.
Alternatively, $G = E/20$ for panels with
exterior glue.)

The nail slip, e_n, depends on the nail size, wood structural panel thickness, load per nail, and type of lumber. Usually, worst-case green lumber is assumed. Nail slip is usually obtained graphically from appropriate sources. Values are typically less than 0.15 in (4 mm), with most values being half that amount.

The maximum deflection of the diaphragm is the acceptable limitation of deflection or drift for the perpendicular walls directly below the diaphragm. This limitation will depend on whether the walls are masonry or concrete and on which authority specifies the limitation. The UBC-97 does not specify actual limitations on diaphragm deflection, but it limits such deflection to amounts that maintain the structural integrity and protect occupants [UBC-97 Sec. 1633.2.9, Item 1].[7]

---

[7]The California *Office of Architecture and Construction* limits deflection to 1/16 in per foot of wall height. SEAOC has developed the following formula for the maximum allowable deflection.

$$\text{deflection (in)} = \frac{75h^2 F_b}{Et}$$

In the above equation, h is the wall height (feet), F_b is the allowable masonry flexural (compressible) stress (psi, increased by the 1/3 allowance for seismic loads), E is the masonry modulus of elasticity (1,500,000 psi for masonry, 2,000,000 psi for concrete), and t is the wall thickness (in inches).

A similar equation has been proposed by the American Institute of Timber Construction, in which the 75 is replaced by 96.

The *Reinforced Masonry Engineering Handbook* suggests the following equation for maximum deflection. This equation can be derived from the SEAOC equation if E = 1,500,000 psi and $F_b = (4/3)(900 \text{ psi})$ is used. 900 psi is appropriate for concrete walls but may be too high for masonry walls.

$$\text{deflection (in)} = \frac{2h^2}{45t}$$

Diaphragm deflection calculations are generally unnecessary and are waived if the UBC-97 provisions (Table 23-II-G) for maximum *diaphragm ratios* for size are followed. (See Sec. 12-4.) The span-to-width (L/b) ratio for most horizontal diaphragms (including edge-nailed wood structural panel diaphragms) is limited to 4:1, except for conventionally constructed diagonal sheathing (as defined by the UBC-97 in Sec. 2315.3.1) for which the limitation is reduced to 3:1.

If the deflection is excessive, it can be reduced by increasing the wood structural panel thickness, decreasing the nail spacing, adding a collector strut, or placing an additional shear wall within the building to reduce the diaphragm span.

7-16 CHORDS

The elements—wall top plates or reinforcement—capable of supporting chord (i.e., compressive and tensile) forces at the edges of the diaphragm along the perpendicular walls are known as *chords*. Chords are generally considered to be tension and compression members, analogous to the flanges of the beam shown in Fig. 7.7. A diaphragm will be constructed with chord elements along all outer edges. The chords that run perpendicular to the applied force, that is, along the perpendicular walls, and are stressed during an earthquake are called the *active chords*. The chord elements in parallel walls are known as the *passive chords*.

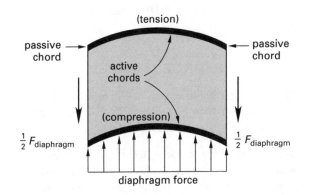

Figure 7.9 Chords

In masonry buildings, reinforcement in the perpendicular walls themselves can (and are intended to) serve as chords if the force is transferred through solidly attached *ledgers* (see Sec. 12-9). This assumes that the masonry walls have adequate tensile reinforcement. Alternatively, a properly spliced wood ledger beam or a bond beam using an embedded reinforcing bar could serve as the chord with masonry walls, depending on the method of attachment. Chords can also be wood (e.g., the double top-plate in conventional wood stud walls), steel, or any other continuous material connected to the diaphragm edge.

The diaphragm may be functionally divided into independent parts, known as *sub-diaphragms*. In this case, chords and struts will run through the diaphragm in addition to around it [UBC-97 Sec. 1633.2.9, Item 4]. One method of supporting internal chord members in masonry construction is with pilasters, as shown in Fig. 7.10. A *pilaster* is constructed as part of the masonry wall and is designed as a column.

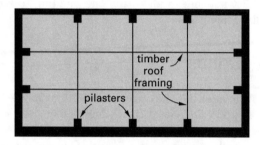

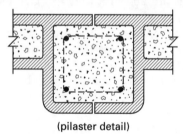

(pilaster detail)

Figure 7.10   Pilaster

7-17 CHORD FORCE

The diaphragm itself is assumed incapable of supporting a normal (bending) stress. Chords are designed to carry all tensile and compressive forces.

The maximum chord force occurs at mid-span ($L/2$). The maximum chord force, C, is calculated as the bending moment of a simple beam under a distributed load (i.e., $wL^2/8$, with w in lbf/ft or N/m) divided by the depth, b, of the diaphragm. The distributed load is $w = F_{\text{diaphragm}}/L$, where $F_{\text{diaphragm}}$ is in pounds (N). (For wind loads, w is the wind load alone.)

$$C = \frac{M}{b} = \frac{wL^2}{8b} = \frac{F_{\text{diaphragm}}L}{8b} \qquad [7.9]$$

Equation 7.9 is derived by equating the applied moment to the resisting moment. The applied moment is $wL^2/8$. The resisting moment comes from two sources: the two chords. One chord is in   compression; the other is in

tension. Both act with a moment arm of $b/2$ (with respect to a neutral axis passing through the midpoint of the diaphragm).

$$\frac{wL^2}{8} = C\left(\frac{b}{2}\right) + C\left(\frac{b}{2}\right) = Cb \qquad [7.10]$$

$$C = \frac{wL^2}{8b} \qquad [7.11]$$

The minimum chord force is zero and occurs at the chord ends. At intermediate locations, the force follows the shape of the bending moment between the two parallel walls.

7-18 CHORD SIZE

The required chord size (cross-sectional area) can be determined from the allowable stress in tension or compression, whichever is less, for the chord material. A one-third increase is permitted when using ASD because the loading is seismic. (See Sec. 6-51.)

$$A_{\text{chord}} = \frac{C}{\frac{4}{3} \times \text{allowable stress}} \qquad [7.12]$$

The chord area may be reduced near the parallel walls, but the reduction must be in accordance with the actual moment distribution.

The allowable tensile stress for steel reinforcing bar depends on its grade (equivalent to its minimum yield strength in ksi (MPa)). Allowable tensile stresses are 20 ksi (137.9 MPa) for grades 40 and 50 steel, and 24 ksi (165.5 MPa) for grade 60 steel [UBC-97 Sec. 1926.3.2, Items 1 and 2].

7-19 OVERTURNING MOMENT

In addition to seismic shear loading, a shear wall (see Fig. 7.11) will be subjected to overturning moments as well. Overturning will not be a problem, however, if there is a larger resisting moment.

The force causing overturning is the seismic force of V_{wall} at the top of the shear wall. If the shear wall has a length of b', the total overturning force parallel to the ground is $V_{\text{wall}} = \bar{v}_{\text{diaphragm}}b'$. ($b'$ and b may be the same if the parallel wall is one piece, or b' may correspond to the length of a tilt-up wall section.) This total roof load acts with a moment arm of h, the height of the force above the ground.

Also contributing to overturning is the seismic force due to self-weight, calculated as either $W_{wall}a/g$ or $4.0C_aI_pW_{wall}$. This seismic force acts halfway up the parallel wall, with a moment arm of $h/2$.

$$M_{overturning} = V_{wall}h + \frac{V_{self}h}{2} = V_{wall}h + \frac{W_{wall}ah}{2g}$$

$$= V_{wall}h + \frac{4.0C_aI_pW_{wall}h}{2} \qquad [7.13]$$

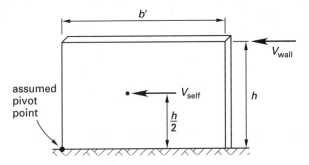

Figure 7.11   Overturning Forces on a
Parallel Wall

The overturning moment is resisted by the weight of the parallel wall and the distributed roof dead load (calculated as the roof load tributary to that panel), both acting with a moment arm of $b'/2$.

$$M_{resisting} = (D_{roof} + W_{wall})\left(\frac{b'}{2}\right) \qquad [7.14]$$

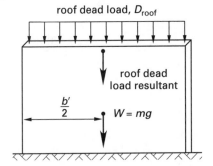

Figure 7.12   Resisting Forces on a Parallel Wall

Measures to resist overturning are well known and include anchoring the parallel wall panels to the foundations and, in the case of tilt-up slab construction, interconnecting adjacent panels.

Example 7.4

A simple four-walled 40 ft by 50 ft (12.2 m by 15.2 m) building is part of a hazardous chemical chlorine storage facility located in seismic zone 4. It is constructed with a wood structural panel sheathed roof on 10-ft-high (3 m) concrete masonry unit (CMU) bearing walls 8 in (203 mm) thick. All walls are reinforced vertically and horizontally. The average weight of the roof diaphragm and mounted equipment is 20 lbf/ft² (0.96 kN/m²). All connections between walls, roof, and foundation are pinned. The building performance is being analyzed for an earthquake acting parallel to the short dimension. Disregard all openings in the walls. The value of the near-source factor, N_a, is 1.0.

(a) Find the unit shear in the diaphragm on line A-B.

(b) Find the required diaphragm edge nailing spacing on line A-B. (Assume 6d nails, case 1 plywood layout, 3/8 in (10 mm) C-D wood structural panel on blocked 2 in (51 mm) frame members.)

(c) Find the maximum chord force on line B-B'.

(d) Find the horizontal shear stress (in psi or kPa) in wall A-B at a point 5 ft (1.5 m) above the foundation.

(e) Determine whether the wall thickness is adequate.

(f) If wall A-B was to be reduced to only 10 ft (3 m) long and the remaining 30 ft (9.1 m) of roof supported by a collector, find the collector force at the end of the wall.

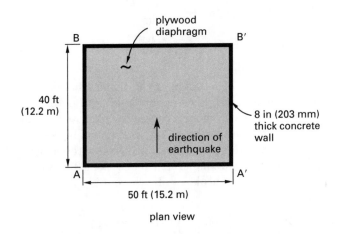

plan view

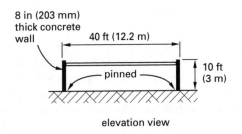

elevation view

Customary U.S. Solution

From Table 6.3, the seismic zone factor for seismic zone 4 is $Z = 0.40$. From Table 6.5 for a hazardous facility, the seismic importance factor, I, is 1.25. From App. K (corresponding to UBC-97 Table 16-K), $I_p = 1.50$. Inasmuch as the soil type is unknown, soil-profile type S_D must be used. (See Sec. 6-15.)

From Table 6.10 for a bearing wall system consisting of concrete shear walls, $R = 4.5$. From Table 6.9, for a soil-profile S_D and $Z = 0.4$, the seismic response coefficient $C_a = 0.44N_a$. With $N_a = 1.0$, the value of $C_a = 0.44$. Thus, the base shear equation (Eq. 6.26) is

$$V = \frac{3.0C_aW}{R} = \frac{(3.0)(0.44)W}{4.5}$$
$$= 0.293W$$

The weight being accelerated by the earthquake consists of the diaphragm weight and a portion of the wall weight. The weight of the diaphragm is

$$W_{\text{diaphragm}} = \left(20 \, \frac{\text{lbf}}{\text{ft}^2}\right)(40 \text{ ft})(50 \text{ ft})$$
$$= 40,000 \text{ lbf}$$

Since concrete has a density of 150 lbf/ft³, the weight of 1 ft² of an 8-in-thick wall is

$$\gamma = \frac{(8 \text{ in})\left(150 \, \frac{\text{lbf}}{\text{ft}^3}\right)}{12 \, \frac{\text{in}}{\text{ft}}}$$
$$= 100 \, \text{lbf/ft}^2$$

For the purpose of determining the diaphragm force, the upper half (i.e., only the upper 5 ft) of both perpendicular walls (i.e., the chords) is used to calculate the wall weight. (See Secs. 6-29 and 7-2.) The remaining seismic force passes directly into the foundation without being carried by the wall-diaphragm connection. The weight of half the perpendicular walls (chords) is

$$W_{\perp\text{walls}} = \left(100 \, \frac{\text{lbf}}{\text{ft}^2}\right)(5 \text{ ft})(2 \text{ walls})\left(\frac{50 \text{ ft}}{\text{wall}}\right)$$
$$= 50,000 \text{ lbf}$$

(a) Equation 6.26 is for calculating the base shear passing through to the foundation and should not strictly be used to calculate the force on connections between elements in the building. Equation 6.38 is for elements of structures and their attachments and must be used. (See Sec. 6-44.) For diaphragms, based on UBC-97 Sec. 1633.2.9, (1) the roof diaphragm is to be designed

to resist a portion of the floor forces above it, as weighted by the floor weights, and (2) the diaphragm force must be within $0.5C_aIW_{px}$ and $1.0C_aIW_{px}$. With the values of $C_a = 0.44$ and $I = 1.25$, the limits on diaphragm force are $0.275W_{px}$ and $0.550W_{px}$.

Inasmuch as this is a simple one-story building with only one diaphragm, all of the inertial load from the accelerating wall and roof masses must be carried by the wall-roof connection, so Eq. 6.26 with the wall and diaphragm weights is ultimately used. Checking, $0.275 < 0.293 < 0.550$ (ok).

$$F_{\text{diaphragm}} = 0.293W$$
$$= (0.293)(40,000 \text{ lbf} + 50,000 \text{ lbf})$$
$$= 26,370 \text{ lbf}$$

The shear per foot of diaphragm width, b, is given by Eq. 7.7.

$$v = \frac{F_{\text{diaphragm}}}{2b} = \frac{26,370 \text{ lbf}}{(2)(40 \text{ ft})}$$
$$= 329.6 \, \text{lbf/ft}$$

(b) Table 12.1 (corresponding to UBC-97 Table 23-II-H) gives the nail spacing directly. The allowable shear is 375 lbf/ft (> 329.6 lbf/ft), with a nail spacing of 2.5 in.

(c) The distributed seismic force, w, across the face of the diaphragm is

$$w = \frac{F_{\text{diaphragm}}}{L} = \frac{26,370 \text{ lbf}}{50 \text{ ft}}$$
$$= 527.4 \, \text{lbf/ft}$$

From Eq. 7.9, the chord force, C, is

$$C = \frac{wL^2}{8b} = \frac{\left(527.4 \, \frac{\text{lbf}}{\text{ft}}\right)(50 \text{ ft})^2}{(8)(40 \text{ ft})}$$
$$= 4120.3 \text{ lbf}$$

(d) The net effect is to include the inertial force for accelerating the diaphragm mass and half of all the walls. The diaphragm and perpendicular wall weights have already been determined. Half the weight of the parallel walls (i.e., the shear walls) is

$$W_{\parallel\text{walls}} = \left(100 \, \frac{\text{lbf}}{\text{ft}^2}\right)(5 \text{ ft})(2 \text{ walls})\left(40 \, \frac{\text{ft}}{\text{wall}}\right)$$
$$= 40,000 \text{ lbf}$$

The total weight is

$$W_{\text{total}} = W_{\text{diaphragm}} + W_{\perp\text{walls}} + W_{\|\text{walls}}$$
$$= 40{,}000 \text{ lbf} + 50{,}000 \text{ lbf} + 40{,}000 \text{ lbf}$$
$$= 130{,}000 \text{ lbf}$$

The seismic force is

$$V = 0.293W = (0.293)(130{,}000 \text{ lbf})$$
$$= 38{,}090 \text{ lbf}$$

Since the perpendicular walls (chords) have no rigidity, all of the seismic force is resisted by the two parallel walls (shear walls). The shear stress is

$$v = \frac{V}{A} = \frac{38{,}090 \text{ lbf}}{(2 \text{ walls})(8 \text{ in})\left(40 \dfrac{\text{ft}}{\text{wall}}\right)\left(12 \dfrac{\text{in}}{\text{ft}}\right)}$$
$$= 4.96 \text{ psi}$$

(e) Part (d) was an analysis problem, so it was not necessary to include the 150% term required by UBC-97 Sec. 2107.1.7. (See Sec. 7-3.) However, in determining adequacy, the increase must be included. At 150% of the seismic load, the shear stress would be

$$v_{\text{design}} = (1.5)(4.96 \text{ psi}) = 7.44 \text{ psi}$$

Since this is less than 35 psi, the limit for in-plane shear reinforcement (see Sec. 11-1(F)), the wall thickness is adequate.

(f) Since the shear load along the wall was calculated in part (a) to be 329.6 lbf/ft, the connection between the 10-ft stub wall and the collector would have to carry 30 ft of compressive or tensile loading.

$$F_{\text{collector}} = (30 \text{ ft})\left(329.6 \dfrac{\text{lbf}}{\text{ft}}\right) = 9888 \text{ lbf}$$

Notice that eliminating 30 ft of parallel wall (shear wall) reduces the accelerating mass but does not reduce the diaphragm force. Only the perpendicular walls affect the diaphragm force.

SI Solution

From Table 6.3, the seismic zone factor for seismic zone 4 is $Z = 0.40$. From Table 6.5 for a hazardous facility, the seismic importance factor, I, is 1.25. From App. K (corresponding to UBC-97 Table 16-K), $I_p = 1.50$. Inasmuch as the soil type is unknown, the soil-profile type S_D must be used. (See Sec. 6-15.)

From Table 6.10 for a bearing wall system consisting of concrete shear walls, $R = 4.5$. From Table 6.9, for a soil-profile S_D and $Z = 0.4$, the seismic response coefficient $C_a = 0.44N_a$. With $N_a = 1.0$, the value of $C_a = 0.44$. Thus, the base shear equation (Eq. 6.26) is

$$V = \frac{3.0C_aW}{R} = \frac{(3.0)(0.44)W}{4.5}$$
$$= 0.293W$$

The weight being accelerated by the earthquake consists of the diaphragm weight and a portion of the wall weight. The weight of the diaphragm is

$$W_{\text{diaphragm}} = \left(0.96 \dfrac{\text{kN}}{\text{m}^2}\right)(12.2 \text{ m})(15.2 \text{ m})$$
$$= 178.0 \text{ kN}$$

Since concrete has a density of 2400 kg/m³, the weight of 1 m² of a 203-mm-thick wall is

$$\gamma = \frac{(203 \text{ mm})\left(2400 \dfrac{\text{kg}}{\text{m}^3}\right)\left(9.81 \dfrac{\text{N}}{\text{kg}}\right)}{\left(1000 \dfrac{\text{mm}}{\text{m}}\right)\left(1000 \dfrac{\text{N}}{\text{kN}}\right)}$$
$$= 4.8 \text{ kN/m}^2$$

For the purpose of determining the diaphragm force, the upper half (i.e., only the upper 1.5 m) of both perpendicular walls (i.e., the chords) is used to calculate the wall weight. (See Secs. 6-29 and 7-2.) The remaining seismic force passes directly into the foundation without being carried by the wall-diaphragm connection. The weight of half the perpendicular walls (chords) is

$$W_{\perp\text{walls}} = \left(4.8 \dfrac{\text{kN}}{\text{m}^2}\right)(1.5 \text{ m})(2 \text{ walls})\left(15.2 \dfrac{\text{m}}{\text{wall}}\right)$$
$$= 219.0 \text{ kN}$$

(a) Equation 6.26 is for calculating the base shear passing through to the foundation and should not strictly be used to calculate the force on connections between elements in the building. Equation 6.38 is for elements of structures and their attachments and must be used. (See Sec. 6-44.) For diaphragms, based on UBC-97 Sec. 1633.2.9, (1) the roof diaphragm is to be designed to resist a portion of the floor forces above it, as weighted by the floor weights, and (2) the diaphragm force must be within $0.5C_aIW$ and $1.0C_aIW_{px}$. With the values of $C_a = 0.44$ and $I = 1.25$, the limits on diaphragm force are $0.275W_{px}$ and $0.550W_{px}$.

Inasmuch as this is a simple one-story building with only one diaphragm, all of the inertial load from the accelerating wall and roof masses must be carried by the wall-roof connection, so Eq. 6.26 with the wall and diaphragm weights is ultimately used. Checking, $0.275 < 0.293 < 0.550$ (ok).

$$F_{\text{diaphragm}} = 0.293W$$
$$= (0.293)(178.0 \text{ kN} + 219.0 \text{ kN})$$
$$= 116.3 \text{ kN}$$

The shear per meter of diaphragm width, b, is given by Eq. 7.7.

$$v = \frac{F_{\text{diaphragm}}}{2b} = \frac{(116.3 \text{ kN})\left(1000 \frac{\text{N}}{\text{kN}}\right)}{(2)(12.2 \text{ m})}$$
$$= 4770 \text{ N/m}$$

(b) Table 12.1 (corresponding to UBC-97 Table 23-II-H) gives the nail spacing. The allowable shear is 5470 N/m (> 4770 N/m), with a nail spacing of 63.5 mm.

(c) The distributed seismic force, w, across the face of the diaphragm is

$$w = \frac{F_{\text{diaphragm}}}{L} = \frac{116.3 \text{ kN}}{15.2 \text{ m}}$$
$$= 7.7 \text{ kN/m}$$

From Eq. 7.9, the chord force, C, is

$$C = \frac{wL^2}{8b} = \frac{\left(7.7 \frac{\text{kN}}{\text{m}}\right)(15.2 \text{ m})^2}{(8)(12.2 \text{ m})}$$
$$= 18.2 \text{ kN}$$

(d) The net effect is to include the inertial force for accelerating the diaphragm mass and half of all the walls. The diaphragm and perpendicular wall weights have already been determined. Half the weight of the parallel walls (i.e., the shear walls) is

$$W_{\parallel\text{walls}} = \left(4.8 \frac{\text{kN}}{\text{m}^2}\right)(1.5 \text{ m})(2 \text{ walls})\left(12.2 \frac{\text{ft}}{\text{wall}}\right)$$
$$= 175.7 \text{ kN}$$

The total weight is

$$W_{\text{total}} = W_{\text{diaphragm}} + W_{\perp\text{walls}} + W_{\parallel\text{walls}}$$
$$= 178.0 \text{ kN} + 219.0 \text{ kN} + 175.7 \text{ kN}$$
$$= 572.7 \text{ kN}$$

The seismic force is

$$V = 0.293W = (0.293)(572.7 \text{ kN})$$
$$= 167.8 \text{ kN}$$

Since the perpendicular walls (chords) have no rigidity, all of the seismic force is resisted by the two parallel walls (shear walls). The shear stress is

$$v = \frac{V}{A} = \frac{(167.8 \text{ kN})\left(1000 \frac{\text{mm}}{\text{m}}\right)}{(2 \text{ walls})(203 \text{ mm})\left(12.2 \frac{\text{m}}{\text{wall}}\right)}$$
$$= 33.9 \text{ kPa}$$

(e) Part (d) was an analysis problem, so it was not necessary to include the 150% term required by UBC-97 Sec. 2107.1.7. (See Sec. 7-3.) However, in determining adequacy, the increase must be included. At 150% of the seismic load, the shear stress would be

$$v_{\text{design}} = (1.5)(33.9 \text{ kPa}) = 50.9 \text{ kPa}$$

Since this is less than 240 kPa, the limit for inplane shear reinforcement (see Sec. 11-1(F)), the wall thickness is adequate.

(f) Since the shear load along the wall was calculated in part (a) to be 4770 N/m, the connection between the 3-m stub wall and the collector would have to carry 9.1 m of compressive or tensile loading.

$$F_{\text{collector}} = \frac{(9.1 \text{ m})\left(4770 \frac{\text{N}}{\text{m}}\right)}{1000 \frac{\text{N}}{\text{kN}}} = 43.4 \text{ kN}$$

Notice that eliminating 9.1 m of parallel wall (shear wall) reduces the accelerating mass but does not reduce the diaphragm force. Only the perpendicular walls affect the diaphragm force.

◇   ◇   ◇   ◇   ◇   ◇   ◇

8

GENERAL STRUCTURAL DESIGN

8-1 DISTRIBUTING STORY SHEARS TO MEMBERS OF UNKNOWN SIZE

Chapter 7 dealt with distributing seismic forces in frames whose resisting elements (fixed columns, shear walls, etc.) were already designed. (The seismic force to be resisted is the sum of story shears for a particular level and all levels above.) If the cross-sectional area and moment of inertia of these resisting elements are not known (as they will not be initially), assumptions must be made about the amount of lateral force each element carries.

There are several approximate methods (e.g., the portal, cantilever, and Spurr methods discussed in the next sections) that distribute the lateral forces to the resisting elements. These methods eliminate the need to use indeterminate solution methods such as moment distribution.[1]

8-2 PORTAL METHOD

The *portal method* is ideal for cases where the framing system is regularly spaced. It assumes that all interior columns carry the same shear, while exterior columns carry half the shear of interior columns.[2] Inflection points are assumed to occur at mid-span in each girder and column. Changes in length due to compression, tension, and deflection are disregarded. Each bay is treated independently of adjacent bays.

---

[1]Since the member geometries (areas and moments of inertia) are initially unknown, a rigorous method cannot be used.

[2]If the bay sizes (i.e., the distances between columns) are not the same, the shear may be distributed to the columns in proportion to the bay sizes.

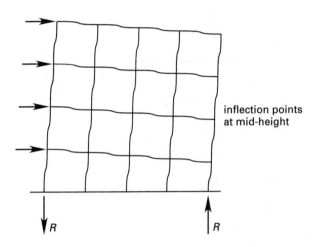

inflection points at mid-height

Figure 8.1   Portal Method Frame Deflection

Example 8.1

The sum of story shears for a particular layer and those above it is 110 kips (490 kN). Use the portal method to find the shears in members 2-3 (i.e., the vertical shear on the horizontal members) and 2-7 (i.e., the horizontal shear on the vertical members).

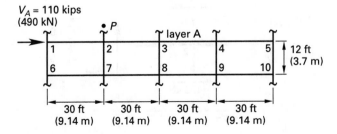

Customary U.S. Solution

Exterior columns 1-6 and 5-10 carry only half the loads of the interior columns. Therefore, there are a total of four full-strength columns. The horizontal shear carried by member 2-7 and all other interior columns is

$$V_{\text{interior}} = \frac{110 \text{ kips}}{4} = 27.5 \text{ kips}$$

Column 2-7 carries no axial load.

Forces are assumed to be applied at the inflection points of the columns and girders. The vertical girder shears, V_{girder}, are obtained by taking a free-body diagram of joint 2 and summing moments about point 2.

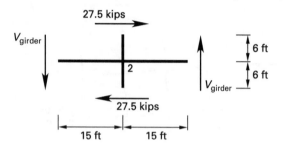

$$\sum M_2 = (2)(27.5 \text{ kips})(6 \text{ ft}) - 2V_{\text{girder}}(15 \text{ ft}) = 0$$
$$V_{\text{girder}} = 11 \text{ kips}$$

SI Solution

Exterior columns 1-6 and 5-10 carry only half the loads of the interior columns. Therefore, there are a total of four full-strength columns. The horizontal shear carried by member 2-7 and all other interior columns is

$$V_{\text{interior}} = \frac{490 \text{ kN}}{4} = 123 \text{ kN}$$

Column 2-7 carries no axial load.

Forces are assumed to be applied at the inflection points of the columns and girders. The vertical girder shears, V_{girder}, are obtained by taking a free-body diagram of joint 2 and summing moments about point 2.

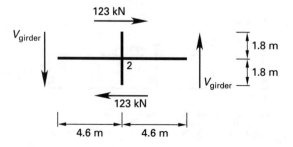

$$\sum M_2 = (2)(123 \text{ kN})(1.8 \text{ m})$$
$$- 2V_{\text{girder}}(4.6 \text{ m}) = 0$$
$$V_{\text{girder}} = 48.1 \text{ kN}$$

◇   ◇   ◇   ◇   ◇   ◇   ◇

8-3 CANTILEVER METHOD

Unlike the portal method, which treats each bay independently of the others, the *cantilever method* assumes the entire floor works together as a unit. Although analysis of the cantilever method must begin at the roof level and work down, this method is preferred for buildings with more than 25 stories. It assumes that the floors remain plane (though not horizontal) and the force in a column is proportional to the distance of the column from the frame's center of gravity. As with the portal method, inflection points of columns and girders are assumed to occur at the mid-lengths.

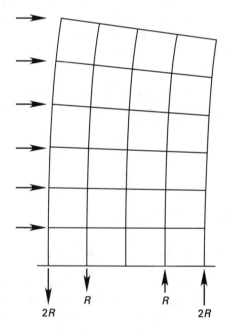

Figure 8.2   Cantilever Method Frame Deflection

◇   ◇   ◇   ◇   ◇   ◇   ◇

Example 8.2

The cumulative lateral force at the roof level of the frame shown is 110 kips (490 kN). Use the cantilever method to determine the shear in members 2-3 and 2-7.

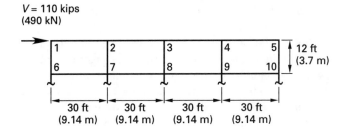

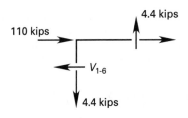

Customary U.S. Solution

The center of gravity of this level is located at point Q. Columns 1-6 and 2-7 are in tension. Columns 4-9 and 5-10 are in compression. Column 3-8 is along the neutral axis and is not stressed axially.

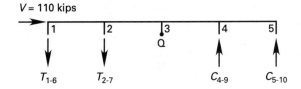

The sum of moments in a member at an inflection point is zero. Summing moments about point Q (halfway between points 3 and 8),

$$\sum M_Q = (6\ \text{ft})(110\ \text{kips}) - (60\ \text{ft})T_{1\text{-}6}$$
$$- (30\ \text{ft})T_{2\text{-}7} - (30\ \text{ft})C_{4\text{-}9}$$
$$- (60\ \text{ft})C_{5\text{-}10}$$
$$= 0$$

Since the column forces are proportional to the distance from column 3-8 and all columns are separated by the same distance, it is apparent that the relationships between the tension, T, and compressive, C, forces are

$$T_{1\text{-}6} = 2T_{2\text{-}7} = 2C_{4\text{-}9} = C_{5\text{-}10}$$

Making these substitutions,

$$T_{1\text{-}6} = C_{5\text{-}10} = 4.4\ \text{kips}$$
$$T_{2\text{-}7} = C_{4\text{-}9} = 2.2\ \text{kips}$$

The vertical shear in member 1-2 is equal to the sum of vertical loads to the left of its inflection point.

$$V_{1\text{-}2} = T_{1\text{-}6} = 4.4\ \text{kips}$$

Taking the free body of the inflection points and summing moments about point 1 gives the horizontal shear in member 1-6.

$$\sum M_1 = (6\ \text{ft})V_{1\text{-}6} - (15\ \text{ft})(4.4\ \text{kips}) = 0$$
$$V_{1\text{-}6} = 11\ \text{kips}$$

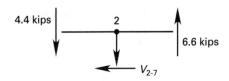

The vertical shear in member 2-3 is

$$V_{2\text{-}3} = 2.2\ \text{kips} + 4.4\ \text{kips} = 6.6\ \text{kips}$$

(This is approximately half the value calculated with the portal method in Ex. 8.1.)

Summing moments about point 2 gives the horizontal shear in member 2-7.

$$\sum M_2 = (6\ \text{ft})V_{2\text{-}7} - (15\ \text{ft})(4.4\ \text{kips})$$
$$- (15\ \text{ft})(6.6\ \text{kips}) = 0$$
$$V_{2\text{-}7} = 27.5\ \text{kips}$$

(This is the same as the value calculated with the portal method in Ex. 8.1.)

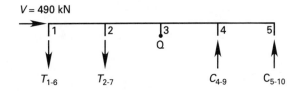

SI Solution

The center of gravity of this level is located at point Q. Columns 1-6 and 2-7 are in tension. Columns 4-9 and 5-10 are in compression. Column 3-8 is along the neutral axis and is not stressed axially.

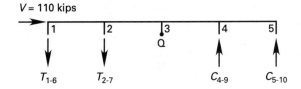

The sum of moments in a member at an inflection point is zero. Summing moments about point Q (halfway between points 3 and 8),

$$\sum M_Q = \left(\frac{3.7\ \text{m}}{2}\right)(490\ \text{kN}) - (18.28\ \text{m})T_{1\text{-}6}$$
$$- (9.14\ \text{m})T_{2\text{-}7} - (9.14\ \text{m})C_{4\text{-}9}$$
$$- (18.28\ \text{m})C_{5\text{-}10}$$
$$= 0$$

Since the column forces are proportional to the distance from column 3-8 and all columns are separated by the same distance, it is apparent that the relationships between the tension, T, and compressive, C, forces are

$$T_{1\text{-}6} = 2T_{2\text{-}7} = 2C_{4\text{-}9} = C_{5\text{-}10}$$

Making these substitutions,

$$T_{1\text{-}6} = C_{5\text{-}10} = 20 \text{ kN}$$
$$T_{2\text{-}7} = C_{4\text{-}9} = 10 \text{ kN}$$

The vertical shear in member 1-2 is equal to the sum of vertical loads to the left of its inflection point.

$$V_{1\text{-}2} = T_{1\text{-}6} = 20 \text{ kN}$$

Taking the free body of the inflection points and summing moments about point 1 gives the horizontal shear in member 1-6.

$$\sum M_1 = (1.83 \text{ m})V_{1\text{-}6} - (4.57 \text{ m})(20 \text{ kN}) = 0$$
$$V_{1\text{-}6} = 50 \text{ kN}$$

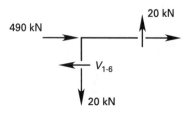

The vertical shear in member 2-3 is

$$V_{2\text{-}3} = 10 \text{ kN} + 20 \text{ kN}$$
$$= 30 \text{ kN}$$

(This is approximately half the value calculated with the portal method in Ex. 8.1.)

Summing moments about point 2 gives the horizontal shear in member 2-7.

$$\sum M_2 = (1.83 \text{ m})V_{2\text{-}7} - (4.57 \text{ m})(20 \text{ kN})$$
$$- (4.57 \text{ m})(30 \text{ kN}) = 0$$
$$V_{2\text{-}7} = 120 \text{ kN}$$

(This is the same as the value calculated with the portal method in Ex. 8.1.)

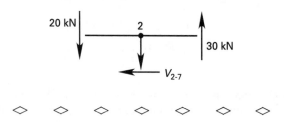

◇   ◇   ◇   ◇   ◇   ◇   ◇

8-4 SPURR METHOD

Although the cantilever method assumes that the floor will remain plane, it does nothing to ensure a plane floor. Irregular lengthening and shortening of the columns produce secondary bending stresses that can become excessive in tall structures and disturb the planar nature of the floor.

If the height-to-bay length ratio is less than 4, the floor will indeed remain fairly plane. However, when the height-to-bay length ratio exceeds 4, interior column elongation and shortening will cause noticeable floor deflection.

The *Spurr method* uses the cantilever method to calculate girder shears, but girders are given specific strengths in order to justify the assumption that inflection points are located at mid-span.[3] Essentially, the girder moments of inertia are made proportional to the girder shear times the square of the span length.

8-5 FRAME MEMBER SIZING

Once the column and girder shears and axial loads are known, these members can be sized by traditional methods. For example, a girder can be considered to be loaded at its mid-points (between two sets of columns) by the vertical girder shear acting in opposite directions.

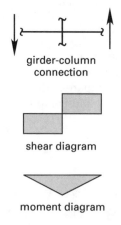

Figure 8.3 Girder Shear and Moment Diagrams

Another method for sizing frame members is to select the columns and girders so that the drift is limited to the UBC-97 maximum [Sec. 1630.10.2] under the action of the horizontal forces. This method, however, requires

---

[3]This method is named for H. V. Spurr, who published the method in his book *Wind Bracing: The Importance of Rigidity in High Towers*, New York: McGraw-Hill, 1930.

making assumptions about the joint rigidity. For example, assuming that a bent consisting of a girder of length L and two columns of height h must limit the drift to 2.5% of the story height ($T < 0.7$ sec) when a total shear force of V is experienced, and assuming the joints are all rigid, the drift criterion is

$$\left(\frac{Vh^2}{12E}\right)\left(\frac{h}{I_{\text{column}}} + \frac{L}{I_{\text{girder}}}\right) = 0.025h \qquad [8.1]$$

There are many possible combinations of column and girder strengths that will satisfy this criterion.

9

DETAILS OF SEISMIC-RESISTANT CONCRETE STRUCTURES

9-1 CONCRETE CONSTRUCTION DETAILS

Concrete itself has poor ductility in shear and is therefore a brittle material. This limited ductility, combined with its higher mass compared to steel (higher mass increases the seismic force) and lower tolerance to errors in design and workmanship, usually makes concrete a second choice for high-rise construction. However, concrete ductile moment-resisting frames now make it possible to design structures with the ductility and energy dissipation capability of steel structures. In some cases (particularly those not requiring seismic provisions), a concrete building may be less expensive to construct.[1] The primary disadvantage of concrete is its weight, which increases the seismic forces. Some advantage can be gained, however, by using lightweight concrete.

Concrete structures can be constructed to behave in a ductile manner. Such structures are loosely referred to as having been constructed of "ductile concrete," "special concrete," or "California concrete." In order to make concrete into a ductile structure, much attention must be given to connection and confinement details.[2] "Ordinary concrete," or concrete construction that does not behave in a ductile way, cannot be used in California.

The two most important concepts in designing ductile concrete are (1) continuity and (2) confinement. Concrete members must not pull apart, and they must not disintegrate when the core becomes cracked or crushed.

---

[1]The assumed economic superiority of steel over concrete is traditional and is a very controversial issue; it is not, by any means, an absolute fact.

[2]It is not always easy, however, to get all the special steel reinforcing and confinement into the beam-column connection area. The joint congestion, called a "tangle" by some, can be considerable.

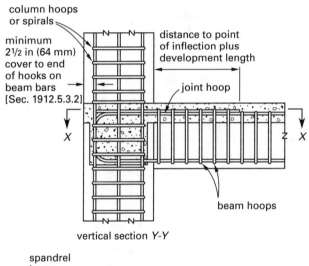

vertical section Y-Y

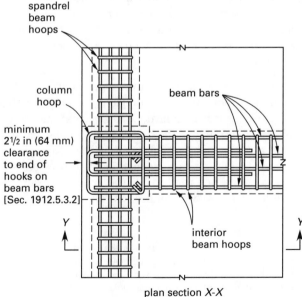

plan section X-X

Figure 9.1 Typical Ductile Frame Joint for Region of High Seismic Risk [UBC-97 Sec. 1921.5]

The UBC-97 discusses the required details of ductile concrete in Secs. 1633 and 1921.[3] The following lettered items list some of the more important provisions. Because all provisions have been greatly simplified in this book, the UBC-97 references should be used for clarification. In almost all cases, other provisions also apply. (In some of the following cases, the accompanying figures contain additional information about special seismic provisions.)

The special provisions in Sec. 1921 of the UBC-97 apply to reinforced concrete frames and shear wall structures that resist earthquake loading. UBC-97 Sec. 1921.7 still requires members that are not part of the lateral force-resisting system to satisfy minimum reinforcement requirements.

The design of concrete structures should be in accordance with the strength design (load and resistance factor design) or allowable stress design method.

A. Orthogonal Effects   [UBC-97 Sec. 1633.1]

Orthogonal effects should be investigated in cases of torsional irregularity, nonparallel structural systems, and where a member is part of two intersecting lateral force-resisting systems (e.g., where the member is a corner column more than lightly loaded by seismic forces).[4]

B. Connections   [UBC-97 Sec. 1633.2.3]

Connections are elements that interconnect two precast members or a precast member and a cast-in-place member. Connections that resist seismic forces must be designed and detailed by an engineer and shown on the drawings.

A strong connection is one that remains elastic while the specified nonlinear action regions (i.e., the member length over which nonlinear action occurs) endure inelastic response under the design basis ground motion. When a connection employs any of the code splicing methods to join precast members and utilizes cast-in-place concrete or grout to fill the splicing closure, it is referred to as a "wet connection." A connection used

between precast members that does not qualify as a wet connection is called a "dry connection."

Based on the UBC-97 [Sec. 1921.2.1.7], in structures having precast gravity systems, all beam-to-column connections that are not part of the lateral-force-resisting system should be designed to resist moments in one direction only, either positive or negative. In designing the connection at the opposite end of the member, it should be noted that the connection should resist moment with the same positive or negative sign. The connections should be allowed to have zero flexural stiffness up to a frame displacement of Δ_s. They should be designed to develop the strength moment (M) that develops at the connection when the frame is displaced by Δ_s, assuming fixity at that connection and a beam flexural stiffness of no less than one-half of the gross section stiffness.

C. Deformation Compatibility
[UBC-97 Sec. 1633.2.4]

Not all members in a structure are part of the lateral-force-resisting system. However, members that are not part of that system and their connections must be adequate to maintain support of design dead plus live loads when subjected to the expected deformations caused by seismic forces. The UBC-97 requires that P-Δ effects on such elements be accounted for. In determining expected deformation, P-Δ effects or the deformation induced by a story drift of 0.0025 times the story height should be considered, and its value should be the greater of the maximum inelastic response displacement ($\Delta_M = 0.7R\Delta_s$) [UBC-97 Sec. 1921.0]. For concrete elements that are not part of the lateral-force-resisting system, provisions of the UBC-97 [Sec. 1921.7] apply.

For concrete elements that are part of the lateral-force-resisting system, the accepted flexural and shear stiffness properties should not exceed one-half of the gross section properties, except on the condition that a rational cracked-section analysis is performed. Foundation flexibility and diaphragm defections may create additional deformation that should also be considered.

D. Ties and Continuity   [UBC-97 Sec. 1633.2.5]

It is important that structural members do not pull apart. As a minimum, all smaller portions of a building must be tied to the rest of the building with elements having at least a strength to resist $0.5C_aI$ times the weight of the smaller portion. Positive connections for beams, girders, and trusses must resist a horizontal force parallel to the member not less than $0.5C_aI$ times the dead plus live loads.

---

[3]The material in UBC-97 Sec. 1921 is equivalent to the material contained in Chapter 21 of the ACI (American Concrete Institute) publication 318. However, the Blue Book commentary also contains valuable information, particularly for seismic zones 3 and 4.

[4]The column is more than lightly loaded if its factored axial load due to seismic forces is greater than 20% of the column axial strength.

E. Collector Elements [UBC-97 Sec. 1633.2.6]

Collector elements transfer lateral forces from a portion of a structure to vertical elements of the lateral-force-resisting system. Concrete collector elements must rely on reinforcing steel to carry drag forces into shear walls.

Based on the UBC-97 [Sec. 1921.6.12], when collector elements in topping slabs are placed over precast floor or roof elements, the slab thickness should not be less than 3 in (76 mm) or six times the diameter of the largest reinforcement ($6d_b$).

F. Concrete Frames [UBC-97 Sec. 1633.2.7]

Concrete frames used in lateral-force-resisting systems in zones 3 and 4 must be special moment-resisting frames (SMRF). In seismic zone 2, as a minimum, these frames should be intermediate moment-resisting frames (IMRF).

G. Anchorage to Concrete and Masonry Walls [UBC-97 Sec. 1633.2.8]

To resist the uplift and sliding forces, the roof should be anchored to the walls and columns, and walls and columns should be anchored to the foundations. The provisions of the UBC-97 [Secs. 1605.2.3, 1611.4, 1633.2.8, and 1633.2.8.1] apply. Applying the greater of the wind or earthquake loads in design, these anchorages should be capable of resisting load combinations using strength design [UBC-97 Sec. 1612.2] and load combinations using allowable stress design [UBC-97 Sec. 1612.3]. As a minimum, these connections to and between walls must be capable of resisting the horizontal force of 280 pounds per foot of wall (4.09 kN/m) substituted for earthquake load (E). When anchor spacing exceeds 4 ft (1219 mm), walls must resist bending between the horizontal anchors.

In seismic zones 3 and 4, when diaphragms are anchored to walls using embedded straps providing out-of-plane lateral support of the wall, the straps must either attach to or hook around the reinforcing steel or terminate so that forces can effectively be transferred to the reinforcing steel. Where flexible diaphragms are used to provide lateral support for the walls (out-of-plane wall anchorage to flexible diaphragms) in seismic zone 4, the value of total design lateral seismic force (F_p) used for the design of the elements of the wall anchorage system should be equal to or greater than 420 pounds per foot of wall (6.1 kN/m) substituted for earthquake load (E).

H. Boundary Elements of Shear Walls and Diaphragms [UBC-97 Sec. 1921.6]

The reinforcement ratio cannot be less than 0.0025 along both the longitudinal and transverse axes of shear walls and diaphragms. Reinforcement spacing cannot exceed 18 in (457 mm). See Fig. 9.2.

Boundary elements of shear walls must not fail due to deformations caused by a major earthquake, particularly when the design is governed by flexure. The design should specify special detailing to prevent tensile fracture as well as compressive crushing and buckling. This is accomplished by requiring two curtains of shear reinforcement for walls with factored shear in excess of $2A_{cv}\sqrt{f_c'}$ (in SI units, $0.166A_{cv}\sqrt{f_c'}$). Additional reinforcement is required at the edges of all shear walls and diaphragms, as well as at the boundaries of all openings.

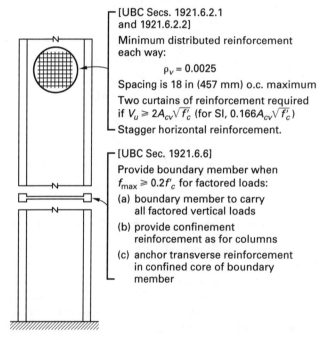

Figure 9.2 Requirements for Shear Walls and Boundary Members

I. Seismic Hooks, Crossties, and Hoops [UBC-97 Sec. 1921.1]

A *seismic hook* is a hook on a stirrup that engages the longitudinal reinforcement and projects into the interior of the stirrup or hoop. For a seismic hook, the bend must be at least 135° with an extension past the bend of at least 6 bar diameters (but not less than 3 in (76 mm)). A *crosstie* is a continuous reinforcing bar. It has a seismic hook at one end and a hook at least 90° with an extension past the bend of at least 6 bar diameters at the other end. A *hoop* is a closed tie made up of several reinforcing elements, each consisting of seismic hooks at both ends. A hoop can also be a continuously wound tie having seismic hooks at both ends. The details of seismic hooks, crossties, and hoops are shown in Fig. 9.3.

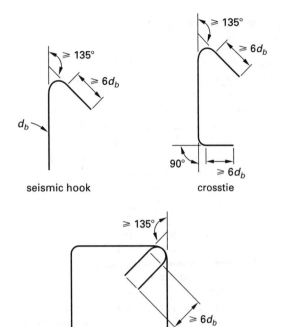

Figure 9.3   Seismic Hook, Crosstie, and Hoop
[UBC-97 Sec. 1921.1]

J. Strength Reduction Factor [UBC-97 Sec. 1921.2.3]

In general, the normal strength reduction factor, ϕ, defined in UBC-97 Sec. 1909.3 is applicable. For seismic zones 3 and 4, provisions of the UBC-97 [Sec. 1909.3.4] apply to strength-reduction factors. The strength-reduction factor is 0.60 for walls, topping slabs used as diaphragms over precast concrete members, and structural framing members (joints are excluded) when their nominal shear strength is less than the shear corresponding to development of their nominal flexural strength. The strength reduction factor is 0.85 for cast-in-place concrete diaphragms and beam-column joints. The nominal flexural strength should concur with the most critical factored axial loads, including earthquake effects. The strength-reduction factor is 0.65 for flexure compression, shear, and bearing of structural plain concrete.

K. Combined Loading Load Factors
[UBC-97 Sec. 1909.2]

To resist dead load and live load, the required minimum strength is

$$U = 1.4D + 1.7L \qquad [9.1]$$

When wind load should be included in design, the greatest required strength is

$$U = 0.75(1.4D + 1.7L + 1.7W) \qquad [9.2]$$

To determine the more severe conditions, load combinations should include both full value and zero value of live load; however, the required strength should not be less than Eq. 9.1.

$$U = 0.9D + 1.3W \qquad [9.3]$$

Based on the UBC-97 [Sec. 1612.2.1, Exception 2], required strength U to resist specified earthquake loads or forces E considering factored load combinations multiplier of 1.1 for concrete should be

$$U = 1.1(1.2D + 1.0E + 0.5L) \qquad [9.4]$$
$$U = 1.1(0.9D - 1.0E) \qquad [9.5]$$

In proportioning concrete structural elements using alternate load-factors combination and strength-reduction factors [UBC-97 Sec. 1928.1.2.3, Div. VIII], the required strength using basic load combinations should be

$$U = 1.4D \qquad [9.6]$$
$$U = 1.2D + 1.6L + 0.5(L_r \text{ or } S \text{ or } R) \qquad [9.7]$$
$$\begin{aligned} U = {} & 1.2D + 1.6(L_r \text{ or } S \text{ or } R) \\ & + (0.5L \text{ or } 0.8W) \end{aligned} \qquad [9.8]$$
$$\begin{aligned} U = {} & 1.2D + 1.3W + 0.5L \\ & + 0.5(L_r \text{ or } S \text{ or } R) \end{aligned} \qquad [9.9]$$
$$U = 1.2D + 1.5E + (0.5L \text{ or } 0.2S) \qquad [9.10]$$
$$U = 0.9D - (1.3W \text{ or } 1.5E) \qquad [9.11]$$

L. Concrete Strength
[UBC-97 Secs. 1921.2.4.1 and 1921.2.4.2]

The compressive strength, f'_c, of normal-weight concrete may not be less than 3000 psi (20.69 MPa). The compressive strength of lightweight concrete may not exceed 4000 psi (27.58 MPa) unless experimental data are submitted to show that it has properties equivalent to normal-weight concrete. However, the compressive strength of lightweight concrete used in calculations cannot exceed 6000 psi (41.37 MPa) under any circumstances.

M. Steel Reinforcement [UBC-97 Sec. 1921.2.5.1]

All reinforcing steel must be of the deformed variety; plain bars are not permitted. Steel used must be low-alloy complying with ASTM A706. ASTM A615 grades 40 or 60 may be used if (1) the actual yield strength is no more than 18,000 psi (124.1 MPa) higher than the specified yield, and (2) the actual ultimate tensile stress is at least 125% of the actual yield strength. Restrictions on welding apply.

N. Frame Flexural Members [UBC-97 Sec. 1921.3]

(This UBC-97 section is specifically for flexural members used in lateral force-resisting systems.) The UBC-97 imposes a geometric restraint (i.e., a width/depth ratio greater than or equal to 0.3) and other constraints [Secs. 1921.3.1.3 and 1921.3.1.4] in order to reduce lateral instability and ensure a transfer of moment during inelastic response. See Fig. 9.4.

Seismic hoops (confinement) are required in all areas of flexural members where yielding is expected (e.g., at the ends of built-in beams and at other plastic hinge points). The first hoop must be within 2 in (51 mm) of the face of the supporting member. Maximum spacing must be less than (a) $d/4$, (b) 8 times the diameter of the smallest longitudinal bar, (c) 24 times the hoop bar diameter, and (d) 12 in (305 mm). Where hoops are not required, stirrup spacing with seismic hooks at both ends must be spaced at no more than $d/2$ throughout the member [Sec. 1921.3.3.4].

O. Top and Bottom Reinforcement
[UBC-97 Secs. 1921.3.2.1 and 1921.3.2.2]

At least two top and two bottom bars must run the entire length of the member. For both top and bottom steel, the amount of reinforcement steel must be greater than $200b_w d/f_y$ (for SI, $1.38b_w d/f_y$), and $\Delta_{r,\max} \leq (\Delta_{f,\max})/5$. The maximum steel is $0.025b_w d$. At the face of a joint, the positive moment strength cannot be less than 50% of the negative moment strength. At any section along the member length, neither the positive nor the negative moment strengths can be less than 25% of the corresponding strength at a joint.

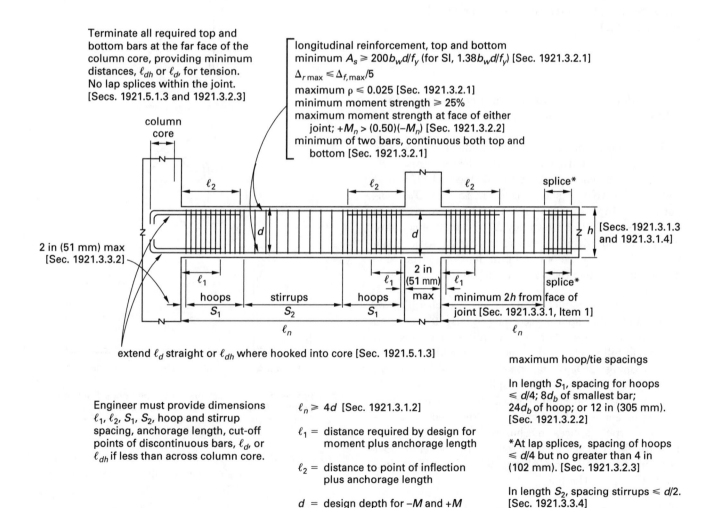

Terminate all required top and bottom bars at the far face of the column core, providing minimum distances, ℓ_{dh} or ℓ_d, for tension. No lap splices within the joint. [Secs. 1921.5.1.3 and 1921.3.2.3]

longitudinal reinforcement, top and bottom
minimum $A_s \geq 200b_w d/f_y$ (for SI, $1.38b_w d/f_y$) [Sec. 1921.3.2.1]
$\Delta_{r\,\max} \leq \Delta_{f,\max}/5$
maximum $\rho \leq 0.025$ [Sec. 1921.3.2.1]
minimum moment strength $\geq 25\%$
maximum moment strength at face of either
  joint; $+M_n > (0.50)(-M_n)$ [Sec. 1921.3.2.2]
minimum of two bars, continuous both top and
  bottom [Sec. 1921.3.2.1]

column core

2 in (51 mm) max [Sec. 1921.3.3.2]

[Secs. 1921.3.1.3 and 1921.3.1.4]

ℓ_2 ℓ_2 ℓ_2 splice*

d d h

ℓ_1 ℓ_1 2 in (51 mm) max ℓ_1 splice*
hoops stirrups hoops minimum $2h$ from face of joint [Sec. 1921.3.3.1, Item 1]
S_1 S_2 S_1
ℓ_n ℓ_n

extend ℓ_d straight or ℓ_{dh} where hooked into core [Sec. 1921.5.1.3]

Engineer must provide dimensions ℓ_1, ℓ_2, S_1, S_2, hoop and stirrup spacing, anchorage length, cut-off points of discontinuous bars, ℓ_d, or ℓ_{dh} if less than across column core.

$\ell_n \geq 4d$ [Sec. 1921.3.1.2]

ℓ_1 = distance required by design for moment plus anchorage length

ℓ_2 = distance to point of inflection plus anchorage length

d = design depth for $-M$ and $+M$

maximum hoop/tie spacings

In length S_1, spacing for hoops $\leq d/4$; $8d_b$ of smallest bar; $24d_b$ of hoop; or 12 in (305 mm). [Sec. 1921.3.2.2]

*At lap splices, spacing of hoops $\leq d/4$ but no greater than 4 in (102 mm). [Sec. 1921.3.2.3]

In length S_2, spacing stirrups $\leq d/2$. [Sec. 1921.3.3.4]

Figure 9.4 Special Flexural Member Reinforcement [UBC-97 Sec. 1921.3]

P. Tension Lap Splices [UBC-97 Sec. 1921.3.2.3]

Lap splices are classified as Class A (with a lap length of one development length) and Class B (with a lap length of 1.3 development lengths) [UBC-97 Sec. 1912.15]. (The Class C splice (1.7 development lengths) was eliminated in 1989, and (in some cases) the basic development length was increased.)

Tension lap splices are prohibited in flexural members (1) within beam-column joints, (2) within a distance from the face of a joint equal to twice the member depth, and (3) where flexural yielding is anticipated.

Lap splices are not permitted where flexural yielding is expected (i.e., at plastic hinge points) because such splices are not reliable when the loading repeatedly exceeds the yield strength of the reinforcement. Where permitted, they are proportioned as tension lap splices. Class B splices must be used unless double (or more) of the required reinforcement steel is provided [UBC-97 Sec. 1912.15.2].

Q. Welded Splices and Mechanical Connections
[UBC-97 Secs. 1912.14.3 and 1921.2.6]

Welded splices and approved mechanical connections can be used but must maintain the clearance and coverage requirements of the UBC-97 [Secs. 1907.6 and 1907.7]. All welding should conform to UBC-97 Standard 19-1. In seismic zones 2, 3, and 4, welded splices on billet steel A615 or low alloy A706 reinforcement are not permitted within (1) an anticipated plastic hinge region, (2) a distance of one beam depth on either side of the plastic hinge region, or (3) a joint.

Splices with mechanical connections are classified as type 1 and type 2 splices according to UBC-97-specified strength capacities. For type 1 splices, a full-welded splice and full mechanical connection should be capable of developing 125% of the specified yield strength (f_y) of the bar. Otherwise, these splices are allowed only for No. 5 bars and smaller when the provided area of reinforcement is at least twice that required by analysis. For type 2 splices, mechanical connections should be capable of developing in tension the lesser of 95% of ultimate tensile strength or 160% of specified yield strength (f_y) of the bar.

Welding of designed reinforcement for any purpose other than approved splicing is prohibited. Welding of stirrups, ties, inserts, and other elements to the longitudinal bars is prohibited [UBC-97 Sec. 1921.2.6.2].

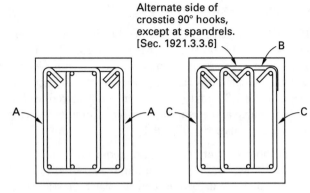

(a) examples of overlapping hoops and stirrups

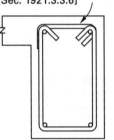

(b) spandrel beam

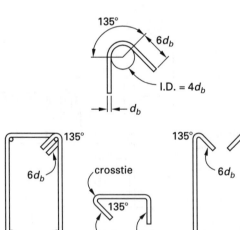

(c) detail A   (d) detail B   (e) detail C

Stirrups required to resist shear shall be hoops. Throughout the length of flexural members where hoops are not required, stirrups shall be spaced at no more than $d/2$ and shall have seismic hooks.

Figure 9.5   Details of Transverse Confinement Reinforcement in Flexural Members [UBC-97 Secs. 1921.1 and 1921.3]

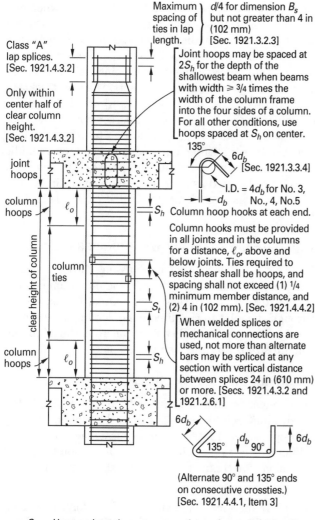

Class "A" lap splices. [Sec. 1921.4.3.2]

Only within center half of clear column height. [Sec. 1921.4.3.2]

joint hoops

column hoops

clear height of column

column ties

column hoops

Maximum spacing of ties in lap length. } $d/4$ for dimension B_s but not greater than 4 in (102 mm) [Sec. 1921.3.2.3]

Joint hoops may be spaced at $2S_h$ for the depth of the shallowest beam when beams with width $\geq 3/4$ times the width of the column frame into the four sides of a column. For all other conditions, use hoops spaced at S_h on center.

135° $6d_b$ [Sec. 1921.3.3.4]

I.D. = $4d_b$ for No. 3, No., 4, No.5 Column hoop hooks at each end.

Column hooks must be provided in all joints and in the columns for a distance, ℓ_o, above and below joints. Ties required to resist shear shall be hoops, and spacing shall not exceed (1) 1/4 minimum member distance, and (2) 4 in (102 mm). [Sec. 1921.4.4.2]

When welded splices or mechanical connections are used, not more than alternate bars may be spliced at any section with vertical distance between splices 24 in (610 mm) or more. [Secs. 1921.4.3.2 and 1921.2.6.1]

$6d_b$ 135° d_b 90° $6d_b$

(Alternate 90° and 135° ends on consecutive crossties.) [Sec. 1921.4.4.1, Item 3]

S_h = Hoop and supplementary crosstie spacing, not to exceed (1) 1/4 minimum member distance, and (2) 4 in (102 mm) [Sec. 1921.4.4.2], $8d_b$ of verticals, $24d_b$ of ties, or $1/2 B_s$ [Sec. 1907.10.5.2]

S_t = Column tie spacing, not to exceed $16d_b$ of verticals, $48d_b$ of ties, or B_s [Sec. 1907.10.5.2]

B_s = Smaller dimension of column cross section.

ℓ_o = Largest column dimension, but not less than one-sixth clear height, or 18 in (457 mm). [Sec. 1921.4.4.4]

Figure 9.6   Special Beam-Column Reinforcement [UBC-97 Sec. 1921.4]

R. Members with Bending and Axial Loads
[UBC-97 Sec. 1921.4]

The minimum member dimension is 12 in (305 mm). The ratio of the shortest-to-longest dimension cannot be less than 0.4. Class A tension lap splices are allowed within the center half of the member's length [Sec. 1921.4.3.2]. The total column moment-resisting strength at a joint must be greater than 6/5 of the total girder moment-resisting strength at a joint [Sec. 1921.4.2.2]. The longitudinal reinforcement ratio must be between 0.01 and 0.06, inclusive. (The 6% limit on column reinforcement ratio is too high for most designs. Considering the difficulty in designing and fabricating a joint as well as the joint congestion caused by additional ductility requirements, a practical upper limit of 4% is more reasonable.) Special provisions for transverse reinforcement (confinement) in the member apply. See Figs. 9.6 and 9.7.

In Fig. 9.6, distance S_h is the maximum spacing of the transverse reinforcement. The UBC-97 [Sec. 1921.4.4.2] specifies this as the smaller of 4 in (102 mm) or one-quarter of the smallest member dimension (B_s in the figure).

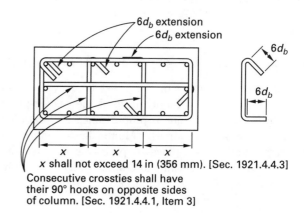

$6d_b$ extension
$6d_b$ extension
$6d_b$
$6d_b$

x x x

x shall not exceed 14 in (356 mm). [Sec. 1921.4.4.3] Consecutive crossties shall have their 90° hooks on opposite sides of column. [Sec. 1921.4.4.1, Item 3]

Figure 9.7   Details of Transverse Confinement Reinforcement in Beam-Columns [UBC-97 Sec. 1921.4]

S. Columns Carrying Discontinuous Walls
[UBC-97 Secs. 1921.4.4.5 and 1921.6.8]

Members supporting discontinuous elements (e.g., two columns supporting a short wall between them) must have special transverse reinforcement for their full height. See Fig. 9.8.

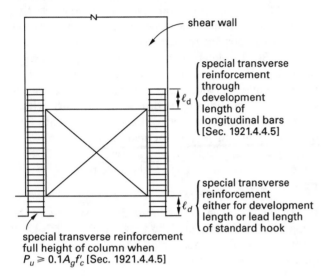

Figure 9.8   Columns Carrying a Discontinuous Wall [UBC-97 Sec. 1921.6.9]

T. Diaphragms   [UBC-97 Secs. 1921.6.11 and 1921.6.12]

A cast-in-place topping on a precast floor system can be considered as the diaphragm as if the cast-in-place topping functioning alone is proportioned and detailed to resist the seismic forces. Monolithic concrete diaphragms and cast-in-place topping over precast elements used to resist seismic forces must be at least 2 in (51 mm) thick [UBC-97 Sec. 1921.6.12, Item 1].

Based on the UBC-97 [Sec. 1921.6.12, Item 2], when mechanical connectors are used to transfer forces between the diaphragm and the lateral system, the anchorage must be able to develop $1.4A_s f_y$. When the collectors and boundary elements in topping slabs are placed over precast floor and roof elements, the slab thickness should not be less than 3 in (76 mm) or $6d_b$ (six times the diameter of the largest reinforcement in the topping slab) [UBC-97 Sec. 1921.6.12, Item 3]. Tendons that can be wire, cable, bar, rod, or strand, or a bundle of such elements used to impart prestress to concrete, should not be used as primary reinforcement in boundaries and collector elements [UBC-97 Sec. 1921.6.12, Item 4].

U. Minimum Column Size   [UBC-97 Sec. 1921.5.1.4]

In order for a horizontal bar passing through a joint to develop its full bond strength, the minimum column size is 20 times the largest longitudinal bar extending through the joint in the direction of loading for normal-weight concrete. For lightweight concrete, the minimum column size is 26 times the bar diameter.

V. Joints   [UBC-97 Sec. 1921.5]

Joints must be able to withstand forces of $1.25f_y$ in the longitudinal reinforcement. Flexural members framing into opposite sides of a column must have continuous reinforcement to pass through the column joint. Special provisions for confinement around the joint apply. See Fig. 9.1.

A 50% reduction in the amount of reinforcement confining a joint is permitted when all four sides of the joint are confined by members (beams) framing into the joint (column). The confinement spacing should be permitted to be increased to 6 in (152 mm) [UBC-97 Sec. 1921.5.2.2].

10

DETAILS OF SEISMIC-RESISTANT STEEL STRUCTURES

10-1 STEEL CONSTRUCTION DETAILS

The design, construction, and quality of steel components (members and connections) in buildings that resist seismic forces should conform to the requirements of the UBC-97 Chap. 22 provisions. The design forces resulting from earthquake motions are determined on the basis of energy dissipation in the nonlinear range of response.

A. Design and Construction Provisions

For design, fabrication, erection, and quality control of structural steel, the UBC-97 permits the use of the load and resistance factor design (LRFD) method as well as the allowable stress design (ASD) method. These methods proportion structural components such as members, connectors, connecting elements, and assemblages. Steel design based on the LRFD method should use the factored load combinations of the UBC-97 Sec. 1612.2 and the applicable requirements of the UBC-97 Sec. 2205. Design standards for the LRFD method are provided in UBC-97 Chap. 22, Div. II. As for steel design based on the ASD method, the factored load combinations of UBC-97 Sec. 1612.3 and the applicable requirements of the UBC-97 Sec. 2205 apply. For the ASD method, UBC-97 Chap. 22, Div. III provides design standards. UBC-97 Sec. 1612.3, Amendments furnish allowable stress increase requirements.

B. Seismic Provisions for Structural Steel Buildings

In addition to the structural steel requirements in UBC-97 Sec. 2205.2, steel structural elements resisting seismic forces should also comply with the UBC-97 seismic provisions in Chap. 22, Divs. IV and V. For steel design based on the ASD method, UBC-97 Chap. 22, Div. V, provides relevant seismic provisions.

Based on the UBC-97 Sec. 2210, Item 2.1, Div. IV, all buildings in seismic zones 0 and 1, and those buildings in seismic zone 2 that are not categorized as essential and hazardous facilities, do not need to be designed or detailed according to the structural steel seismic provisions.

C. Loads and Load Combinations

The appropriate critical combination of factored loads, as required by UBC-97 Sec. 1612.2 at load and resistance factor limits or UBC-97 Sec. 1612.3 at allowable stress limits with stress increases allowed by UBC-97 Sec. 1612.3.2, should be used in determining the required strength of the steel structure and its elements. Where the amplified horizontal earthquake is required, the following additional load combinations apply.

$$1.2D + f_1 L + E \qquad [10.1]$$
$$0.9D \pm E \qquad [10.2]$$

(f_1 is equal to 1.0 for floors in places of public assembly, for live loads in excess of 100 lbf/ft^2 (4.79 kN/m^2), and for garage live loads. For other live loads, f_1 is 0.5.)

D. Steel Frames

The most common structural system for high-rise buildings in seismically active areas is the ductile steel frame. Ductile frame behavior is much easier to achieve with steel construction than with concrete construction because steel is intrinsically a ductile material. Design emphasis is therefore shifted away from ensuring that the material itself will behave in a ductile manner to ensuring that the structural frame will behave in a ductile manner.

Ductile frame operation is accomplished by extensive use of moment-resisting connections known as *type I connections.*[1]   These connections transmit column moments to beams and girders, forcing those members to carry the moments. It is easier for beams and girders to resist moments because their lengths (and, hence, their moment arms) are long. Without moment-resisting joints, the columns would most likely fail (by web crippling, for example) before the beams and girders became fully stressed. Therefore, a beam-column connection must be able to transmit without failure a moment equal to the plastic capacity of the beam. Frames that use joints that do this are known as *special moment-resisting frames* (SMRF).[2]

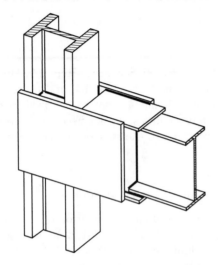

Figure 10.1   Moment-Resisting Joint for New Construction Incorporating SidePlate™ Connection Technology

SidePlate™ is the trademark of SidePlate Systems, Inc., of Long Beach, CA.

E. Steel Properties [UBC-97 Sec. 2213.4.1]

With limited exceptions, only certain steels (i.e., A36, A500, A501, A572 (Grades 42 and 50), A588, and A913 (Grades 50 and 65)) are permitted in seismic force-resisting frames. Structural steel conforming to A283 (Grade D) can be used for base plates and anchor bolts. Other steels not listed and those with yield strengths in excess of 50 ksi (345 MPa) cannot be used.

Based on the statistical data, A36 steel (Grade 36) has an actual yield strength of approximately 49,000 psi

(345 MPa). ASTM A913 (Grades 50 and 65), which was recently added to a list of structural steels permitted for lateral-force-resisting systems, has higher yield strength, increased toughness, and greatly improved weldability due to a lower carbon equivalent. Under high seismic demands, use of A913 is recommended.

F. Member Strength [UBC-97 Sec. 2213.4.2]

The term *strength* (as in "full strength") means that the members must be capable of developing the following forces or moments.

◇ moment in flexural members $M_s = ZF_y$
  Z represents plastic modulus of the section, and F_y is specified minimum yield stress of the steel.

◇ shear in flexural members $V_s = 0.55F_y dt$
  d and t denote overall depth and thickness of the member, respectively.

◇ axial compression in
  flexural members $P_{sc} = 1.7F_a A$
  F_a signifies axial design strength, and A symbolizes cross-section area of the member.

◇ axial tension in
  flexural members $P_{st} = F_y A$

◇ bolts $1.7 \times$ allowable

◇ full-penetration welds $F_y A$

◇ partial-penetration welds $1.7F_s$

◇ fillet welds $1.7F_s$
  F_s represents the applicable allowable stress values. Based on the UBC-97 [Sec. 2213.4.2], the one-third allowable stress increase is not permitted for the purpose of determining member or connection strength.

G. Column Requirements [UBC-97 Sec. 2213.5]

Load combinations required by UBC-97 Secs. 1612.2 and 1612.3 must be supported at load and resistance factor limits or at allowable stress limits. For ASD, a one-third stress increase is permitted according to the UBC-97 [Sec. 1612.3.2]. Because the integrity of a structural steel system that has experienced ductile yielding is strongly dependent on the axial capacity of the columns, with limited exceptions, columns in seismic zones 3 and 4 must have the strength (as defined in UBC-97 Sec. 2213.5.1) to support the load combinations given.[3]

$$1.0D + 0.7L + E \qquad \text{[axial compression]} \qquad [10.3]$$
$$0.85D \pm E \qquad \text{[axial tension]} \qquad [10.4]$$

---

[1] *Type II joints* do not develop any appreciable moments. Steel structures in any seismic zone can use type II joints. This is typical of braced-frame designs.

[2] The term *ductile moment-resisting space frame* (DMRSF), once used extensively, is no longer commonly used.

[3] Despite appearances, this is not ultimate strength (plastic or limit) design.

Other UBC-97 specifications relating to column splices [Sec. 2213.5.2] and slenderness [Sec. 2213.5.3] must also be met.

H. Ordinary Moment Frames [UBC-97 Sec. 2213.6]

Ordinary moment frames (OMFs) are permitted, but the Blue Book commentary discourages such use. It is likely, in any case, that OMFs will be heavier and thus costlier than SMRFs, since OMFs resist seismic forces elastically. OMFs may have applications where seismic forces are low and the design is controlled by wind.

OMFs should be designed to resist the load combinations in UBC-97 Sec. 1612.3. In resisting earthquake forces, UBC-97 Sec. 2213.6, Items 1–3, list requirements for the beam-to-column connections in OMFs.

I. Special Moment-Resisting Frames
[UBC-97 Sec. 2213.7]

This UBC-97 section is designed to ensure that joints and members behave in a ductile manner in SMRFs. The areas covered are the beams and columns in which plastic hinges can form and in the joint *panel zones* in which shear yielding can occur. The panel zone is normally the column web central to the beam-column connection. (SidePlate™ Connection Technology eliminates reliance on panel zone deformation by providing three panel zones: the two side plates, plus the column's own web.) With specific limitations, inelastic behavior in these areas is permitted.

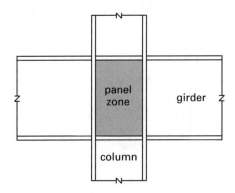

Figure 10.2   Panel Zone

For special moment-resisting frames, the adequacy of the beam-column connections is crucial to the integrity of the entire system. Previously, the beam-column connections were required to have minimum strength (i.e., strength equal to the beam in flexure). The code said this strength was achieved when beam flanges were full-penetration welded to columns and the beam web-to-column portion of the connection alone was able to support the gravity and seismic shear force. Bolted connections and other methods were also permitted if justified.[4]

After the 1994 Northridge earthquake in California, new design guidelines were adopted regarding beam-column connection configurations, and the current UBC-97 reflects these changes (see Sec. 10-1(H)).

Doubler plates can be used, with restrictions, to reduce panel zone shear stress or web depth/thickness ratio. Doubler plates should be welded across the plate width top and bottom. The minimum fillet weld is 3/16 in (4.7 mm). *Continuity plates* (column stiffeners at top and bottom of the panel zone) must satisfy a certain width/thickness ratio [Secs. 2213.7.2.3 and 2213.7.4].[5]

A minimum *strength ratio* is given in order to prevent column failures at joints when the beams are "stronger" than the columns. This is the "strong column-weak beam" test. Some exceptions are permitted, however [Sec. 2213.7.5].

Within strength requirements, *trusses* (see Sec. 10-1(J)) may be used as horizontal members in SMRFs [Sec. 2213.7.6].

In order to withstand reverse bending moments, intersecting beams or bottom flange diagonal bracing is required.

To reduce the earthquake hazards of welded steel moment frame structures in seismic zones 3 and 4, it is implied that (1) plastic hinge location should be designed to develop within the beam span at a minimum distance of half the beam depth from the column face, (2) all elements of frame and their connections should be designed to develop plastic hinge moment strength, (3) weld filler metals should be specified with rated toughness values, (4) connections should be detailed with beam flange continuity plates, (5) column panel zones should be designed without doubler plate consideration where practical, (6) backing bars and weld tabs should be eliminated, (7) connection design should be proficient through prototype testing or by reference to tests of comparable connection configuration, and (8) the quality of material and construction is essential and should be considered crucial to total frame behavior.

---

[4] Full-penetration welding of beam flanges to columns is expensive and is not the standard method preferred by erectors.

[5] The Blue Book commentary strongly suggests the use of continuity plates.

J. Special Truss Moment-Frames
[UBC-97 Sec. 2213.11]

A special truss moment frame STMF is a combination of a strong column and a ductile girder. STMF is a new design concept based on extensive analytical and experimental research at the University of Michigan that provides ductile behavior of truss girders. STMF should be designed in accordance with UBC-97 Sec. 2213.11. In Fig. 10.3, each horizontal truss that is part of the moment frame is shown with a *special segment* located within the middle one-half length of the truss.

Special segments perform inelastically when induced by earthquake forces. In the fully yielded case, these special segments should develop vertical shear strength through flexure of the chord members and through axial tension and compression of diagonal web members.

The maximum span length between columns that include such trusses is 50 ft (15.24 m), and the overall depth should not exceed 6 ft (183 cm). The length of the special segment ranges from 0.1 to 0.5 times the truss span length. All panels within the special segment should be either Vierendeel trusses or have bracing, and their length-to-depth ratio should be equal to or greater than 0.67, and less than or equal to 1.5.

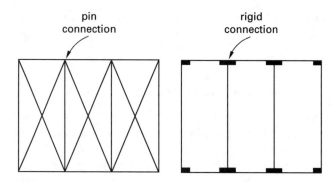

Figure 10.4   X and Vierendeel Trusses (elevations)

Within special segments, when diagonal members with identical sections are used, they should be arranged in an X pattern and be separated by vertical members. The UBC-97 requires that such vertical members be interconnected at points of crossing and that the connections have the strength to resist a minimum force equal to 0.25 times diagonal member tension strength. Connections of all elements in the truss frames, including those within the truss, should be designed according to UBC-97 Sec. 2213.11.5. Bolted connections and splicing of chord members within the special segment are not permitted by the UBC-97.

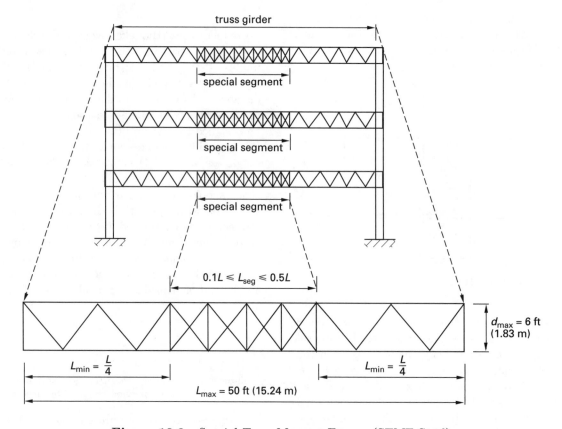

Figure 10.3   Special Truss Moment Frames (STMF-Steel)

The maximum axial stress in diagonal web members due to concentrated dead plus live loads acting within the special segment is $0.03F_y$.

K. Welded and Bolted Connections

In response to fractures in beam-column connections of steel moment frames during the 1994 Northridge earthquake in California, the UBC-97 retains the "emergency code change" provisions as adopted on September 14, 1994.

The Northridge earthquake damaged a considerable number of welded beam-column connections in moment-resisting frames. The most common failure involved cracking of the welded connection from the beam bottom flange to the column flange. In some instances, the crack propagated into the column flange and/or beam web. In response, UBC-97 Sec. 2213.7.1.2 specifies that the connection configurations (e.g., beam-column) should demonstrate, by approved cyclic load test results or calculation, the capability to uphold inelastic rotation and develop the UBC-97 strength criterion for ASD [Sec. 2213.7.1.1, Items 1 and 2] and the UBC-97 criterion for LRFD [Sec. 2210, Amendment 8.2.c]. Additionally, the effect of steel overstrength (expected value of yield strength) and strain hardening should be considered.

Reasons that contributed to the damage of welded steel moment frames can be summarized: (1) In resisting earthquake demands, the number of frame bays were limited. (2) The actual strengths versus code specified strengths were conflicting. (3) At beam-column connections, vast inelastic demands and stress concentrations occurred as a result of inappropriate detailing practice. (4) Column panel zones were weak. (5) Weld metal materials that were low in toughness were utilized. (6) In the welding procedures, levels of quality control were inadequate.

For welded steel moment frames that are expected to experience significant postelastic earthquake demands, or in seismic zones 3 and 4 where highly reliable postelastic seismic performance is expected, connections should be designed so that the plastic hinge location develops within the beam span at a minimum distance of half the beam depth from the column face, as shown in Fig. 10.5. This can be ensured either by reinforcing the beam at the connection or by decreasing the cross section of the beam a distance of half the beam depth from the column face.

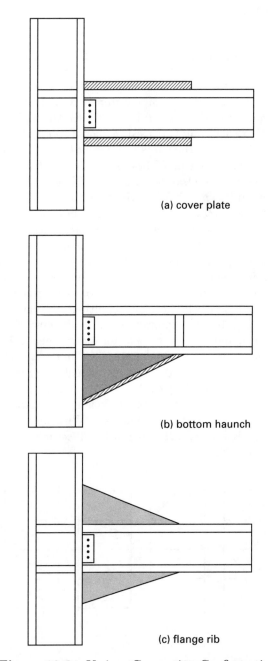

(a) cover plate

(b) bottom haunch

(c) flange rib

Figure 10.6 Various Connection Configurations

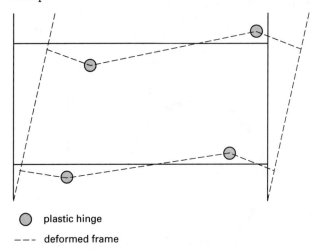

⬤ plastic hinge

--- deformed frame

Figure 10.5 Desired Plastic Hinge Behavior

L. Girder-Column Joint Restraints
[UBC-97 Sec. 2213.7.7]

Restrictions are given to ensure that moment frames are capable of reaching their seismic design capacity. Specifically, if the columns remain elastic, column flanges require lateral support only at the level of the top girder flange. (The UBC-97 describes the conditions under which the column can be assumed to remain elastic.) If the column does not remain elastic, column flange support is required at the tops and bottoms of the girder flanges. The UBC-97 [Sec. 2213.7.7.1, Items 1–4] provides acceptable conditions for columns to be assumed to remain elastic.

M. Braced Frames   [UBC-97 Sec. 2213.8 and 2213.9]

Various requirements, including slenderness ratios, are given for *concentrically braced frames* (CBF) and *special concentrically braced frames* (SCBF). Different systems of braced frames are used: *diagonal bracing* (where diagonals connect the joints in adjacent levels); *chevron bracing* (where a pair of braces terminate at a single point within the clear beam span); *V bracing* (a form of chevron bracing that intersects a beam from above); *inverted V bracing* (a form of chevron bracing that intersects a beam from below); *K bracing* (where a pair of braces located on one side of a column terminate at a single point within the clear column height); and *X bracing* (where a pair of diagonal braces cross near mid-length of the bracing members). Chevron bracing can be used when certain conditions are met [Sec. 2213.8.4.1], but the lateral seismic load must be increased by multiplying by a factor of 1.5 when designing bracing members [Sec. 2213.8.4.1, Item 1]. When chevron bracing is used in special concentrically braced frames, the UBC-97 [Sec. 2213.9.1] requires that any member intersected by a brace (beam) be continuous through the connection (columns). With limited exceptions, K bracing is not permitted in seismic zones 3 and 4 [Sec. 2213.8.4.2].

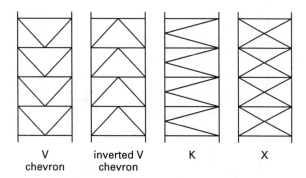

Figure 10.7   Types of Braced Frames

The provisions of UBC-97 Secs. 2213.8 and 2213.9 apply to ordinary and special concentrically braced frames, respectively. Based on UBC-97 Table 16-N, the response modification factors (R) for these frames are 5.6 and 6.4. The higher R value for special concentrically braced frames implies a more ductile system than ordinary concentrically braced frames. In special concentrically braced frames, all members and connections should be designed and detailed to resist shear and flexure. Also, any member that is intersected by a brace should be continuous through the connection.

Based on the UBC-97 [Secs. 2213.8.2.5 and 2213.9.2.4], it is required that compression elements in special and ordinary concentrically braces frames be compact sections in order to minimize the potentiality of local buckling.

The width-thickness ratio (w/t) of angle sections should be limited to

$$\frac{w}{t} \leq \frac{52}{\sqrt{F_y}} \qquad \text{[U.S.]} \quad \text{[10.5(a)]}$$

$$\frac{w}{t} \leq 0.31\sqrt{\frac{E}{F_y}} \qquad \text{[SI]} \quad \text{[10.5(b)]}$$

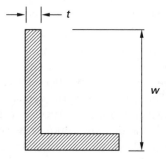

Figure 10.8   Angle Sections

The outside diameter-wall thickness ratio (d/t) of circular sections should be limited to

$$\frac{d}{t} \leq \frac{1300}{F_y} \qquad \text{[U.S.]} \quad \text{[10.6(a)]}$$

$$\frac{d}{t} \leq \frac{7.63E}{F_y} \qquad \text{[SI]} \quad \text{[10.6(b)]}$$

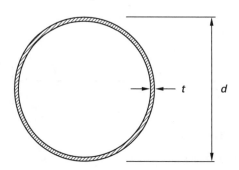

Figure 10.9   Circular Sections

The outside width-wall thickness ratio (w/t) of rectangular tubes should be limited to

$$\frac{w}{t} \le \frac{110}{\sqrt{F_y}} \qquad \text{[U.S.]} \quad \text{[10.7(a)]}$$

$$\frac{w}{t} \le 0.65\sqrt{\frac{E}{F_y}} \qquad \text{[SI]} \quad \text{[10.7(b)]}$$

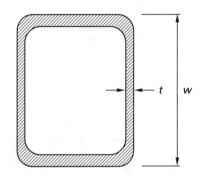

Figure 10.10   Rectangular Tubes

10-2 ECCENTRICALLY BRACED FRAMES
[UBC-97 SEC. 2213.10]

Figure 10.11 illustrates the basic types of *eccentrically braced frames* (EBF) in which at least one end of each bracing member connects to a beam a short distance from a beam-to-column connection or from another beam-to-brace connection.[6] The short section of girder between the brace end and the column is known as the *link beam*. Types (a) and (b) are known as *end link EBFs*, and type (c) is known as the *center link EBF*. EBFs may be inverted, and different EBF systems can be mixed.

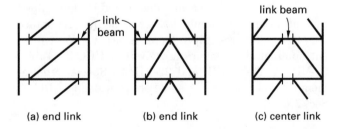

Figure 10.11   Types of Eccentrically Braced Frames

---

[6]This is exactly the type of design that engineers are taught to avoid in traditional classes on structures. However, this is an area pioneered by the works of Egor P. Popov. One of the earliest California buildings to use this type of feature is the 16-floor Bank of America Regional Office and Branch Bank Building in San Diego, which Popov assisted in designing.

The concept behind EBFs is that in low-to-moderate ground shaking, a frame using them performs as a braced frame rather than as a moment frame. Therefore, the structure experiences small drifts, little if any architectural damage, and no structural damage. The link beam is specifically designed to yield in a major seismic event, thereby absorbing large quantities of seismic energy and preventing buckling of the other bracing members. To limit yielding to the link beam requires attention to detail at the connection. Figure 10.12 illustrates typical details. (Notice the use of web stiffeners in the links to keep the web from buckling. These may not be needed in every instance.)

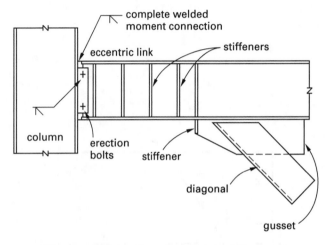

Figure 10.12   Detail of Original Eccentric Link Design

There is also some evidence that the beam-to-column connection should be at the lower end of the diagonal rather than at the top. Tests seem to show that the high ends are always stressed to yielding while the links at the low end remain elastic. This is illustrated in Fig. 10.13.

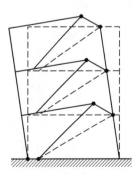

Figure 10.13   Typical Deflections of Eccentric Links

Section 2213.10 of the UBC-97 covers the design of EBFs. Shear in the link beam and bay drift are the primary design variables. Shear stress in the link beam is limited to $(0.80)(0.55)F_y = 0.44F_y$ [UBC-97 Secs. 2213.4.2 and 2213.10.5]. The braces impose large axial forces in the beams between two brace points. Therefore, critical bending/axial stress loads usually occur outside the link beam.

The shear force on a link beam—either end link or center link—is equal to the story shear times the story height divided by the distance between the column center lines (i.e., the *bay length*). For a given story height and distance between columns, frames with the same story shear will have the same link beam shear, regardless of their geometries (i.e., end link EBF or center link EBF). The load in each column is equal to the link beam shear. (There are two columns to carry the two link beam shears.)

$$V_{\text{link beam}} = \frac{F_x h}{L} \qquad [10.8]$$

$$P_{\text{column}} = V_{\text{link beam}} \qquad [10.9]$$

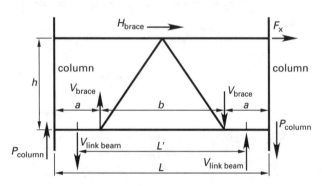

Figure 10.14   End Link Beam

Figure 10.14 illustrates an *end link EBF*, also known as a *Type (a) EBF*. The location of the link beam's inflection point initially must be assumed until all member sizes have been determined.[7] A location at mid-length of the link beam is a reasonable initial assumption. The fraction of the story shear carried by the inclined braces compared to that carried by direct moment frame action (approximately 0.75) is equal to the ratio of the distance between the points of inflection on the link beams (shown in Fig. 10.14 as distance L') to the bay length, b.

$$V_{\text{brace}} = \frac{V_{\text{link beam}} L'}{b} \qquad [10.10]$$

$$H_{\text{brace}} = \frac{F_x L'}{L} \qquad [10.11]$$

For a *center link beam*, also known as a *Type (b) EBF*, the beam-to-column connection is normally "pinned" in the frame's plane, and all of the lateral shear is taken by the braces. The link beam point of inflection is at the (vertical) centerline of the bay.[8]

$$V_{\text{link beam}} = \frac{F_x h}{L}$$

$$V_{\text{brace}} = \frac{V_{\text{link beam}} L}{2a} \qquad [10.12]$$

$$H_{\text{brace}} = \frac{F_x}{2} \qquad [10.13]$$

$$P_{\text{column}} = V_{\text{link beam}}$$

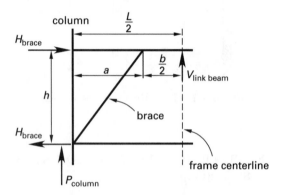

Figure 10.15   Center Link Beam

10-3 STEEL STUD WALL SYSTEMS

Steel stud wall systems used to resist lateral loads generated by wind or earthquake should comply with UBC-97 Sec. 2219. For ASD, the allowable shear value should be determined by dividing the nominal shear value by a factor of safety. The factor of safety is either 2.5 or 3.0. For LRFD, the design shear value should be determined by multiplying the nominal shear value by a resistance factor (ϕ). The resistance factor is either 0.45 or 0.55. All nominal shear values are given in UBC-97 Tables 22-VIII-A, 22-VIII-B, and 22-VIII-C. Special requirements for seismic zones 3 and 4 are provided in UBC-97 Sec. 2220.

---

[7]This is a direct result of the need to weld the beam-column connection in order to obtain a moment-resisting condition.

[8]This is a direct result of the fact that the beam-column connection can be pinned.

11

DETAILS OF SEISMIC-RESISTANT MASONRY STRUCTURES

11-1 MASONRY STRUCTURES

The materials, design, construction, and quality assurance of masonry should be in accordance with UBC-97 Chap. 21.

A. Design and Construction Provisions

Masonry structures should be designed by one of the following methods.

1. Working stress design—For this method, masonry-designed structures should comply with the provisions of UBC-97 Secs. 2106 and 2107.

2. Strength design—For this method, masonry-designed structures should comply with the provisions of UBC-97 Secs. 2106 and 2108.

3. Empirical design—For this method, masonry-designed structures should comply with the provisions of UBC-97 Secs. 2106 and 2109.

Masonry should also comply with the requirements of UBC-97 Secs. 2101 through 2105, which relate to materials and construction. Masonry units should be dry at the time of placement. Specifically, the UBC-97 prohibits wet or frozen masonry units from being laid. For various temperature ranges, there are special requirements outlined in UBC-97 Sec. 2104.3.3. For reinforcement, metal reinforcement should be placed prior to grouting. Positioning bolts for reinforcement requires templates or approved equivalent means to prevent dislocation during grouting.

B. Seismic Provisions for Structural Masonry Buildings

Special provisions for seismic resistance are given in UBC-97 Sec. 2106.1.12. In particular, UBC-97 Sec.

2106.1.12.4 is concerned with provisions for seismic zones 3 and 4.

C. Loads and Load Combinations

Masonry buildings and structures should be designed to resist the load combinations specified in UBC-97 Chap. 16. Where strength (load and resistance) factor design is used, structures and all portions should resist the most critical effects from the combinations of factored loads provided in the UBC-97 [Sec. 1612.2]. When allowable stress design is used, the most critical effects from the combinations of factored loads provided in the UBC-97 [Sec. 1612.3] should be resisted by the masonry structures. In addition, for seismic design regardless of the applied design method, the special load combinations of the UBC-97 [Sec. 1612.4] should also be considered.

D. Masonry Construction Details

The strength of masonry structures is sensitive to materials, design, construction, and quality control. Quality assurance ensures that materials, construction, and workmanship comply with the plans and specifications and with UBC-97 Sec. 2105. Masonry that is constructed under the watchful eye of an expert or other qualified person (as defined in UBC-97 Sec. 1701.2) is entitled to higher design stresses. In particular, when using working stress design, the allowable stresses for masonry in UBC-97 Sec. 2107 must be reduced by 50% if *special inspection* is not provided [UBC-97 Sec. 2107.1.2].

The following important points must be observed in masonry design for structures in seismic zones 3 and 4:

1. For columns, all longitudinal bars must be encircled by lateral ties. The lateral ties give support to the longitudinal bars. Based on the working

stress design and strength design requirements for reinforced masonry, the minimum and maximum distance from the surface of the column to the lateral ties and longitudinal bars are 1.5 in (38 mm) and 5 in (127 mm), respectively [UBC-97 Sec. 2106.3.6].

In columns that are stressed by tensile or compressive axial overturning forces from earthquake loading, column ties should be spaced a maximum of 8 in (203 mm) for the full height of such columns. In all other columns, column ties should be used only in the tops and bottoms of the column with the same maximum spacing of 8 in (203 mm) for a distance of one-sixth of the clear column height, 18 in (457 mm), or the maximum column cross-sectional dimension, whichever is greater [UBC-97 Sec. 2106.1.12.4, Item 1].

2. Masonry shear walls must be reinforced with both vertical and horizontal reinforcement. Reinforcement should be continuous around wall corners and through intersections.

3. For masonry shear walls, the minimum requirement for the sum of areas of vertical and horizontal reinforcement is 0.002 times the gross cross-sectional areas of wall. Also, the minimum area of reinforcement in either direction must be less than 0.0007 times the gross cross-sectional area of the wall [UBC-97 Sec. 2106.1.12.4, Item 2.3].

4. #4 bars are the smallest that may be used for vertical reinforcement at each corner or wall intersection and at the edge of any openings. Maximum bar spacing is 4 ft (122 cm) [UBC-97 Secs. 2106.1.12.3, Item 2, and 2106.1.12.4, Item 2.3].

5. #4 bars are the smallest that may be used for horizontal reinforcement at the tops and bottoms of walls and at the tops and bottoms of any openings. Maximum bar spacing is 10 ft (305 cm) [UBC-97 Secs. 2106.1.12.3, Item 3, and 2106.1.12.4, Item 2.3].

6. The minimum steel ratio (based on area bt) for vertical and horizontal reinforcement is 0.0007 for stack bond (open-end) *concrete masonry units* (CMUs) and 0.0015 for closed-cell CMUs [UBC-97 Sec. 2106.1.12.4, Item 2.4].

7. Types O and N mortar are not permitted [UBC-97 Secs. 2106.1.12.3, Item 5, and 2106.1.12.4, Item 3].

E. Masonry Moment-Resisting Wall Frames

UBC-97 Sec. 2108.2.6.1 provides general requirements for the design of fully grouted moment-resisting wall frames constructed of reinforced, open-end hollow-unit concrete or hollow-unit clay masonry. In designing a *masonry moment-resisting wall frame* (MMRWF), the primary requirement is to ensure a ductile response to lateral forces.

Fig. 11.1 illustrates piers (columns) and beams in a CMU wall. Beams interconnect vertical elements of the lateral-load resisting system. For the beams, clear span should not be less than twice the depth. The minimum nominal depth of the beam should be two units or 16 in (406 mm), whichever is greater. The ratio of nominal beam depth to nominal beam width should be less than six. As for piers, the maximum nominal depth is 96 in (2438 mm). The minimum nominal depth is two full units or 32 in (813 mm), whichever is greater. These dimensions ensure that flexural yielding will be limited to the beams at the face of the piers and to the bottom of the columns at the base of the structure.

F. Allowable Shear Stress

The *allowable shear stress* in a masonry wall depends on the compressive strength of the masonry, f'_m. When in-plane flexural reinforcement is present, the maximum allowable shear stress is 35 psi (240 kPa). Similarly, with shear reinforcement (i.e., the horizontal steel) taking all the shear, the maximum shear stress is 75 psi (520 kPa) [UBC-97 Sec 2107.2.9]. The allowable shear stress on reinforced walls without special inspection is always half of what is permitted with special inspection [UBC-97 Sec. 2107.1.2]. Horizontal steel must be provided to carry all shear stress higher than the allowable value. While the horizontal steel is used for shear, it is normal practice to have the same reinforcement ratio horizontally and vertically. Unreinforced masonry walls are not permitted in seismic zones 3 and 4 [UBC-97 Sec. 2106.1.12.4, Item 2.3]. Therefore, the empirical design method [UBC-97 Sec. 2109] and UBC-97 Table 21-M ("Allowable Compressive Stresses for Empirical Design of Masonry") cannot be used in those zones.

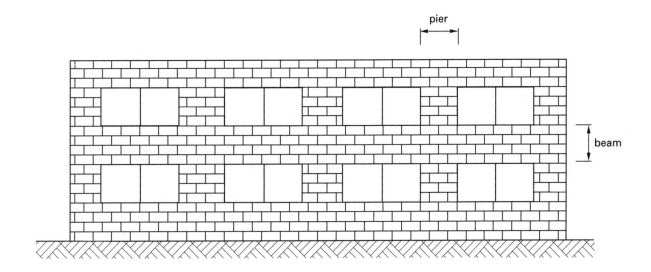

Figure 11.1   Masonry Moment-Resisting Wall Frame (MMRWF)

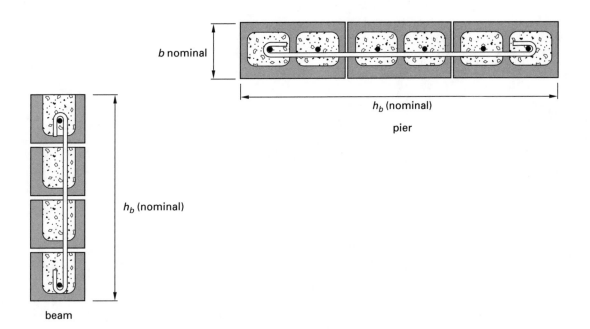

Figure 11.2   Masonry Moment-Resisting Wall Frame Details

12

DETAILS OF SEISMIC-RESISTANT WOOD STRUCTURES

12-1 WOOD STRUCTURES

The quality and design of wood members and their fastenings should be in accordance with the UBC-97 Chap. 23.

A. Design and Construction Provisions

Wood structures should be designed by one of the following design methods.

1. *Allowable stress design.* For this method, wood designed structures should comply with the provisions of UBC-97 Sec. 2305 and Chap. 23, Div. I; Div. II, Part I; and Div. III.

2. *Conventional light-frame construction.* For this method, wood designed structures should comply with the provisions of UBC-97 Sec. 2305 and Chap. 23, Div. IV.

The Load and Resistance Factor Design (LRFD) method for wood construction is adopted by reference in UBC-97 2303.5.54. LRFD currently is an alternate design procedure to the traditional allowable stress design method as prescribed in the National Design Specification (NDS). Provisions for the design of wood buildings in UBC-97 Chap. 23, Div. III are based on the 1991 NDS. (NDS, published by the American Forest and Paper Association, is a nationally recognized guide for wood structural design.) The NDS method incorporates factored loads and design provisions and guidelines for structural lumber, glued-laminated timber, poles and piles, connections, I-joists, structural composite lumber, trusses, structural panels, shear walls and diaphragms, and structural framing connections.

B. Seismic Provisions for Structural Wood Buildings

Special provisions for seismic load-resisting systems for all engineered wood structures are provided in UBC-97

Chap. 23, Div. II, Part II. In particular, the UBC-97 Secs. 2315.5, 2319, 2320.4, and 2320.5 are concerned with provisions for seismic zones 3 and 4.

C. Loads and Load Combinations

Wood buildings and structures should be designed to resist the load combinations specified in UBC-97 Chap. 16. Where allowable stress design is used, the most critical effects from the combinations of factored loads provided in the UBC-97 Sec. 1612.3 should be resisted by the wood structures. When load and resistance factor design is applied, structures and all portions should resist the most critical effects from the combinations of factored loads provided in the UBC-97 Sec. 1612.2. For seismic load-resisting systems for all engineered wood structures, the special load combinations of the UBC-97 Sec. 1612.4 should also be accounted.

12-2 WOOD SHEAR WALL AND WOOD STRUCTURAL PANEL DIAPHRAGM DESIGN CRITERIA

Design of wood shear walls and wood structural panel diaphragms requires consideration of diaphragm ratios, horizontal and vertical diaphragm shears, and connector/fastener values.[1]

Figure 12.1 illustrates the types of wood structural panel sheathing used to construct diaphragms. *Diagonal sheathing*, consisting of 1-in (25-mm) (nominal) sheathing boards laid at an angle of approximately 45 degrees to the supports, can also be used [UBC-97 Sec. 2315.3.1].

---

[1]The UBC-97 also specifies in Sec. 2315.1 that diaphragm deflection must be limited to the "permissible deflection" of attached distributing and resisting elements such as walls. *Permissible deflection* is defined as a deflection that maintains the structural integrity when loaded, that is, deflection that can support the loads without endangering the occupants. (See Sec. 7-15.)

In addition, there are special requirements for seismic zones 3 and 4, introduced in the following subsections.

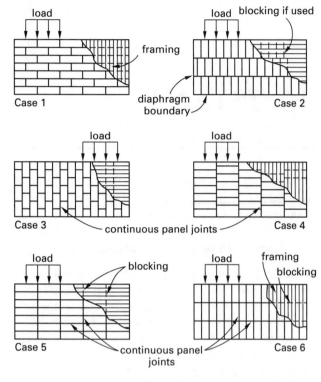

Figure 12.1   Types of Wood Structural Panel Diaphragm Sheathing

A. Framing   [UBC-97 Sec. 2315.5.2]

Collectors (i.e., drag struts) are required. Openings in diaphragms require perimeter framing.k Such perimeter framing must be detailed to distribute shear along its length. Diaphragm plywood cannot be used to splice the perimeter members. Chords must be in the plane of the diaphragm unless it can be shown that chords in other locations of the walls will work. (See Fig. 12.2.)

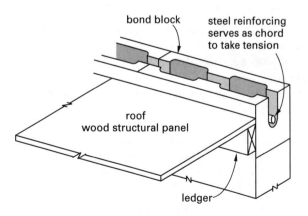

Figure 12.2   Chord in the Plane of a Diaphragm

B. Wood Structural Panels   [UBC-97 Sec. 2315.5.3]

For types of *wood structural panel* diaphragm sheathing, see Fig. 12.1.

Wood structural panels must be of the exterior glue type. For wood structural panel diaphragms and shear walls, sheets must be at least 4 ft by 8 ft (122 cm by 244 cm), except at edges of the diaphragm where pieces with a minimum dimension of 2 ft (61 cm) may be used.[2] Wood structural panel sheathing can be used to splice members that are not placed in cross-grain tension and bending from the splicing nails.

Particleboard with type M or better exterior glue can be used in shearwall construction [UBC-97 Sec. 2315.5.5].

Though wood structural panel can be used for splicing members in shear [UBC-97 Sec. 2315.5.3], it cannot be used to splice collector and perimeter members that carry tension and compression [UBC-97 Sec. 2315.5.2].

Per UBC-97 [Sec. 2319.3], in resisting horizontal forces, horizontal and vertical diaphragms may be sheathed with wood structural panels; however, the horizontal forces should be equal to or less than those established in Table 12.1 (corresponding to UBC-97 Table 23-II-H) for horizontal diaphragms and App. J of this book (corresponding to UBC-97 Table 23-II-I-1) for vertical diaphragms.

When both faces of shear walls are sheathed with the wood structural panels in accordance with Table 12.1, allowable shear for the wall may be taken as twice the tabulated shear for one side assuming the shear capacities are equal. Otherwise, the allowable shear should be taken as either the shear for the side with the higher capacity or twice the shear for the side with lower capacity, whichever is greater.

C. Requirements for Resisting Horizontal Forces in Concrete/Masonry Buildings
[UBC-97 Sec. 2315.2]

In buildings with concrete or masonry walls, wood horizontal floor and roof trusses and diaphragms can be used to resist seismic and wind forces as long as such forces are not resisted by intended torsion or rotation of the trusses or diaphragms.[3] Vertical shear walls can be used to resist horizontal shear in concrete or masonry buildings up to two stories in height as long as certain requirements are met. For example, wall heights

---

[2]Pieces even smaller than 2 ft (61 cm) may be used if all edges are properly blocked.

[3]UBC-97 Sec. 2315.1 states that diaphragms are to be considered incapable of supporting rotation (torsion). Also, see Sec. 7-12.

Table 12.1
Allowable Shear for Horizontal Wood Structural Panel Diaphragms with Douglas Fir-Larch
or Southern Pine Framing[1] (pounds per foot)
(See Fig. 12.1 for cases.)
[UBC-97 Table 23-II-H]

PANEL GRADE	COMMON NAIL SIZE	MINIMUM NAIL PENETRATION IN FRAMING (inches) × 25.4 for mm	MINIMUM NOMINAL PANEL THICKNESS (inches)	MINIMUM NOMINAL WIDTH OF FRAMING MEMBER (inches)	BLOCKED DIAPHRAGMS Nail spacing (in.) at diaphragm boundaries (all cases), at continuous panel edges parallel to load (Cases 3 and 4) and at all panel edges (Cases 5 and 6) × 25.4 for mm				UNBLOCKED DIAPHRAGMS Nails spaced 6 in (152 mm) max. at supported edges	
					6	4	2½[2]	2[2]	Case 1 (No unblocked edges or continous joints parallel to load)	All other configurations (Cases 2,3,4,5 and 6)
					Nail spacing (in.) at other panel edges × 25.4 for mm					
					6	6	4	3		
					× 0.0146 for N/mm					
Structural I	6d	1¼	5/16	2	185	250	375	420	165	125
				3	210	280	420	475	185	140
	8d	1½	3/8	2	270	360	530	600	240	180
				3	300	400	600	675	265	200
	10d[3]	1⅝	15/32	2	320	425	640	730	285	215
				3	360	480	720	820	320	240
C-D, C-C, Sheathing, and other grades covered in UBC Standard 23-2 or 23-3	6d	1¼	5/16	2	170	225	335	380	150	110
				3	190	250	380	430	170	125
			3/8	2	185	250	375	420	165	125
				3	210	280	420	475	185	140
	8d	1½	3/8	2	240	320	480	545	215	160
				3	270	360	540	610	240	180
			7/16	2	255	340	505	575	230	170
				3	285	380	570	645	255	190
			15/32	2	270	360	530	600	240	180
				3	300	400	600	675	265	200
	10d[3]	1⅝	15/32	2	290	385	575	655	255	190
				3	325	430	650	735	290	215
			19/32	2	320	425	640	730	285	215
				3	360	480	720	820	320	240

[1]These values are for short-time loads due to wind or earthquake and must be reduced 25 percent for normal loading. Space nails 12 in (305 mm) on center along intermediate framing members.

Allowable shear values for nails in framing members of other species set forth in Part III of Division III shall be calculated for all other grades by multiplying the shear capacities for nails in Structural I by the following factors: 0.82 for species with specific gravity greater than or equal to 0.42 but less than 0.49, and 0.65 for species with a specific gravity less than 0.42.

[2]Framing at adjoining panel edges shall be 3-in (76-mm) nominal or wider and nails shall be staggered where nails are spaced 2 in (51 mm) or 2½ in (64 mm) on center.

[3]Framing at adjoining panel edges shall be 3-in (76-mm) nominal or wider and nails shall be staggered where 10d nails having penetration into framing of more than 1⅝ in (41 mm) are spaced 3 in (76 mm) or less on center.

must be less than 12 ft (366 cm), and deflection/drift cannot exceed 0.5% (0.005 times story height as defined in Sec. 1630.10). Per UBC-97 [Sec. 1633.2.9], design seismic forces for flexible diaphragms that provide lateral support for shear walls of masonry or concrete should be determined using a structure response modification factor (R) of 4 or smaller. Other requirements apply.

12-3 WOOD STRUCTURAL PANEL SHEAR WALLS

The design of wood structural panel-sheathed shear walls is covered in UBC-97 Sec. 2315, and in particular, Sec. 2315.5. Appendix J of this book, "Allowable Shear for Wind or Seismic Forces for Wood Structural Panel Shear Walls with Framing of Douglas Fir-Larch or Southern Pine" (corresponding to UBC-97 Table 23-II-I-1), gives the nailing schedule and other details for the design and analysis of such walls.

In seismic zone 4, unblocked shear walls are prohibited. Shear walls may suffer major damage or complete failure in the event of severe earthquakes due to inadequate nail edge distance into 2-in (51-mm) nominal framing members. For seismic zones 3 and 4, where allowable shear for shear walls exceeds 350 lbf/ft (5.11 N/mm), foundation sill plates and all framing members receiving edge nailing from abutting panels should be 3-in

(76-mm) nominal members [UBC-97 Table 23-II-1-1, Ftn. 3].

12-4 DIAPHRAGM RATIOS

Maximum aspect ratios of diaphragms are limited to the values listed in Table 12.2 (corresponding to UBC-97 Table 23-II-G).

Table 12.2
Maximum Diaphragm Dimension Ratios
[UBC-97 Table 23-II-G]
(This table is not applicable to subfloor-underlayment.)

MATERIAL	HORIZONTAL DIAPHRAGMS	SHEAR WALLS
	Maximum Span-Width Ratios	Maximum Height-Width Ratios
1. Diagonal sheathing, conventional	3:1	1:1[1]
2. Diagonal sheathing, special	4:1	2:1[2]
3. Wood structural panels and particleboard, nailed all edges	4:1	2:1[2,3]
4. Wood structural panels and particleboard, blocking omitted at intermediate joints	4:1	[4]

[1]In Seismic Zones 0, 1, 2 and 3, the maximum ratio may be 2:1.
[2]In Seismic Zones 0, 1, 2 and 3, the maximum ratio may be 3½:1.
[3]In Seismic Zone 4, the maximum ratio may be 3½:1 for walls not exceeding 10 ft (3048 mm) in height on each side of the door to a one-story Group Occupancy.
[4]Not permitted.

12-5 WOOD STRUCTURAL PANEL DIAPHRAGM CONSTRUCTION DETAILS

A wood structural panel must be nailed continuously along its edges (*edge nailing*) and throughout its interior (*field nailing*) to achieve full development of the sheet. Table 12.1 (UBC-97 Table 23-II-H) specifies the edge nail spacing required to carry a particular shear load. The plywood is nailed to joists, blocking, and ledgers, as shown in Fig. 7.5.

All force elements must have a full transmission path across the diaphragm. Collectors must frame into suitable walls or other collector elements. *Continuity ties* are required between adjacent edges of sheathing where edge nailing from the sheathing places the framing member below in cross-grain tension (see Sec. 12-9).

Figure 12.3 illustrates how ties can be used to transmit tension and compression forces through a perpendicular girder.

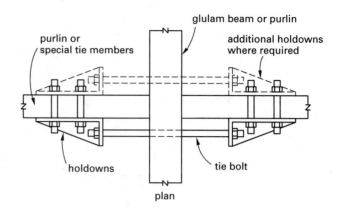

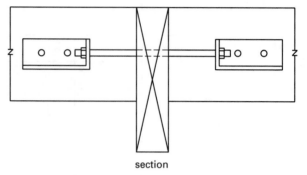

Figure 12.3   Typical Seismic Tie

12-6 BLOCKING

Beams, rafters, and joists should be supported laterally to prevent rotation or lateral displacement. The shear transfer action is provided by blocking (see Fig. 12.1).

Blocking in accordance with UBC-97 Sec. 2320.12.8 provides lateral support and prevents long joists from buckling—that is, "rolling out" or rotating out from under the members the joists support. When a joist buckles, its moment of inertia in the plane of support decreases significantly. Blocking is typically toe-nailed to the joists.

A blocked diaphragm is one in which all edges of the wood structural panels not falling on structural framing members are supported on and connected to blocking.

12-7 SUBDIAPHRAGMS

A *subdiaphragm*, as defined in UBC-97 Sec. 2302, is "a portion of a larger wood diaphragm designed to anchor and transfer local forces to primary diaphragm struts and the main diaphragm." For example, in Fig. 12.4, the lateral forces from the masonry wall are transferred to the subdiaphragm through the anchor ties. The subdiaphragm span is the distance between the end shear wall and the center diaphragm strut. The anchor ties run the full depth of the subdiaphragm. The subdiaphragm should be designed as if it acts alone in transferring the tie forces developed in the anchor to the shear wall and strut.

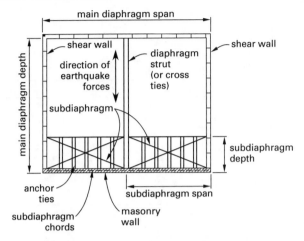

Figure 12.4 Subdiaphragm

The maximum length-to-width ratio for a wood structural subdiaphragm is 2.5:1 [UBC-97 Sec 1633.2.9]. Therefore, the subdiaphragm depth limits the length of anchor ties required. The lengths of diaphragm struts and cross ties at diaphragm discontinuities are similarly limited, as shown in Fig. 12.5.

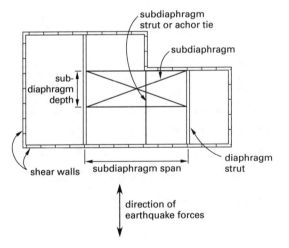

Figure 12.5 Subdiaphragm Limiting Strut Length

12-8 CONNECTOR STRENGTHS

Connectors, or fasteners, for wood are typically nails, lag bolts (i.e., pointed bolts installed in pilot holes from one side), and machine bolts (i.e., nutted bolts). Connectors can fail in one of several ways. Connectors, particularly nails, can pull out; wood in shear connections can fail in bearing; connectors can fail in shear or bending. (Since they are much stronger than the wood pieces they connect, the connectors seldom fail in tension.)

It is generally unnecessary to deal with properties of the connectors such as yield strength in shear, longitudinal friction factor, and so on. The UBC-97 provides tables of allowable loads for various types of connectors.[4] (See Tables 12.3 and 12.4.)

A. Nails [UBC-97 Sec. 2318.3]

Tables 12.3 and 12.4 (corresponding to UBC-97 Tables 23-III-C-1 and 23-III-C-2) give the allowable lateral force design values, Z, when a nail (box or common) is driven the specified distance perpendicular to the grain. These tables apply to *sinkers*, smooth framing nails coated for easy penetration, only when values are adjusted for wire diameter and penetration. Note that values may be increased 30% for use in wood diaphragm calculations, although it is conservative not to take the increase. The allowable lateral loads are two-thirds of the table's values when nails are driven parallel to the grain. *Toe-nails* may be used to support up to five-sixths of the lateral load. However, toe-nailing may not be used to connect diaphragms supporting concrete or masonry walls to ledgers in seismic zones 2, 3, and 4. In seismic zones 3 and 4, toe-nails may not transfer lateral forces in excess of 150 lbf/ft (2188 N/m) from diaphragms to shear walls, drag struts (collectors), or other elements [UBC-97 Sec. 1633.2.9, Item 5, and Sec. 2318.3.1].

Table 12-5 (corresponding to UBC-97 Table 23-III-D) gives the maximum withdrawal loads for nails driven perpendicular to the grain. Nails driven parallel to the grain of the wood are not permitted to support withdrawal loads. Other requirements for spacing and edge distances are given.

Nails are to be driven so that their heads are flush with, but do not fracture, the surface of the sheathing [UBC-97 Sec. 2315.1].

B. Bolts [UBC-97 Sec. 2318.2]

Allowable lateral design values for bolts loaded in shear in wood-to-wood connections are given in Tables 12.6 and 12.7 (corresponding to UBC-97 Tables 23-III-B-1 and 23-III-B-2). The shear force permitted in members used to connect wood to concrete or masonry is one-half of the double-shear values in the table for a wood member that is considered twice as thick as its actual thickness. The values may be increased by one-third for seismic loadings.

---

[4]It is also important to recognize that connector forces used in wood-to-concrete and wood-to-masonry are limited by the strength of the concrete and masonry. Inasmuch as the wood is the weaker material, it seems logical that the wood provisions would determine the design, but there is no guarantee of this. Limitations are covered in UBC-97 Sec. 1923 for concrete and Secs. 2106.2.14 and 2107.1.5 for masonry.

Table 12.3
Box Nail Design Values (Z) for
Single-Shear (Two-Member) Connections[1,2,3]
(with both members of identical species)
[UBC-97 Table 23-III-C-1]

side member thickness t_s (inches)	nail length L (inches)	nail diameter D (inches)		$G = 0.55$ southern pine Z lbs.	$G = 0.50$ Douglas-fir larch Z lbs.	$G = 0.42$ spruce-pine-fir Z lbs.
	×25.4 for mm		penny-weight		×4.45 for N	
$1/2$	2	0.099	6d	55	48	38
	$2^1/_2$	0.113	8d	67	59	47
	3	0.128	10d	82	73	59
	$3^1/_4$	0.128	12d	82	73	59
	$3^1/_2$	0.135	16d	89	79	65
	4	0.148	20d	101	90	73
	$4^1/_2$	0.148	30d	101	90	73
	5	0.162	40d	117	105	87
$3/4$	2	0.099	6d	61	55	47
	$2^1/_2$	0.113	8d	79	72	57
	3	0.128	10d	101	87	68
	$3^1/_4$	0.128	12d	101	87	68
	$3^1/_2$	0.135	16d	108	94	74
	4	0.148	20d	121	105	83
	$4^1/_2$	0.148	30d	121	105	83
	5	0.162	40d	138	121	96
1	$2^1/_2$	0.113	8d	79	72	61
	3	0.128	10d	101	93	79
	$3^1/_4$	0.128	12d	101	93	79
	$3^1/_2$	0.135	16d	113	103	86
	4	0.148	20d	128	118	96
	$4^1/_2$	0.148	30d	128	118	96
	5	0.162	40d	154	141	109
$1^1/_2$	$3^1/_4$	0.128	12d	101	93	79
	$3^1/_2$	0.135	16d	113	103	88
	4	0.148	20d	128	118	100
	$4^1/_2$	0.148	30d	128	118	100
	5	0.162	40d	154	141	120

[1]Tabulated lateral design values (Z) for nailed connections shall be multiplied by all applicable adjustment factors (see Division III, Part I).

[2]Tabulated lateral design values (Z) are for box nails inserted in side grain with nail axis perpendicular to wood fibers and with the following nail bending yield strengths (F_{yb}):

   F_{yb}=100,000 psi (690 N/mm^2) for 0.099- (2.5-mm), 0.113- (2.9-mm), 0.128- (3.3-mm) and 0.135-inch-diameter (3.4-mm) box nails.

   F_{yb}=90,000 psi (621 N/mm^2) for 0.148- (3.8-mm) and 0.162-inch-diameter (4.1-mm) box nails.

[3]For other species and configurations, see Division III, Part I.

Table 12.4
Common Wire Nail Design Values (Z) for
Single-Shear (Two-Member) Connections[1,2,3]
(with both members of identical species)
[UBC-97 Table 23-III-C-2]

side member thickness t_s (inches)	nail length L (inches)	nail diameter D (inches)		$G = 0.55$ southern pine Z lbs.	$G = 0.50$ Douglas-fir larch Z lbs.	$G = 0.42$ spruce-pine-fir Z lbs.
	×25.4 for mm		penny-weight		×4.45 for N	
$1/2$	2	0.113	6d	67	59	47
	$2^1/2$	0.131	8d	85	76	61
	3	0.148	10d	101	90	73
	$3^1/4$	0.148	12d	101	90	73
	$3^1/2$	0.162	16d	117	105	87
	4	0.192	20d	137	124	103
	$4^1/2$	0.207	30d	148	134	112
	5	0.225	40d	162	147	123
	$5^1/2$	0.244	50d	166	151	127
	6	0.263	60d	188	171	144
$3/4$	$2^1/2$	0.131	8d	104	90	70
	3	0.148	10d	121	105	83
	$3^1/4$	0.148	12d	121	105	83
	$3^1/2$	0.162	16d	138	121	96
	4	0.192	20d	157	138	111
	$4^1/2$	0.207	30d	166	147	119
	5	0.225	40d	178	158	129
	$5^1/2$	0.244	50d	182	162	132
	6	0.263	60d	203	181	149
1	3	0.148	10d	128	118	96
	$3^1/4$	0.148	12d	128	118	96
	$3^1/2$	0.162	16d	154	141	109
	4	0.192	20d	183	159	124
	$4^1/2$	0.207	30d	192	167	131
	5	0.225	40d	202	177	140
	$5^1/2$	0.244	50d	207	181	143
	6	0.263	60d	227	199	159
$1^1/2$	$3^1/2$	0.162	16d	154	141	120
	4	0.192	20d	185	170	144
	$4^1/2$	0.207	30d	203	186	158
	5	0.225	40d	224	205	172
	$5^1/2$	0.244	50d	230	211	175
	6	0.263	60d	262	240	191

[1]Tabulated lateral design values (Z) for nailed connections shall be multiplied by all applicable adjustment factors (see Division III, Part I).
[2]Tabulated lateral design values (Z) are for common wire nails inserted in side grain with nail axis perpendicular to wood fibers and with the following nail bending yield strengths (F_{yb}):
   F_{yb}=100,000 psi (690 N/mm^2) for 0.113- (2.9-mm), 0.131- (3.3-mm) and 0.135-inch-diameter (3.4-mm) common wire nails.
   F_{yb}=90,000 psi (621 N/mm^2) for 0.148- (3.8-mm) and 0.162-inch-diameter (4.1-mm) common wire nails.
   F_{yb}=80,000 psi (552 N/mm^2) for 0.192- (4.9-mm), 0.207- (5.3-mm) and 0.225-inch-diameter (5.7-mm) common wire nails.
   F_{yb}=70,000 psi (482 N/mm^2) for 0.244- (6.2-mm) and 0.263-inch-diameter (6.7-mm) common wire nails.
[3]For other species and configurations, see Division III, Part I.

Table 12.5
Nail and Spike Withdrawal Design Values (W)[1,2]
[UBC-97 Table 23-III-D]

Tabulated withdrawal design values (W) are in pounds per inch of penetration into side grain of main member.

| | specific gravity, G | common wire nails, box nails and common wire spikes diameter, D | | | | | | | | | | | | | |
		0.099″	0.113″	0.128″	0.131″	0.135″	0.148″	0.162″	0.192″	0.207″	0.225″	0.244″	0.263″	0.283″	0.312″	0.375″
southern pine	0.55	31	35	40	41	42	46	50	59	64	70	76	81	88	97	116
Douglas-fir larch	0.50	24	28	31	32	33	36	40	47	50	55	60	64	69	76	91
spruce-pine-fir	0.42	16	18	20	21	21	23	26	30	33	35	38	41	45	49	59

[1]Tabulated withdrawal design values (W) for nail or spike connections shall be multiplied by all applicable adjustment factors (see Division III, Part I).
[2]For other species and configurations, see Division III, Part I.

Table 12.6
Bolt Design Values (Z) for Single-Shear
(Two-Member) Connections[1,2,3]
(for sawn lumber with both members of identical species)
[UBC-97 Table 23-III-B-1]

| main member lm (inches) | side member ls (inches) | bolt diameter D (inches) | $G = 0.55$ southern pine | | | $G = 0.50$ Douglas fir-larch | | | $G = 0.42$ spruce-pine-fir | | |
			$Z_\parallel$ lbs.	$Z_{s\perp}$ lbs.	$Z_{m\perp}$ lbs.	$Z_\parallel$ lbs.	$Z_{s\perp}$ lbs.	$Z_{m\perp}$ lbs.	$Z_\parallel$ lbs.	$Z_{s\perp}$ lbs.	$Z_{m\perp}$ lbs.
×25.4 for mm			×4.45 for N								
$1^1/_2$	$1^1/_2$	$1/_2$	530	330	330	480	300	300	410	240	240
		$5/_8$	660	400	400	600	360	360	510	290	290
		$3/_4$	800	460	460	720	420	420	610	340	340
		$7/_8$	930	520	520	850	470	470	710	380	380
		1	1060	580	580	970	530	530	810	430	430
$3^1/_2$	$1^1/_2$	$1/_2$	660	400	470	610	370	430	540	320	370
		$5/_8$	940	560	620	880	520	540	780	410	430
		$3/_4$	1270	660	690	1200	590	610	1080	450	480
		$7/_8$	1680	720	770	1590	630	680	1340	490	540
		1	2010	770	830	1830	680	740	1530	530	590
$5^1/_2$	$1^1/_2$	$5/_8$	940	560	640	880	520	590	780	410	520
		$3/_4$	1270	660	850	1200	590	790	1080	450	690
		$7/_8$	1680	720	1090	1590	630	980	1440	490	760
		1	2150	770	1190	2050	680	1060	1760	530	830
$7^1/_2$	$1^1/_2$	$5/_8$	940	560	640	880	520	590	780	410	520
		$3/_4$	1270	660	850	1200	590	790	1080	450	690
		$7/_8$	1680	720	1090	1590	630	1010	1440	490	890
		1	2150	770	1350	2050	680	1270	1760	530	1110

[1]Tabulated lateral design values (Z) for bolted connections shall be multiplied by all applicable adjustment factors (see Division III, Part I).
[2]Tabulated lateral design values (Z) are for "full diameter" bolts with a bending yield strength (F_{yb}) of 45,000 psi (310 N/mm^2).
[3]For other species and configurations, see Division III, Part I.

Table 12.7

Bolt Design Values (Z) for Double-Shear
(Three-Member) Connections[1,2,3]
(for sawn lumber with all members of identical species)
[UBC-97 Table 23-III-B-2]

main member lm (inches)	side member ls (inches)	bolt diameter D (inches)	$G = 0.55$ southern pine			$G = 0.50$ Douglas fir-larch			$G = 0.42$ spruce-pine-fir		
			$Z_\parallel$ lbs.	$Z_{s\perp}$ lbs.	$Z_{m\perp}$ lbs.	$Z_\parallel$ lbs.	$Z_{s\perp}$ lbs.	$Z_{m\perp}$ lbs.	$Z_\parallel$ lbs.	$Z_{s\perp}$ lbs.	$Z_{m\perp}$ lbs.
×25.4 for mm			×4.45 for N								
$1^1/_2$	$1^1/_2$	$1/_2$	1150	800	550	1050	730	470	880	640	370
		$5/_8$	1440	1130	610	1310	1040	530	1100	830	410
		$3/_4$	1730	1330	660	1580	1170	590	1320	900	450
		$7/_8$	2020	1440	720	1840	1260	630	1540	970	490
		1	2310	1530	770	2100	1350	680	1760	1050	530
$3^1/_2$	$1^1/_2$	$1/_2$	1320	800	940	1230	730	860	1080	640	740
		$5/_8$	1870	1130	1290	1760	1040	1190	1570	830	960
		$3/_4$	2550	1330	1550	2400	1170	1370	2160	900	1050
		$7/_8$	3360	1440	1680	3280	1260	1470	2880	970	1130
		1	4310	1530	1790	4090	1350	1580	3530	1050	1230
$5^1/_2$	$1^1/_2$	$5/_8$	1870	1130	1290	1760	1040	1190	1570	830	1040
		$3/_4$	2550	1330	1690	2400	1170	1580	2160	900	1380
		$7/_8$	3360	1440	2170	3180	1260	2030	2880	970	1780
		1	4310	1530	2700	4090	1350	2480	3530	1050	1930
$7^1/_2$	$1^1/_2$	$5/_8$	1870	1130	1290	1760	1040	1190	1570	830	1040
		$3/_4$	2550	1330	1690	2400	1170	1580	2160	900	1380
		$7/_8$	3360	1440	2170	3180	1260	2030	2880	970	1780
		1	4310	1530	2700	4090	1350	2530	3530	1050	2240

[1]Tabulated lateral design values (Z) for bolted connections shall be multiplied by all applicable adjustment factors (see Division III, Part I).
[2]Tabulated lateral design values (Z) are for "full diameter" bolts with a bending yield strength (F_{yb}) of 45,000 psi (310 N/mm^2).
[3]For other species and configurations, see Division III, Part I.

C. Ties

Allowable loads on proprietary structural straps and
ties, as shown in Fig. 12.6, must be given by the man-
ufacturer of those ties.

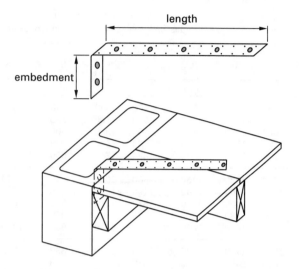

Figure 12.6   Typical Tie Installation

12-9 FLEXIBLE DIAPHRAGM TO WALL CONNECTION DETAILS

There are many acceptable ways that flexible diaphragms can be connected to shear walls. (There are
even more methods of making connections to wood-
framed walls.) All acceptable methods provide an un-
broken path for the force to follow from the diaphragm
to the foundation. (Any detail where the joists, ledgers,
or diaphragm merely sit on supports without positive
connection is unsatisfactory. Toe-nailing and nailing
subject to withdrawal are not permitted in seismic zones
2, 3, and 4 [UBC-97 Sec. 1633.2.9, Item 5]).

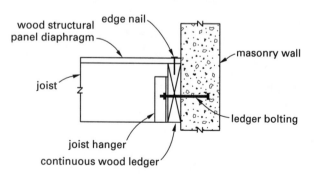

Figure 12.7   Inadequate Joist-Wall Framing

Figure 12.7 shows a detail for a wall into which the joist
(which may also be called a purlin or other framing
member term) ends the frame. The lateral force from

the diaphragm travels from the plywood through the
edge nails into the *ledger*. (A ledger may also be called
a *nailer* or *sill*.) A diaphragm shear force parallel to
the masonry wall will continue through the ledger bolts
into the masonry wall and through the parallel wall to
the foundation and ground.

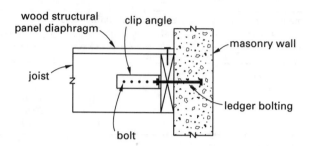

Figure 12.8   Adequate Joist-Wall Framing for
Tension Forces

The framing method shown in Fig. 12.7 is not adequate
(nor is it permitted [UBC-97 Sec. 1633.2.9, Item 5]) for
forces perpendicular to the wall because of cross-grain
bending. (See Sec. 12-10.) Additional tension connection straps, as shown in Figs. 12.8 and 12.9, must
be added [UBC-97 Sec. 1633.2.9, Item 4]. The tension connection should be continuously nailed back (i.e.,
strapped) a considerable distance into the diaphragm to
eliminate the high local tensile stress that would otherwise occur.

Figure 12.9 illustrates the typical details of a connection using both anchor bolts (for the ledger) and an
embedded tie (for the joist[5]). Figure 12.6 illustrates
yet another option.

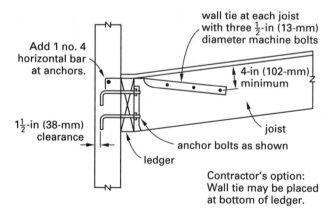

Figure 12.9   Ledger and Joist Tie

---

[5]As Sec. 7-7 implies, the term *joist* may be replaced by *purlin* or
other framing member term.

The framing for a wall parallel to the joists is shown in Fig. 12.10. The design (i.e., using a ledger) is basically the same as for the connections at the ends of the joists. (Notice that the detail as shown places the ledger in cross-grain bending. (See Fig. 12.11.) Ties would also be required between the diaphragm and wall.)

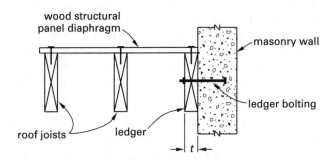

Figure 12.10   Joist-Parallel Wall Framing

The UBC-97 [Secs. 1605.2.3, 1611.4, and 1633.2.8.1] require that (1) the connection between concrete and masonry walls and floor and roof diaphragms must be designed to resist the load combination of UBC-97 Secs. 1612.2 or 1612.3, using the greater of wind or seismic-induced forces or a minimum horizontal force of 280 pounds per foot (4.09 kN/m) of wall, (2) the anchor spacing cannot exceed 4 ft (122 cm) unless the wall is designed to resist bending between anchors, and (3) the anchors must be grouted in place when they are embedded in hollow masonry blocks or cavity walls.[6] For flexible diaphragms (the usual case with wood-framed diaphragms) in seismic zone 4, the minimum horizontal force is increased from 280 to 420 pounds per foot (6.1 kN/m) of wall. (See Sec. 6-44 for special provisions regarding the connector design force in the center of the diaphragm.)

12-10 CROSS-GRAIN LOADING

Cross-grain bending and *cross-grain tension* in ledgers and joists, as illustrated in Fig. 12.11, are not permitted [UBC-97 Sec. 1633.2.9, Item 5].

---

[6]It is usually easier to specify a 4-ft (122-cm) anchor spacing than to design the wall connection for bending.

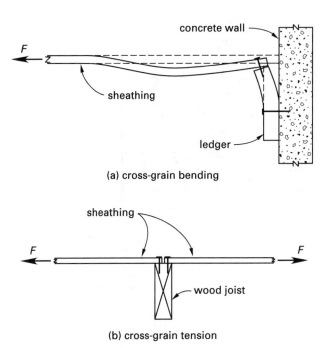

Figure 12.11   Cross-Grain Bending and Tension

12-11 FRAMING FOR DIAPHRAGM OPENINGS

It is not uncommon for a diaphragm to have openings for skylights, furnace flues, stairwells, and so on. In Sec. 2315.5.2, the UBC-97 requires that such openings be completely blocked around the opening edges, with such blocking framing into joists and other structural members. The purpose of the framing is to redistribute shears from areas adjacent to the openings around, or past, the openings to other collection elements.

An opening in the diaphragm imposes the following two requirements if the diaphragm is to operate correctly.

◇ The blocking around the perimeter of the opening must transfer the unequal loads on each side of the opening.

◇ The members around the perimeter of the opening must run from wall to wall, with tension straps used at corner connections to maintain continuity of the members. (See Fig. 12.12.) It is common to use double members for perimeter blocking. This eliminates the prohibited practice of using the diaphragm wood structural panel to splice perimeter wood.

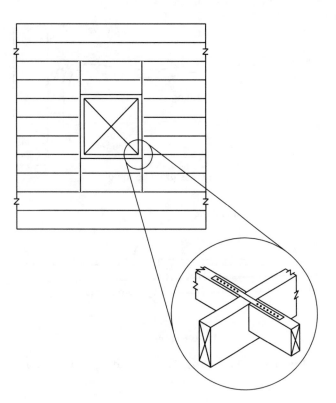

Figure 12.12   Framing an Opening

12-12 CRIPPLE WALLS

When the first level of a structure is constructed above the original ground (i.e., raised foundation), the space between the first level and the ground is referred to as *crawl space*. The created crawl space varies in size depending on the length of wooden stud walls used on the top of an exterior foundation to support a building. These stud walls are termed *cripple walls*. The UBC-97 provisions for the cripple walls are given in Sec. 2320.11.5. Bracing these short walls in accordance with UBC-97 Sec. 2320.11.3 prevents their failures during earthquakes, which would cause damage to the entire structure.

Solid blocking or wood structural panel sheathing may be used to brace cripple walls having a cripple stud height of 14 in (356 mm) or less. For those cripple walls with stud heights exceeding 14 in (356 mm), the UBC-97 requires that they be braced in accordance with UBC-97 Table 23-IV-C-2.

13
TILT-UP CONSTRUCTION

A *tilt-up building* typically uses precast structural panels, a wood structural panel diaphragm roof supported on wood joists or purlins, and either steel or glued-laminated (glulam) wood girders.

Analysis of tilt-up concrete shear walls is essentially the same as for cast-in-place concrete walls except that the panel-to-panel, panel-to-ceiling, and panel-to-floor details become critical. Shear between two panels must be developed by shear keys, dowels, or welded inserts. Contact joints are assumed to develop no strength in shear or in tension.

Tilt-up wall construction used in one-story industrial and commercial buildings fared poorly in the 1971 San Fernando and 1987 Whittier earthquakes. The main weaknesses were found in the connections, or *anchorages*, between the roof and walls, particularly between main girders, purlins, and joists and the walls, the connection of the perimeter wood *ledger* to the wall, and the nailing of the wood structural panel diaphragm to the ledger. Basically, the walls moved outward, the girders and purlins detached from the ledgers, and the roofs fell to the ground.

Another problem with tilt-up construction occurs because each panel in a line (i.e., as part of a wall) is separate from the other panels and therefore resists a seismic load parallel to the panel in proportion to its relative rigidity. Since all of the panels are connected at their tops and bottoms, all will deflect the same amount. However, solid panels will resist the seismic load in shear (i.e., as a shear wall) while panels with large openings such as windows or doors will resist the seismic load in bending (i.e., as a beam).

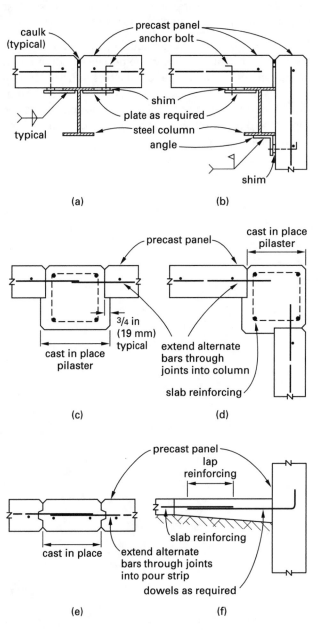

Figure 13.1   Details of Tilt-Up Construction
Connections

Connections, such as those at weld plates, to shear walls should be numerous and regular, with a maximum spacing of approximately 4 ft (122 cm) along the top of the shear wall, and such connections should be capable of transferring three times the expected lateral load. It is also necessary for each panel to be attached to the floor to counteract the overturning moment. (See Sec. 7-19.) These details will prevent the poor performance that has been experienced in some previous earthquakes.

Openings in tilt-up walls have become so numerous that a wall, with its openings and spandrels, is sometimes more like a frame. The UBC-97 [Sec. 1921.1] defines a *wall pier* as "a wall segment with a horizontal length-to-thickness ratio between 2.5 and 6, and whose clear height is at least two times its horizontal length."

In the 1989 Loma Prieta earthquake, a number of "well-designed" tilt-up buildings experienced differential movement between the panel pilasters and the glulam roof beams that sat atop the pilasters. Greater attention should be given to detailing the tie spacing at the top of the pilaster to keep the anchor bolts from spalling the pilaster concrete.

14

SPECIAL DESIGN FEATURES

14-1 ENERGY DISSIPATION SYSTEMS

Various active and passive devices that reduce the magnitude or duration (or both) of the seismic force are in use or evaluation. These devices include active mass systems, passive visco-elastic dampers, tendon devices, and base isolation. Such devices may be incorporated into a design when approved by the building official [UBC-97 Sec. 1629.10.2].

14-2 BASE ISOLATION

The base shear experienced by a structure is, simplistically, the product of the structure mass and the acceleration (i.e., $F = ma$). Little can be done to reduce the mass of a structure in an earthquake, but the acceleration can be reduced if the structure is not attached rigidly to its foundation. Application of this concept is known as *base isolation* or *decoupling*, and the connections between the structure and the foundation are known as *isolation bearings*. Base isolation is applicable to bridges as well as to buildings.[1]

---

[1]It is generally accepted that only bridges supported on box-girders, less than 300 ft (91.4 m) long, and whose superstructures are supported at every pier (rather than being monolithic) are true candidates for base isolation. Longer monolithic bridges are already so flexible that their natural period is long enough to reduce stresses significantly. It is also usually cheaper to design the bridge pier foundations as moment-resisting members than to specify base isolation in new construction (as opposed to retrofitting old bridges). The first new bridge to use base isolation was the Sexton Creek Bridge near Cairo, Illinois, installed by the Illinois Department of Transportation. In 1986, the Metropolitan Water District (MWD) of Southern California used base isolation on its Santa Ana River crossing of the Upper Feeder pipeline. California's Department of Transportation (CALTRANS) has retrofitted several bridge structures in this manner.

In effect, the ground is allowed to move back and forth under a building during an earthquake, leaving the building "stationary." Since the building theoretically does not accelerate, it does not experience a seismic force. In most cases of base isolation, the building is partially constrained, but the concept is the same. This can be done by "skewering" the base isolator with a vertical rod surrounded by a clearance hole. It may be necessary to excavate a trench (or "moat") around the building to allow for differential movement.[2] Other nonstructural considerations include attaching utility service with flexible pipes and cables, as well as suspending elevator pits from the basement.

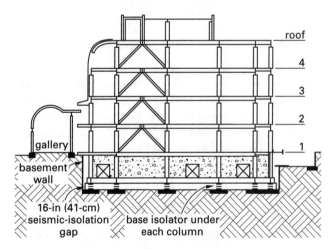

Figure 14.1   Base Isolation (Foothill Center, San Bernardino County, California)

---

[2]Every building is different, but a seismic isolation gap of approximately 12 to 16 in (31 to 41 cm) should be used, although differential motion may need to be limited to less than the full gap size. After that, the deflection control element takes over and the building follows the earthquake motion. Refer to UBC-97 Secs. 1633.2.11 and 1661.3.2 (Appendix Chapter 16, Div. IV).

Bearings consisting of elastomeric (e.g., neoprene) alone may be suitable for absorbing horizontal (thrusting) loads, but metal is incorporated into designs that support vertical loads. There are three primary methods of base isolation. These are supporting the building on (1) large ball bearings sandwiched between plates, (2) elastomeric bearings consisting of alternating layers of steel (or lead) and rubber, and (3) traditional structural expansion joints consisting of a layer of Teflon and a layer of rubber sandwiched between two steel plates. Combinations of these three methods (e.g., some bearings and some expansion joints) are desirable from a cost standpoint, since true bearings are costly. Some combinations work nearly as well as "pure" base isolation systems.

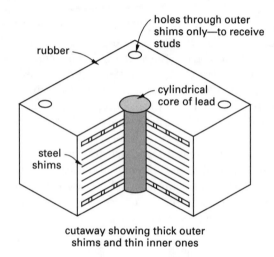

Figure 14.2   One Type of Isolation Bearing

Most base isolators consist of alternate layers of some elastomer (i.e., rubber) and steel plate. Another type of isolator is a Teflon-coated slider. Teflon sliders, though capable of isolation, do not provide significant damping, so it must be provided elsewhere.

The number of isolators necessary in an existing building depends on the type of foundation. A structure with a continuous foundation around its periphery may require hundreds of small isolators to spread out the building weight. However, a structure with a column-like support system might require only one isolator per column.

While base isolation is not yet in widespread routine use, there is significant evidence that the technique is successful.[3] Natural building periods of structures using base isolation of the earliest designs have been more

than doubled, moving the structures into areas on the response spectrum of lower acceleration. (See Sec. 5-1.) In several recent earthquakes, most buildings experienced maximum accelerations at the roof well in excess of the ground acceleration. However, buildings using base isolation experienced accelerations from 25% to 50% lower than the ground acceleration. Building and bridge periods were 2 to 10 times larger than their original values.

It has been suggested that a building fitted with base isolation will experience a Richter magnitude 8 earthquake as a magnitude 5 or 5.5 earthquake.[4] Instrumentation in the four-story Foothill Center illustrated in Fig. 14.1 was operational during the April 1990 magnitude 5.5 earthquake epicentered approximately 7 mi away. (See App. C.) The base isolation system reduced the seismic forces by almost half. The foundation acceleration was 0.15 g but was only 0.08 g above the isolators. Acceleration was 0.16 g at the roof. The acceleration at two similar buildings approximately 6 mi from the epicenter was 0.13 g at the ground and 0.39 g at the roof.

One concern is that no base-isolated building has experienced a truly significant earthquake. Some engineers are concerned that the elastomer will deteriorate, perhaps due to atmospheric ozone or other contaminants in the building, or that the bearing will "freeze up" after many years.

Another possible problem with base isolation derives from its very benefit—that of lengthening the natural period of the building. A natural soil period on the order of 3 or more seconds (as in cases where the soil is "soft") may coincide with the lengthened period of the building. This will produce resonance effects, such as those that occurred in the 1985 Mexico City and 1989 Loma Prieta earthquakes.

Base isolation is not suitable, economically or practically, for every building. One reason is that the base isolation bearings must all, within reason, be located at the same elevation. Buildings with footings that step down hillsides, for example, are poor candidates for base isolation.

Only a short time ago, seismic isolators were "outside" the provisions of the UBC and were considered exotic and experimental. Now they are installed routinely as

---

[3]The first new building to use base isolation was the 1983 Foothill Communities Law and Justice Center in San Bernardino,

California. (See Fig. 14.1.) Base isolation is commonly used in the "seismic-proofing" of existing historical structures, such as the Salt Lake City and County Building (which opened in 1989), since it is one of the few methods that does not require extensive exterior work.

[4]Time will tell.

retrofit devices to add seismic protection to buildings and bridges. Base isolation is now specifically permitted (subject to the approval of the building official) [UBC-97 Sec. 1629.10.2].

14-3 SEISMIC-ISOLATED DESIGN DETAILS

UBC-97 App. 16, Div. IV, "Earthquake Regulations for Seismic-Isolated Structures," provides requirements for every seismic-isolated structure. These requirements are partly parallel to UBC-97 Chap. 16 [Sec. 1629], "Earthquake Design Criteria Selection." This appendix defines the design lateral force for buildings and elements of structures based on a design displacement. Design displacement is described as the design-basis earthquake (i.e., having a 10% probability of being exceeded in 50 years) lateral displacement, excluding additional displacement due to actual and accidental torsion required for design of the isolation system. In addition, the seismic-isolated structures should be designed to withstand the maximum capable earthquake. In seismic zones 3 and 4, the maximum capable earthquake has a 10% probability of being exceeded in a 100-year time period.

For the design of seismic-isolated structures, there are some exceptions to the procedures and limitations provided by the UBC-97 [Sec. 1629] when considering seismic zones, occupancy, site characteristics, structural configuration, structural system, vertical acceleration, and structural height. These exceptions are given in UBC-97 App. 16, Div. IV [Sec. 1657].

For a seismic-isolated structure, the importance factor, I, is always 1.0 regardless of its occupancy category [UBC-97 Sec. 1657.3]. The stability of the vertical load-carrying elements of the isolation system should be properly substantiated by analysis and tests. The structural designation (regular or irregular) should be based on the structural configuration above the isolation system in accordance with the UBC-97 [Sec. 1629.5]. The response modification factor for the seismic-isolated structures that primarily ensures elastic response of the structure is denoted as R_1. The values of R_1 are provided in UBC-97 Table A-16-E of App. 16, Div. IV. The R values of the fixed-based structures are greater than the R_1 values of the seismic-isolated structures.

As for the selection of a lateral response procedure, any seismic-isolated structures can be designed by the dynamic lateral response procedure of the UBC-97 [Sec. 1659]. The static lateral response procedure of the UBC-97 [Sec. 1658] can only be used for design of seismic-isolated structures that are listed in the UBC-97 [Sec. 1657.5.2]. Mainly, these listed seismic-isolated structures are required (1) to be structurally regular

in configuration, (2) not to be located within 6.2 mi (10 km) of any active seismic source, (3) not to be constructed on soil-profile types S_E or S_F, (4) to be limited to a maximum of four stories or 65 ft (19.8 m) in height, and (5) to have an isolated effective period, T_D, greater than three times the elastic fixed-based period of the structure above the isolation system, while not exceeding the maximum three-second limit of isolated effective period, T_M.

Based on the UBC-97 [Sec. 1658.4.3], the total design base shear of the seismic-isolated structures, V_s, determined by the static lateral response procedure should be equal to or greater than the base shear for a fixed-based structure having the same weight and period, the design wind load, or the lateral seismic force required to completely activate the isolated system factored by 1.5. The UBC-97 [Sec. 1658.5] provides for the distribution of the total design base shear over the height of the seismic-isolated structure above the isolation interface in accordance with UBC-97 Formula 58-9.

The maximum interstory drift ratio of the structure above the seismic-isolated structure should not exceed $0.010/R_1$.

14-4 DAMPING SYSTEMS

Passive and active damping systems, like base isolation, are in their infancy, at least in terms of large-scale use. These systems increase the damping ratio of the building and, in so doing, decrease the amplitude of swaying. Some involve moving blocks and counterweights, while others, such as *passive visco-elastic dampers* or *friction dampers*, are not much more than large shock absorbers.[5] Those that require power for motors and information from sensors for computers are known as *active systems*; those that do not are *passive systems*.

Already used in some high-rise buildings to reduce wind drift, *active mass dampers* are nothing more than multi-ton blocks, usually of concrete or steel, suspended like a pendulum by a cable or mounted on tracks in one of the building's upper stories. When the wind or an earthquake makes the building sway, a computer sensing the motion signals a motor to move the weight in the opposite direction, thereby minimizing or neutralizing the motion.

Only a specific size of block will work in a building, because the weight of the block depends on the building's weight, the location of the block, the lag time,

---

[5]Friction dampers have more in common with devices used to absorb coupling shocks in railway rolling stock than with automobile shock absorbers.

and the mode to be counteracted. Therefore, the mass is "tuned" to the structure, and the systems are also known as *tuned mass dampers* (TMD).

Active tendon and *active pulse systems* are similar, except that the building is moved by hydraulic pistons in the foundation or between stories instead of by a mass at the top. The energy pulse usually only needs to be applied once or twice each building motion cycle.

These devices typically reduce the lateral forces by one-third to one-half while increasing the building weight approximately 1%. They are also suitable for torsion control when placed off-center in a structure.

The most significant drawback to active systems is the fact that they require external power not only for the computer but also for the motors driving the masses. Further, the large masses currently in use ride on oil bearings that take up to four minutes to pressurize. Thus, while active systems are useful in reducing drift during a predicted and slowly increasing windstorm, such devices are not yet substitutes for proper seismic design.

Active damping systems remain largely experimental. An exception is located in Tokyo in the Kyobashi Seiwa Building known for its extraordinary shape (11 stories high and only 13 ft (396 cm) wide).

Most damping devices installed in buildings are passive. (One of the best-known passive systems is located on the top floor of the 59-story Citicorp Building in Manhattan, where a 400-ton concrete-tuned block is located.)

An *Added Damping and Stiffness* (ADAS) element is a passive damping system that generally consists of a combination of steel plates and spacers. The plates are bolted to structural bracing at the plate tops and bottoms. As the top and bottom structural bracing members displace relative to one another, the ADAS plates bend (i.e., yield) and dampen vibrations. The advantage of ADAS elements over conventional damping systems or shock absorbers is that ADAS elements contain no moving parts and require no maintenance.

14-5 ARCHITECTURAL CONSIDERATIONS

Due to planned yielding, the inter-story deflections in a major earthquake will be several times larger than the elastic deflections that are calculated from the base shear equation in the UBC-97. Therefore, damage to architectural (i.e., nonstructural) items is likely. Even for lesser-magnitude events, however, nonstructural items must be properly detailed.[6]

Proper architectural detailing means providing proper clearances for exterior cladding, glazing (i.e., glass), wall finishes (e.g., marble veneers), interior partitions, and wall panels. Chimneys in residential buildings must be properly reinforced internally and securely strapped to the building at the roof line. Some elements can be free-floating—that is, they can move independently of the building. Proper attention must be given to the connection of these elements to the building.

To ensure that the occupants are able to get out of the building, doors should be designed to remain functional.

Floor coverings must be capable of three-dimensional movements.

All elements capable of falling and causing damage or injury must be rigidly attached to structural members. For example, suspended ceilings (e.g., tee-bar) and lights in drop-in ceilings must be tied to ceiling members above. Partitions, particularly those that do not run floor-to-ceiling, require special attention.

Columns in traditional moment-resisting frames are typically spaced 10 to 20 ft (3 to 7 m) apart. This spacing presents a challenge to architects when they try to provide unobstructed occupant space in high-rise buildings. Designers who want exterior column spacings greater than this must use other techniques to increase the strength of their buildings. Use of high-strength concrete, braced cores, ductile frames, ductile outrigger framing, and bandages are some of the techniques used.

Most trusses contain numerous axial members that are arranged in triangular sections. These triangular sections, though effective, make it difficult to include corridors (when in the core of the building) and windows (when on the perimeter of the building) in the design.

With the ability to create efficient moment-resisting joints, an increasing number of engineers are designing trusses comprised of rectangular sections. Bridge-like trusses comprised of rectangular sections with moment-resisting joints are known as *quadrangular girders, open-web girders* or *trusses, ductile frames*, or *Vierendeel girders*. In addition to providing unobstructed access through the openings in the truss, these structures are economical to build.

---

[6]This sentence probably should begin "Particularly for lesser-magnitude events" because nobody wants architectural damage, even in the more common small events.

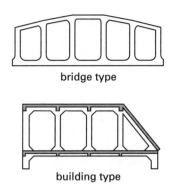

Figure 14.3   Vierendeel (Quadrangular) Girders

In-fill panels can either be given sufficient clearance so that they are not crushed by the deflection of adjacent structural elements, as in a panel forming a wall between two columns, or short sections of the panel (usually at its vertical edges) can be designed to be flexible or much weaker and replaceable.[7] Buttresses—short panel tees or wall returns—should be used to prevent the panel from falling over.

Computer room floors that are raised on pedestals and placed on unbraced stringers have fared poorly in past earthquakes.

Equipment such as air-conditioning devices, motors, pumps, tanks, piping, and air ducts must be securely bolted to their foundations to prevent sliding and overturning. Mere use of clamps or clips to prevent overturning is inadequate because the equipment can slide out from under the clips during an earthquake.

14-6 UPGRADING EXISTING CONSTRUCTION

Structures built in prior years can be upgraded with new features to make them more resistant to seismic forces. Such an activity is known as *retrofitting*. Bridges (some of which were heavily damaged in the 1971 San Fernando and 1989 Loma Prieta earthquakes) can be fitted with cable restraints and increased-capacity shear keys to restrict longitudinal motion.

Nonductile concrete columns with rectangular cross sections, built before the ductility requirements were added to the UBC in 1973, have been upgraded by being wrapped in steel plates. Columns with round or nonrectangular cross sections can be wrapped with steel or composite fiber wire to give them greater ductility. Carbon fibers are probably too brittle to be used to wrap columns as a means of achieving spiral strand confinement. However, Kevlar fibers (the same material used in most bullet-proof vests) show promise to wrap columns.

Steel jackets are comprised of two semicircular portions welded up the seams. Grout is injected into the space between the jacket and column. Thickness of the steel jacket depends on the loads expected.

The proper application of flat steel plates ("Vierendeel bandages," "Vlasovian bandages," or just "bandages") at selected exterior locations on a structure can increase resistance to seismic forces. Bandages do this by increasing the torsional moment of inertia of the structure (when the bandages are connected to core framing members) and by providing alternate vertical load paths (when the bandages are connected to load-carrying members at different levels). Inasmuch as only portions of a structure are covered, bandages are not effective in containing concrete. Therefore, this technique should not be confused with the encasement of concrete columns in steel jackets.

Belting, as a means of upgrading an unreinforced masonry building, is a technique that extends long steel rods horizontally from corner to corner, where the rods are attached to plates on the building's corners. The resulting "girdled" nature of the building is supposed to keep the building's walls from pulling away from the building core during an earthquake. For this technique to work, the rods must placed relatively close together, and they must extend all the way up the building.

Although California has adopted seismic retrofit standards for hospitals, there are no national codes specifically governing the methods of retrofitting or upgrading existing structures. For that matter, no one really knows how successfully retrofitted structures will fare in an earthquake. Approximately 40% of the unreinforced masonry buildings that had been retrofitted sustained damage in the 1987 Whittier earthquake. During the 1989 Loma Prieta earthquake, a four-story unreinforced masonry building in San Francisco that had been seismically retrofitted by belting suffered major seismic damage.

---

[7]Care must be taken when using flexible materials that they remain flexible indefinitely. Foamed polyethylene and polysulfide products appear *not* to meet this requirement.

15

PRACTICE PROBLEMS

15-1 EARTHQUAKES IN GENERAL

1. What is the lowest acknowledged numerical Richter magnitude that would identify a major earthquake?

Answer

This is an ambiguous question since different people would interpret the word "major" differently. In general, a moderate earthquake would have a Richter magnitude of 5, a strong earthquake would have a magnitude of 7, and a great earthquake would have a magnitude of 8 or higher. (See Sec. 2-3.)

2. What is the theoretical upper numerical limit on the Richter magnitude scale?

Answer

There is no theoretical upper limit on the Richter magnitude scale. (See Sec. 2-3.)

3. What does the Richter magnitude scale measure?

Answer

It is not clear what is intended by the word *measure*. The Richter apparatus *detects* earth movement. The numerical magnitude *describes* earthquake strength. The numerical value *represents* a measure of energy release on a logarithmic scale. (See Sec. 2-3.)

15-2 BUILDINGS

1. What is the difference between *stiffness* and *rigidity* as used in seismic consideration?

Answer

Stiffness is the force that will deflect a structure elastically a unit amount in a given direction. *Rigidity*—strictly *relative rigidity*—is a normalized stiffness. Whereas the stiffness of a single member can be used in numerical calculations, rigidities can only be used when forces are being distributed among several members. (See Secs. 4-3 and 4-4.)

2. What is the difference between *ductility* and *flexibility* as used in seismic consideration?

Answer

Flexibility is the reciprocal of *stiffness*. It is the elastic deflection obtained when a unit force is applied. (See Sec. 4-3.) *Ductility* is the ability of a material to distort and yield without fracture or collapse. (See Sec. 5-5.) Since flexibility deals with elastic deformation and ductility deals with inelastic deformation, there is little connection between the two concepts.

3. What is the relationship between *rigidity* and the variables of pier height, depth, and thickness?

Answer

Roughly, *rigidity* is proportional to the first power of thickness and to the cube of pier depth and is inversely proportional to the cube of pier height. (See Sec. 4-4.)

4. What is *ductility*?

Answer

Ductility is the ability of a material to distort and yield without fracture or collapse. (See Sec. 5-5.)

5. What is the *ductility factor*?

Answer

The *ductility factor* of a material is the ratio of its strain energy at fracture to its strain energy at yield. There are other similar and related definitions. (See Sec. 5-6.)

6. What factors influence the ductility factor?

Answer

From a metallurgical perspective, temperature and previous stress-strain history influence the ductility of a ductile material such as steel. The higher the temperature, the greater the ductility. The more the material has been worked or stressed in previous cycles or events, the more brittle (the opposite of ductile) it becomes. From a structural perspective, ductility depends on the type of construction (i.e., steel or concrete), the structural system, the quality of construction, the detailing, and the redundancy. (See Secs. 5-5 and 5-6.)

The strain rate can embrittle certain steels and welds, as can the welding procedure and post-weld heat treatments. Member orientation during rolling processes can alter stress-strain behavior, particularly in rolled plates.

7. What is the minimum recommended ductility factor?

Answer

It is not possible to specify a minimum recommended ductility factor exactly because it depends on the type of structure, construction material used, intended use of the structure, and many other factors. However, the ductility factor should be well in excess of 1.0 and seems to be no less than 2.5 for modern structures. (See Secs. 5-5 and 5-6.)

8. What is *ductile framing*?

Answer

In its simplest interpretation, a structure with ductile framing will not collapse even though its structural frame has sustained significant distortion, misalignment, and other yielding damage. (See Sec. 9-1.)

9. What is the principle reason for specifying a minimum ductility factor?

Answer

The principle reason for specifying a minimum ductility factor is to obtain a *ductility margin* (i.e., the ductility between yield and collapse) sufficient to ensure survivability in a design earthquake.

10. Why will a theoretical analysis of elastic response of a structure usually overestimate the stresses resulting from an earthquake?

Answer

A structure will not behave totally elastically during an earthquake. Local yielding at high stress locations reduces the seismic energy (i.e., the energy of oscillation) initially present in the structure.

11. Describe the two components of *drift*.

Answer

Shear drift is the sideways deflection of a building due to lateral (sideways) loads. *Chord drift* is the sideways deflection due to axial (vertical) loads. (See Sec. 5-12.)

12. What is the P-Δ effect?

Answer

The P-Δ effect is an additional column bending stress caused by eccentric vertical loads. (See Sec. 5-13.)

13. How are drift and the P-Δ effect related?

Answer

When a structure drifts, its vertical loads become eccentric. The eccentric loading increases the column stress, and the stress increase is called the P-Δ effect. (See Secs. 5-12 and 5-13.)

14. What is the *natural period* of a building?

Answer

The *natural period* of a building is the time it takes the building to complete one full swing in its primary mode of oscillation. (See Sec. 3-8.)

15. What does the term *redundancy* mean as it is used in the context of modern high-rise buildings?

Answer

Redundancy, as used in a seismic context, is synonymous with *distributed excess capacity* and *multiple load paths*. A *redundant design* has a safety factor, but the converse statement is not necessarily true. For example, if a vertical 100-kip (445 kN) load is supported by a single column having a 120-kip (534 kN) capacity, the design will have excess capacity but no redundancy since the structure will collapse if the column fails. If the 100-kip (445 kN) load is supported by 12 columns, each with a 10-kip (45 kN) capacity, the design will have both redundancy and excess capacity.

16. Has the recent trend in high-rise buildings been toward increased or decreased redundancy? Why?

Answer

Redundancy is increasingly seen as a crucial characteristic of high-rise designs. Multiple redundant load paths greatly increase the reliability of a structure.

Since building members (e.g., columns, girders, and shear walls) and details (i.e., column-girder joints) do not always behave as intended (due to our meager knowledge of the behavior of so-called ductile designs, design or construction errors, and higher-than-expected loading), the design should allow for the unintended loss of all capacity in a small fraction of the members. The tolerable loss of this excess capacity is the principle of redundant design.

17. What causes *torsional shear*?

Answer

Torsional shear occurs when an earthquake acts on a structure whose centers of mass and rigidity do not coincide. (See Sec. 5-14.)

18. What is *negative torsional shear*?

Answer

Negative torsional shear is the torsional shear on one side of a structure that is opposite in sign to the shear force, or direct shear, induced by the base shear. (See Sec. 5-15.)

19. How should negative torsional shear be treated?

Answer

Inasmuch as the direction of an earthquake is not known in advance, negative torsional shear should be disregarded—it may not be used to decrease the size of a wall or other member. (See Sec. 5-15.)

20. What is the *Rayleigh method* and where would it be used?

Answer

The Rayleigh method is a method determining the mode shape of a multiple-degree-of-freedom system through an iterative process. (See Sec. 4-19.)

21. Explain *critical damping*.

Answer

Critical damping is the amount of structural damping that causes oscillation to die out and return to the equilibrium position faster than any other amount of damping. (See Sec. 4-7.)

22. What is the *damping ratio*?

Answer

The *damping ratio* is the ratio of the actual damping coefficient to the critical damping coefficient. (See Sec. 4-7.)

23. What is the practical range of damping ratios?

Answer

Damping ratios of typical buildings range from approximately 0.02 for steel-frame construction to around 0.15 for wood-frame construction. (See Sec. 4-9.)

24. To what extent does damping affect the natural period of vibration of a structural frame?

Answer

Damping increases the actual period of vibration slightly, compared to the natural period of vibration. However, even with highly damped structures, the increase is usually 1% or less. Therefore, the natural period is used in the UBC-97 calculations and the effect of damping is disregarded. (See Sec. 4-10.)

25. What is a *response spectrum*?

Answer

A *response spectrum* is a graph of the maximum response (acceleration, velocity, or displacement) to a specified excitation of a single degree of freedom system plotted as a function of the SDOF system's natural period. (See Sec. 5-1.)

26. How does the *portal method* deal with the effects of column lengthening and shortening?

Answer

The *portal method* disregards changes in column length. (See Sec. 8-2.)

15-3 STRUCTURAL SYSTEMS

1. Distinguish between a *moment-resisting frame* and a *special moment-resisting frame*.

Answer

A *moment-resisting frame* has rigid connections between members (e.g., between girders and columns) such that moments applied to columns are partially resisted by girder bending and vice versa. With sufficiently large moments, however, even the elastic capacity of such structural systems that share loads between girders and columns can be exceeded. The integrity and load-carrying ability of a *special moment-resisting frame* (previously known as a *ductile moment-resisting frame*) will remain intact even after yielding has been experienced. (See Sec. 6-21.)

2. What are *bearing wall systems* and *box systems*?

Answer

Box system is another name for *bearing wall system*. A bearing wall system relies on shear and load-bearing walls to carry dead, live, and seismic loads. (See Sec. 6-21.)

3. Generally speaking, which is more likely to have a smaller natural frequency of vibration, a steel moment-resisting frame or a concrete moment-resisting frame, given equal heights and moments of inertia?

Answer

While the performance of both steel and concrete frames are similar in this regard, some theoretical generalizations are possible. The steel frame may be slightly more flexible (smaller stiffness), producing a smaller frequency and longer period. Also, the concrete frame may have greater mass, producing a smaller frequency and longer period. (The effect of damping on the period is minimal.) The UBC-97 equation for period (see Eq. 6.8) clearly indicates that steel buildings generally are expected to have longer periods (smaller frequencies). (See Sec. 6-26.)

4. Which is more likely to have a larger damping ratio, a steel or concrete moment-resisting frame?

Answer

Concrete construction generally has a greater damping ratio. (See Sec. 4-9.)

5. Are plastic hinges designed in columns, in girders, or in both?

Answer

A *plastic hinge* forms when a member yields. The yielding of a girder or of a girder-column joint may produce distortion, floor and roof sagging, and misalignment without collapse. The yielding of a column, however, may lead to structural collapse. Therefore, unlike girders, columns should not be designed to form plastic hinges.

6. What is the structural system called that does not have a complete vertical load-carrying space frame?

Answer

Bearing wall systems, or box systems, use walls, not frame members, to carry the vertical loads. (See Sec. 6-21.)

15-4 UBC

1. Describe how base shear is calculated according to the UBC-97 equation.

Answer

The base shear is calculated as an equivalent static loading based on a fraction of the structure weight. Terms are included to account for the seismic zone, building occupancy, building period, soil-profile type, proximity to known faults, and structural system. (See Sec. 6-30.)

2. Which of the terms in the base shear equation can be equal to 1.0?

Answer

The importance factor (I) can have a value of 1.0. It is also possible for the natural structure period (T) to have a calculated value of 1.0 sec. The near-source factors N_a and N_v can have values of 1.0, but these are not in the base shear equation. However, they are needed for calculating the site-specific velocity and acceleration-controlled seismic base shear coefficients (C_v and C_a). (See Secs. 6-14, 6-17, 6-18, and 6-26.)

3. In the application of the base shear formula, what factor of R should apply if a structure is a steel eccentrically braced frame in one direction and a concrete shear-wall building in the orthogonal direction?

Answer

Since the structure's performance is analyzed independently for earthquakes in the two orthogonal directions, the structure is treated as an eccentrically braced frame building ($R = 7.0$) for an earthquake in one direction and as a shear-wall building ($R = 4.5$) for the orthogonal direction. (See Sec. 6-21.)

4. Explain in general terms how the base shear is distributed in a horizontal plane to the various resisting elements.

Answer

The base shear is distributed to the resisting elements in proportion to their rigidities. (See Sec. 4-4.)

5. Draw the seismic force diagram acting on a multi-story building.

Answer

Refer to Fig. 6.15(a), Sec. 6-35, or Fig. 6-20, Sec. 6-42.

6. Draw the cumulative shear diagram acting on a multi-story building.

Answer

Refer to Fig. 6.15(b), Sec. 6-35.

7. What is the UBC-97 building height limit for ordinary shear-wall construction in a bearing-wall structural system located in seismic zones 3 and 4?

Answer

Shear wall construction may not be used in buildings higher than 160 ft (49 m). (See Table 6.10.)

8. Which has a smaller R value, a steel or concrete special moment-resisting frame?

Answer

Steel and concrete structures with special moment-resisting frames have the same R value, 8.5. (See Table 6.10.)

9. What possible values can R take on for a moment-resisting frame?

Answer

R can have values of 8.5 (SMRF of steel or concrete), 5.5 (concrete IMRF), 4.5 (steel OMRF), 3.5 (concrete OMRF), 6.5 (masonry MMRWF), and 6.5 (steel STMF). (See Table 6.10.)

10. What R value would be used for (a) a large football grandstand with bleachers and (b) a tall vertical tank supported on a raised platform supported by a braced framework?

Answer

(a) A football grandstand with bleachers is a self-supporting nonbuilding structure falling under the jurisdiction of the local building official. Its mass can be considered to be lumped at the various spectator levels, and the supporting system continues between floors. Therefore, it is covered by the UBC-97. However, it is not specifically mentioned in Table 6.15. (See Sec. 6-46.) It would probably have an R value of 2.9.

(b) The framework-supported vertical tank is specifically mentioned in the Blue Book commentary as a case not covered at all by the code provisions. The combined tall structure and frame base does not meet the requirement of a distributed supporting system, and mass cannot be considered lumped at various levels. (See Sec. 6-46.) Therefore, no R can be assigned.

11. What is the absolute UBC-97 limitation on drift?

Answer

The key word in this question is "absolute." There is no absolute limitation on drift in the UBC-97, as any drift that can be shown to be "tolerable" is permitted. Other limitations apply, however, when drift is not tolerable. (See Sec. 6-40.)

12. What percentage of the seismic load should be carried by a special moment-resisting frame in a building taller than 160 ft (49 m)?

Answer

All of the seismic load is carried by a special moment-resisting frame. (See Sec. 6-21.) There is no height limit for special moment-resisting frames.

13. What percentage of the live load in a warehouse should be added to the dead load when calculating base shear?

Answer

A minimum of 25% of the warehouse live load should be added. (See Sec. 6-29.)

14. What are the various seismic zones in California?

Answer

California has two seismic zones: 3 and 4. (See Sec. 3-2.)

15. What are the approximate maximum design accelerations in the California seismic zones?

Answer

Referring to Table 3.1, the approximate maximum accelerations in zones 3 and 4 are 0.3 g and 0.4 g, respectively. These correspond to the zone factors, Z, in Table 6.3.

15-5 CONCRETE AND MASONRY STRUCTURES

1. What are the possible modes of failure due to seismic forces if the lateral force-resisting system of a high-rise building is constructed of reinforced concrete?

Answer

(See Sec. 9-1.) This question does not specify whether the concrete is specially reinforced or whether the concrete is used in a frame or shear-wall structure. In general, a reinforced concrete frame will have failed if the concrete spalls or crushes before plastic yielding of the steel reinforcing occurs, or if the steel reinforcing is stressed plastically. Failure can be expected to occur:

(a) at the ends of well-designed columns when there is insufficient shear resistance (i.e., such that the column breaks out of its supports)

(b) in poorly designed columns with insufficient confinement

(c) in shear walls due to inadequate vertical reinforcing

(d) at construction joints due to inadequate bonding between members

(e) in beams due to inadequate shear reinforcing

(f) in columns due to excessive drift and overturning moment

2. What are the most important considerations in achieving ductility in concrete frames?

Answer

The most important considerations are confinement and continuity. (The steel in confined, or specially reinforced, concrete should yield before the concrete crushes.) Adequate bonding between steel and concrete must be ensured. Steel must be capable of developing its full tensile strength. Members must be adequately tied together at joints. (See Sec. 9-1.)

3. What is meant by *confined concrete?*

Answer

Confined concrete is also called *ductile concrete* or *specially reinforced concrete.* The steel in ductile concrete will yield before the concrete crushes. This enables the concrete member to develop its full compressive strength without yielding. (See Sec. 9-1.)

4. Why is concrete confined at joints and in members?

Answer

The confining steel in ductile concrete enables the concrete to develop its full compressive strength without yielding. (See Sec. 9-1.)

5. What are some of the construction methods used to ensure ductile behavior of concrete?

Answer

To confine concrete, columns are spiral wrapped at closer intervals and additional hoops are used at joints and other locations. Continuity of reinforcement is achieved by special attention to splices. Special attention is given to reinforcement of shear walls. Hooks, ties, stirrups, and hoops are detailed to prevent pull-out. (See Sec. 9-1.)

6. With regard to resistance to seismic forces, which is better in steel-reinforced concrete columns, spiral ties or horizontal hoops? Why?

Answer

Spiral transverse reinforcement is the most efficient confinement, but it may not be possible to use it. From a construction standpoint, spiral reinforcement for smaller columns is easier to form in the field. From a seismic standpoint, spiral reinforcement provides slightly better confinement. Larger columns, however, cannot be wrapped in the field, and extending spirals into beam-column joints is difficult, so individual factory-fabricated hoops must be used.

7. What is the function of the spiral and individual ties used in a concrete column?

Answer

Ties confine the concrete and keep it from crushing. (See Sec. 9-1.)

8. Do special reinforcement hoops replace regular ties in beams and columns?

Answer

Special reinforcement (primarily in the form of additional hoops) is used in addition to regular beam stirrups and column ties. In beams, special reinforcement is required at points of expected yielding (i.e., at plastic hinges). In columns, hoops are required at column-girder connections. (See Sec. 9-1.)

9. What is the effectiveness of stirrups in deep concrete beams?

Answer

After inclined cracks form at the ends of deep beams, the load is carried in a "tied arch" configuration that has considerable remaining strength. Stirrups in the center of a deep beam are not particularly effective.

10. What is the maximum permitted shear stress (psi or kPa) for a masonry wall with special inspection?

Answer

The maximum shear stress is 35 psi (240 kPa) when in-plane flexural reinforcement is present. The maximum shear stress is 75 psi (520 kPa) when steel takes all of the shear stress. (See Sec. 11-1(F).)

15-6 STEEL STRUCTURES

1. What are the possible modes of failure due to seismic forces if the lateral force-resisting system of a high-rise building is constructed of steel?

Answer

In general, steel will be considered to have "failed" if it yields. (See Sec. 10-1.)

2. Where is a steel-framed building with a properly designed special moment-resisting frame most likely to yield in an earthquake?

Answer

Yielding and formation of plastic hinges in a steel structure can be expected at points where the moments are greatest, such as at girder ends and at column-girder connections. Columns can buckle due to bending and eccentric effects. Flanges and webs of members can buckle from local stresses and fail from fatigue loading. (Girder-column connections should not fail, however, through weld and bolt failure. All connections should be able to sustain the full plastic moment of connected members.)

15-7 SOILS AND FOUNDATIONS

1. Given a soil engineering report, what would you propose in order to improve the seismic response of the building?

Answer

Devastating resonance effects can be avoided if the natural building period does not coincide with the site period. (See Secs. 3-6, 3-8, and 4-13.) If the site has fine sand, soil liquefaction effects must be considered. (See Sec. 3-7.)

15-8 DESIGN PROBLEMS

1. The roof of the structure shown is rigid. A 10-kip (44.48-kN) load is applied at the roof. What are the resisting shears in each column?

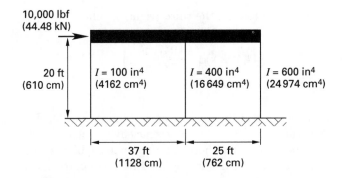

Customary U.S. Solution

The total rigidity is proportional to the sum of the moments of inertia. The total is

$$I_{\text{total}} = I_1 + I_2 + I_3$$
$$= 100 \text{ in}^4 + 400 \text{ in}^4 + 600 \text{ in}^4$$
$$= 1100 \text{ in}^4$$

The load carried by the first (left) column is

$$V_1 = (10 \text{ kips}) \left(\frac{100 \text{ in}^4}{1100 \text{ in}^4} \right)$$
$$= 0.91 \text{ kips}$$

The load carried by the second (middle) column is

$$V_2 = (10 \text{ kips}) \left(\frac{400 \text{ in}^4}{1100 \text{ in}^4} \right)$$
$$= 3.64 \text{ kips}$$

The load carried by the third (right) column is

$$V_3 = (10 \text{ kips}) \left(\frac{600 \text{ in}^4}{1100 \text{ in}^4} \right)$$
$$= 5.45 \text{ kips}$$

SI Solution

Since column lengths and materials are the same, the total rigidity is proportional to the sum of the moments of inertia. The total is

$$I_{\text{total}} = I_1 + I_2 + I_3$$
$$= 4162 \text{ cm}^4 + 16\,649 \text{ cm}^4 + 24\,974 \text{ cm}^4$$
$$= 45\,785 \text{ cm}^4$$

The load carried by the first (left) column is

$$V_1 = (44.48 \text{ kN}) \left(\frac{4162 \text{ cm}^4}{45\,785 \text{ cm}^4} \right) = 4.04 \text{ kN}$$

The load carried by the second (middle) column is

$$V_2 = (44.48 \text{ kN}) \left(\frac{16\,649 \text{ cm}^4}{45\,785 \text{ cm}^4} \right) = 16.17 \text{ kN}$$

The load carried by the third (right) column is

$$V_3 = (44.48 \text{ kN}) \left(\frac{24\,947 \text{ cm}^4}{45\,785 \text{ cm}^4} \right) = 24.24 \text{ kN}$$

2. A power transformer is mounted on a pedestal. The structural adequacy for the seismic environment is to be demonstrated. The horizontal component of the seismic force is to be defined by the El Centro north-south elastic response spectra. If the lateral frequency of the system exceeds 30 Hz, the response is defined as being equivalent to a 0.5 g static load. The vertical response is defined as 50% of the horizontal response, and both occur simultaneously. Free lateral vibration tests on the structure show that the ratio of successive swings of the structure is 0.882. Assume that the pedestal mass and shear flexibility, foundation stiffness, and torsion can be neglected. Specific properties are shown in the figure.

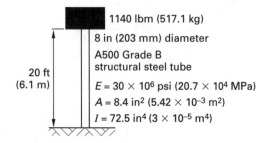

1140 lbm (517.1 kg)

8 in (203 mm) diameter
A500 Grade B
structural steel tube

20 ft
(6.1 m)

$E = 30 \times 10^6$ psi (20.7×10^4 MPa)
$A = 8.4$ in^2 (5.42×10^{-3} m^2)
$I = 72.5$ in^4 (3×10^{-5} m^4)

(a) Calculate the undamped lateral frequency and period.

(b) Calculate the undamped vertical frequency and period.

(c) Explain how the assumptions affect these frequencies and periods.

(d) What is the logarithmic decrement for damping of lateral vibrations?

(e) What is the damping ratio?

(f) What is the lateral acceleration?

(g) What is the vertical acceleration?

(h) What is the maximum shear force at the base using a quasistatic solution method?

(i) Are the assumptions used to calculate the frequencies conservative with respect to the lateral load?

(j) Assume that the pedestal fails by bending. Does the structure have high ductility?

Customary U.S. Solution

(a) From Table 4.1, the lateral stiffness of a vertical cantilever is

$$k = \frac{3EI}{h^3} = \frac{(3) \left(3 \times 10^7 \dfrac{\text{lbf}}{\text{in}^2} \right) (72.5 \text{ in}^4)}{(20 \text{ ft})^3 \left(12 \dfrac{\text{in}}{\text{ft}} \right)^2}$$

$$= 5664 \text{ lbf/ft}$$

The mass in slugs (equivalent to lbf-sec^2/ft) can be calculated from the mass in pounds by dividing by the gravitational constant, g_c.

$$m_{\text{slugs}} = \frac{m_{\text{lbm}}}{g_c} = \frac{1140 \text{ lbm}}{32.2 \dfrac{\text{ft-lbm}}{\text{lbf-sec}^2}}$$

$$= 35.4 \text{ slugs}$$

From Eqs. 4.11 and 4.12, the natural period for lateral vibrations is

$$T_{\text{lateral}} = \frac{2\pi}{\omega} = 2\pi \sqrt{\frac{m}{k}}$$

$$= 2\pi \sqrt{\frac{35.4 \text{ slugs}}{5664 \dfrac{\text{lbf}}{\text{ft}}}}$$

$$= 0.497 \text{ sec (say 0.5 sec)}$$

From Eq. 4.10, the frequency is

$$f = \frac{1}{T} = \frac{1}{0.5 \text{ sec}} = 2 \text{ Hz}$$

(b) From Table 4.1, the deflection due to a compressive load is

$$x = \frac{Fh}{AE}$$

The stiffness is the force per unit deflection.

$$k = \frac{F}{x} = \frac{AE}{h}$$

$$= \frac{(8.4 \text{ in}^2) \left(30 \times 10^6 \dfrac{\text{lbf}}{\text{in}^2} \right)}{20 \text{ ft}}$$

$$= 1.26 \times 10^7 \text{ lbf/ft}$$

Again from Eqs. 4.11 and 4.12, the natural period for vertical vibrations is

$$T_{\text{vertical}} = \frac{2\pi}{\omega} = 2\pi \sqrt{\frac{m}{k}}$$

$$= 2\pi \sqrt{\frac{35.4 \text{ slugs}}{1.26 \times 10^7 \dfrac{\text{lbf}}{\text{ft}}}}$$

$$= 1.05 \times 10^{-2} \text{ sec}$$

From Eq. 4.10, the frequency is

$$f = \frac{1}{T} = \frac{1}{1.05 \times 10^{-2} \text{ sec}} = 95 \text{ Hz}$$

(c) It is assumed that all flexibility comes from the tube. The flexibility of other elements in the structural system decreases the stiffness, which increases the period. It is also assumed that the tube is massless. Increasing the vibrating mass also increases the period.

(d) From Eq. 4.17, the logarithmic decrement, δ, is

$$\delta = \ln\left(\frac{x_n}{x_{n+1}}\right) = \ln\left(\frac{1}{0.882}\right) = 0.126$$

(e) From Eq. 4.17, the damping ratio is solved directly.

$$\delta = 0.126 = \frac{2\pi\xi}{\sqrt{1-\xi^2}}$$
$$\xi = 0.02 \ (2\%)$$

(f) From Fig. 5.4 with 2% damping (determined in part (e)),

$$S_d = 2.2 \text{ in}$$
$$S_v = 28 \text{ in/sec}$$
$$S_a = (0.9 \text{ g})\left(32.2 \frac{\text{ft}}{\text{sec}^2\text{-g}}\right)$$
$$= 29 \text{ ft/sec}^2$$

(g) The vertical response is 50% of the horizontal response.

$$S_d = (0.5)(2.2 \text{ in}) = 1.1 \text{ in}$$
$$S_v = (0.5)\left(28 \frac{\text{in}}{\text{sec}}\right) = 14 \text{ in/sec}$$
$$S_a = (0.5)\left(29 \frac{\text{ft}}{\text{sec}^2}\right) = 14.5 \text{ ft/sec}^2$$

(h) The maximum shear by quasistatic approach is given by Eq. 3.1.

$$V_{\text{horizontal}} = mS_a$$
$$= (35.4 \text{ slugs})\left(29 \frac{\text{ft}}{\text{sec}^2}\right)$$
$$= 1027 \text{ lbf}$$
$$V_{\text{vertical}} = (35.4 \text{ slugs})\left(14.5 \frac{\text{ft}}{\text{sec}^2}\right)$$
$$= 513 \text{ lbf}$$

(i) As determined in part (c), the assumptions result in a smaller period. Since lower values of period give a higher acceleration (see Fig. 5.1), the assumptions result in the structure being designed for higher forces. Thus, the assumptions are conservative.

(j) Failing by inelastic bending, as opposed to fracture or collapse, is one of the indications of a ductile structure.

SI Solution

(a) From Table 4.1, the lateral stiffness of a vertical cantilever is

$$k = \frac{3EI}{h^3} = \frac{(3)(20.7 \times 10^4 \text{ MPa})(3 \times 10^{-5} \text{ m}^4)}{(6.1 \text{ m})^3}$$
$$= 8.2 \times 10^{-2} \text{ MN/m}$$

From Eqs. 4.11 and 4.12, the natural period for lateral vibration is

$$T_{\text{lateral}} = \frac{2\pi}{\omega} = 2\pi\sqrt{\frac{m}{k}}$$
$$= 2\pi\sqrt{\frac{517.1 \text{ kg}}{\left(8.2 \times 10^{-2} \dfrac{\text{MN}}{\text{m}}\right)\left(1 \times 10^6 \dfrac{\text{N}}{\text{MN}}\right)}}$$
$$= 0.499 \text{ s (say 0.5 s)}$$

From Eq. 4.10, the frequency is

$$f = \frac{1}{T} = \frac{1}{0.5 \text{ s}} = 2 \text{ Hz}$$

(b) From Table 4.1, the deflection due to a compressive load is

$$x = \frac{Fh}{AE}$$

The stiffness is the force per unit deflection.

$$k = \frac{F}{x} = \frac{AE}{h}$$
$$= \frac{(5.42 \times 10^{-3} \text{ m}^2)(20.7 \times 10^4 \text{ MPa})}{6.1 \text{ m}}$$
$$= 1.84 \times 10^2 \text{ MN/m}$$

Again from Eqs. 4.11 and 4.12, the natural period for vertical vibrations is

$$T_{\text{vertical}} = \frac{2\pi}{\omega} = 2\pi\sqrt{\frac{m}{k}}$$
$$= 2\pi\sqrt{\frac{517.1 \text{ kg}}{\left(1.84 \times 10^2 \dfrac{\text{MN}}{\text{m}}\right)\left(1 \times 10^6 \dfrac{\text{N}}{\text{MN}}\right)}}$$
$$= 1.05 \times 10^{-2} \text{ s}$$

From Eq. 4.10, the frequency is

$$f = \frac{1}{T} = \frac{1}{1.05 \times 10^{-2} \text{ s}} = 95 \text{ Hz}$$

(c) It is assumed that all flexibility comes from the tube. The flexibility of other elements in the structural system decreases the stiffness, which increases the period. It is also assumed that the tube is massless. Increasing the vibrating mass also increases the period.

(d) From Eq. 4.17, the logarithmic decrement, δ, is

$$\delta = \ln\left(\frac{x_n}{x_{n+1}}\right) = \ln\left(\frac{1}{0.882}\right) = 0.126$$

(e) From Eq. 4.17, the damping ratio is solved directly.

$$\delta = 0.126 = \frac{2\pi\xi}{\sqrt{1-\xi^2}}$$

$$\xi = 0.02 \ (2\%)$$

(f) From Fig. 5.4 with 2% damping (determined in part (e)),

$$S_d = (2.2 \text{ in})\left(0.0254 \frac{\text{m}}{\text{in}}\right) = 5.6 \times 10^{-2} \text{ m}$$

$$S_v = \left(28 \frac{\text{in}}{\text{s}}\right)\left(0.0254 \frac{\text{m}}{\text{s}}\right) = 0.71 \times 10^{-2} \text{ m/s}$$

$$S_a = (0.9 \text{ g})\left(9.81 \frac{\text{m}}{\text{s}^2 \cdot \text{g}}\right)$$

$$= 8.83 \text{ m/s}^2$$

(g) The vertical response is 50% of the horizontal response.

$$S_d = (0.5)(5.6 \times 10^{-2} \text{ m}) = 2.8 \times 10^{-2} \text{ m}$$

$$S_v = (0.5)\left(0.71 \times 10^{-2} \frac{\text{m}}{\text{s}}\right) = 0.36 \times 10^{-2} \text{ m/s}$$

$$S_a = (0.5)\left(8.83 \frac{\text{m}}{\text{s}^2}\right) = 4.42 \text{ m/s}^2$$

(h) The maximum shear by quasistatic approach is given by Eq. 3.1.

$$V_{\text{horizontal}} = mS_a$$

$$= (517.1 \text{ kg})\left(8.83 \frac{\text{m}}{\text{s}^2}\right)$$

$$= 4570 \text{ N}$$

$$V_{\text{vertical}} = (517.1 \text{ kg})\left(4.42 \frac{\text{m}}{\text{s}^2}\right)$$

$$= 2290 \text{ N}$$

(i) As determined in part (c), the assumptions result in a smaller period. Since lower values of period give a higher acceleration (see Fig. 5.1), the assumptions result in the structure being designed for higher forces. Thus, the assumptions are conservative.

(j) Failing by inelastic bending, as opposed to fracture or collapse, is one of the indications of a ductile structure.

3. A two-story jail uses concrete shear walls as shown. None of the shear walls has openings, but only the east wall covers the entire length. The story height is 12 ft (366 cm). The thickness of the roof, wall, and floor slabs are 5, 10, and 5 in (127, 254, and 127 mm), respectively. Assume all walls shown can be considered fixed piers.

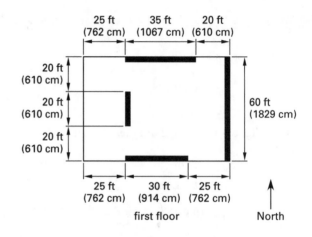

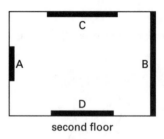

(a) What are the relative rigidities of the second story walls A, B, C, and D?

(b) Where is the center of rigidity?

(c) How would the shears in the second floor be determined? (Do not actually calculate the shears.)

(d) What is the effect on the first floor of offsetting wall A as shown?

Customary U.S. Solution

(a) As stated in the problem, the walls are fixed piers. Use App. D to determine the rigidities of the fixed piers.

The rigidities of perpendicular walls are taken as zero without regard to the h/d values.

wall	$\left(\dfrac{h}{d}\right)_{\text{E-W}}$	$\left(\dfrac{h}{d}\right)_{\text{N-S}}$	$R_{\text{E-W}}$	$R_{\text{N-S}}$
A	–	0.6	0	4.96
B	–	0.2	0	16.447
C	0.34	–	9.44	0
D	0.40	–	7.911	0

(b) The center of rigidity is located at (x_R, y_R). Distances are measured from the southwest corner.

$$x_R = \frac{(0\text{ ft})(4.96) + (80\text{ ft})(16.447)}{4.96 + 16.447} = 61.5\text{ ft}$$

$$y_R = \frac{(0\text{ ft})(7.911) + (60\text{ ft})(9.44)}{7.911 + 9.44} = 32.6\text{ ft}$$

(c) The wall shears are distributed in proportion to the second-floor wall rigidities.

(d) The second-story shear from the outside (west) wall will have to be transferred through the second-story floor (first-story ceiling) slab to wall A below. Failure may occur in the slab if it is not properly detailed.

SI Solution

(a) As stated in the problem, the walls are fixed piers. Use App. D to determine the rigidities of the fixed piers. The rigidities of perpendicular walls are taken as zero without regard to the h/d values.

wall	$\left(\dfrac{h}{d}\right)_{\text{E-W}}$	$\left(\dfrac{h}{d}\right)_{\text{N-S}}$	$R_{\text{E-W}}$	$R_{\text{N-S}}$
A	–	0.6	0	4.96
B	–	0.2	0	16.447
C	0.34	–	9.44	0
D	0.40	–	7.911	0

(b) The center of rigidity is located at (x_R, y_R). Distances are measured from the southwest corner.

$$x_R = \frac{(0\text{ cm})(4.96) + (2439\text{ cm})(16.447)}{4.96 + 16.447} = 1874\text{ cm}$$

$$y_R = \frac{(0\text{ cm})(7.911) + (1830\text{ cm})(9.44)}{7.911 + 9.44} = 996\text{ cm}$$

(c) The wall shears are distributed in proportion to the second-floor wall rigidities.

(d) The second-story shear from the outside (west) wall will have to be transferred through the second-story floor (first-story ceiling) slab to wall A below. Failure may occur in the slab if it is not properly detailed.

4. The structure shown is subjected to a lateral loading of 0.3 g. The columns are set in hard rock concrete footings, but the tops can be considered to be pinned to the slab. The reinforced concrete slab is 6 in (152 mm) thick and has a finished density of 150 lbm/ft^3 (2400 kg/m^3). The columns are rectangular A500 Grade B steel tubing, TS $5 \times 5 \times \frac{5}{16}$ in. Disregard the dead load of the columns, all live load, slab rotation, axial column compression, combined stresses, and column deflection. Determine if the columns are adequate.

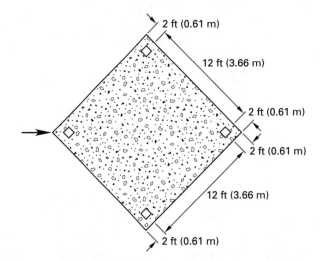

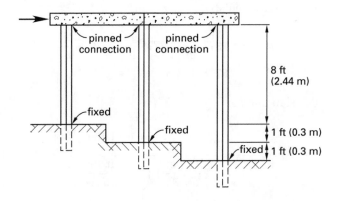

Customary U.S. Solution

The properties of A500 Grade B TS $5 \times 5 \times \frac{5}{16}$ in square structural tubing are found in the AISC manual. The modulus of elasticity is approximately 29×10^6 psi.

$$A = 5.61\text{ in}^2$$

$$I_x = I_y = 20.1\text{ in}^4$$

Although the tubing is oriented 45° from the plane of the principal axis, the moment of inertia per tube is still 20.1 in^4 in the direction of bending. This is because the tube is symmetrical and the product of inertia is zero.

Although the lower ends of the columns are fixed, the tops are pinned. Therefore, simple cantilever curvature occurs. From Table 4.1, the stiffness is

$$k = \frac{3EI}{h^3}$$

There are three different column lengths, so there are three stiffnesses.

$$k_{8\text{-ft}} = \frac{(3)\left(29 \times 10^6 \dfrac{\text{lbf}}{\text{in}^2}\right)(20.1 \text{ in}^4)}{(8 \text{ ft})^3 \left(12 \dfrac{\text{in}}{\text{ft}}\right)^3}$$

$$= \frac{1.75 \times 10^9 \text{ in}^2\text{-lbf}}{8.847 \times 10^5 \text{ in}^3}$$

$$= 1977 \text{ lbf/in}$$

$$k_{9\text{-ft}} = \frac{(2 \text{ columns})(1.75 \times 10^9 \text{ in}^2\text{-lbf})}{(9 \text{ ft})^3 \left(12 \dfrac{\text{in}}{\text{ft}}\right)^3}$$

$$= 2778 \text{ lbf/in}$$

$$k_{10\text{-ft}} = \frac{1.75 \times 10^9 \text{ in}^2\text{-lbf}}{(10 \text{ ft})^3 \left(12 \dfrac{\text{in}}{\text{ft}}\right)^3} = 1013 \text{ lbf/in}$$

The total stiffness consists of the sum of these three terms. (Notice that the stiffnesses for the two 9-ft columns were calculated as one sum.)

$$k_{\text{total}} = 1977 \frac{\text{lbf}}{\text{in}} + 2778 \frac{\text{lbf}}{\text{in}} + 1013 \frac{\text{lbf}}{\text{in}}$$

$$= 5768 \text{ lbf/in}$$

The slab mass is

$$m = \frac{V\rho}{g_c} = \frac{(16 \text{ ft})(16 \text{ ft})(6 \text{ in})\left(150 \dfrac{\text{lbm}}{\text{ft}^3}\right)}{\left(12 \dfrac{\text{in}}{\text{ft}}\right)\left(32.2 \dfrac{\text{ft-lbm}}{\text{lbf-sec}^2}\right)}$$

$$= 596 \text{ slugs}$$

The seismic force is

$$F = ma = (596 \text{ slugs})(0.3 \text{ g})\left(32.2 \frac{\text{ft}}{\text{sec}^2\text{-g}}\right)$$

$$= 5760 \text{ lbf}$$

The resisting force in each tube is proportional to its relative rigidity (stiffness). The shortest tube has the highest stiffness, hence the shortest tube experiences the highest stress. The prorated portion of the force carried by the shortest tube is

$$F_{8\text{-ft}} = \left(\frac{1977 \dfrac{\text{lbf}}{\text{in}}}{5768 \dfrac{\text{lbf}}{\text{in}}}\right)(5760 \text{ lbf}) = 1974 \text{ lbf}$$

The moment at the base of the shortest tube is

$$M = FL = (1974 \text{ lbf})(8 \text{ ft})\left(12 \frac{\text{in}}{\text{ft}}\right)$$

$$= 1.895 \times 10^5 \text{ in-lbf}$$

The bending stress is

$$f_b = \frac{Mc}{I} = \frac{(1.895 \times 10^5 \text{ in-lbf})\left(\dfrac{5 \text{ in}}{\sqrt{2}}\right)}{20.1 \text{ in}^4}$$

$$= 33{,}333 \text{ lbf/in}^2 \text{ (psi)}$$

The minimum yield strength of normally stocked A500 Grade B square steel tubing is 42 ksi. AISC Sec. F3.1 specifies the maximum allowable bending stress as 0.66 times the yield stress. Using the one-third increase in allowable stress from seismic loads, the maximum allowable seismic bending stress is

$$F_b = (1.33)(0.66)F_y = (1.33)(0.66)(42{,}000 \text{ psi})$$

$$= 36{,}868 \text{ psi}$$

Since $f_b < F_b$, the design is acceptable.

SI Solution

The properties of A500 Grade B TS $5 \times 5 \times \frac{5}{16}$ in square structural tubing are found in the AISC manual. The modulus of elasticity is approximately 2×10^5 MPa.

$$A = 3.62 \times 10^{-3} \text{ m}^2$$

$$I_x = I_y = 8.37 \times 10^{-6} \text{ m}^4$$

Although the tubing is oriented 45° from the plane of the principal axis, the moment of inertia per tube is still 8.37×10^{-6} m^4 in the direction of bending. This is because the tube is symmetrical and the product of inertia is zero.

Although the lower ends of the columns are fixed, the tops are pinned. Therefore, simple cantilever curvature occurs. From Table 4.1, the stiffness is

$$k = \frac{3EI}{h^3}$$

There are three different column lengths, so there are three stiffnesses.

$$k_{2.44\text{-m}} = \frac{(3)(2 \times 10^5 \text{ MPa})(8.37 \times 10^{-6} \text{ m}^4)}{(2.44 \text{ m})^3}$$

$$= \frac{5.022 \text{ MPa·m}^4}{14.527 \text{ m}^3}$$

$$= 0.346 \text{ MN/m}$$

$$k_{2.74\text{-m}} = \frac{(2 \text{ columns})(5.022 \text{ MPa·m}^4)}{(2.74 \text{ m})^3}$$

$$= 0.488 \text{ MN/m}$$

$$k_{3.05\text{-m}} = \frac{5.022 \text{ MPa·m}^4}{(3.05 \text{ m})^3}$$

$$= 0.177 \text{ MN/m}$$

The total stiffness consists of the sum of these three terms. (Notice that the stiffnesses for the two 2.74-m columns were calculated as one sum.)

$$k_{\text{total}} = 0.346 \frac{\text{MN}}{\text{m}} + 0.488 \frac{\text{MN}}{\text{m}} + 0.177 \frac{\text{MN}}{\text{m}}$$

$$= 1.011 \text{ MN/m}$$

The slab mass is

$$m = V\rho$$

$$= (4.88 \text{ m})(4.88 \text{ m})(152 \text{ mm}) \left(\frac{1}{1000 \frac{\text{mm}}{\text{m}}} \right)$$

$$\times \left(2400 \frac{\text{kg}}{\text{m}^3} \right)$$

$$= 8687 \text{ kg}$$

The seismic force is

$$F = ma = (8687 \text{ kg})(0.3 \text{ g}) \left(9.81 \frac{\text{m}}{\text{s}^2 \cdot \text{g}} \right)$$

$$= 25\,566 \text{ N} \quad (25.6 \text{ kN})$$

The resisting force in each tube is proportional to its relative rigidity (stiffness). The shortest tube has the highest stiffness, hence the shortest tube experiences the highest stress. The prorated portion of the force carried by the shortest tube is

$$F_{2.44\text{-m}} = \left(\frac{0.346 \frac{\text{MN}}{\text{m}}}{1.011 \frac{\text{MN}}{\text{m}}} \right) (25.6 \text{ kN}) = 8.76 \text{ kN}$$

The moment at the base of the shortest tube is

$$M = FL = (8.76 \text{ kN})(2.44 \text{ m})$$

$$= 21.37 \text{ kN·m}$$

The bending stress is

$$f_b = \frac{Mc}{I}$$

$$= \frac{(21.37 \text{ kN·m})\sqrt{2} \left(\dfrac{127 \text{ mm}}{2} \right)}{(8.37 \times 10^{-6} \text{ m}^4) \left(1000 \dfrac{\text{mm}}{\text{m}} \right)}$$

$$= 229{,}300 \text{ kPa} \quad (229.3 \text{ MPa})$$

The minimum yield strength of normally stocked A500 Grade B square steel tubing is 290 MPa. AISC Sec. F3.1 specifies the maximum allowable bending stress as 0.66 times the yield stress. Using the one-third increase in allowable stress from seismic loads, the maximum allowable seismic bending stress is

$$F_b = (1.33)(0.66)F_y = (1.33)(0.66)(290 \text{ MPa})$$

$$= 255 \text{ MPa}$$

Since $f_b < F_b$, the design is acceptable.

5. An eight-story (including penthouse) office building is supported by an ordinary steel chevron-braced frame intended to carry all vertical loads and resist all seismic loads. There are no other load-supporting walls. The weights and heights of each floor are as given in the illustration. The building is located in seismic zone 4 on S_A soil-profile. Perform a seismic analysis consistent with the UBC-97, and determine (a) the base shear and (b) the seismic force at the third floor. The site is positioned 4.65 mi (7.5 km) from a type A potential seismic source.

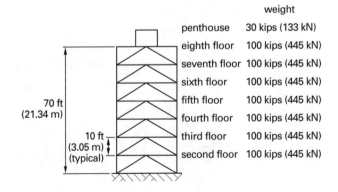

Customary U.S. Solution

(a) The natural period (T) can be determined from Method A of the UBC-97 [Sec. 1630.2.2, Item 1]. From Eq. 6.8, the natural building period is

$$T = C_t(h_n)^{\frac{3}{4}}$$

For an ordinary steel-braced frame system, $C_t = 0.020$. $h_n = 70$ ft. Thus,

$$T = (0.020)(70 \text{ ft})^{\frac{3}{4}}$$
$$= 0.48 \text{ sec}$$

Assuming the penthouse is not part of the structural system and is more an "element" than a building, this building is really a seven-story office building. Therefore, the building weight is

$$W = (7 \text{ stories})\left(100 \text{ } \frac{\text{kips}}{\text{story}}\right) + 30 \text{ kips}$$
$$= 730 \text{ kips}$$

From Table 6.3, $Z = 0.40$. From Table 6.5, $I = 1.00$. From Table 6.10 for an ordinary steel-braced frame, $R = 5.6$. From Table 6.9, for soil-profile type S_A and for seismic zone factor $Z = 0.40$, the applicable velocity and acceleration-controlled seismic response coefficients are

$$C_v = 0.32 N_v$$
$$C_a = 0.32 N_a$$

From Table 6.8, for a site being positioned 4.65 mi from a potential seismic source type A, the applicable near-source factors N_v and N_a are 1.4 and 1.1, respectively, based on the linear interpolation of given values for distances other than those indicated in the table. Therefore,

$$C_v = (0.32)(1.4)$$
$$= 0.45$$
$$C_a = (0.32)(1.1)$$
$$= 0.35$$

Thus, from Eq. 6.22, the total design base shear is

$$V = \left(\frac{C_v I}{RT}\right) W$$
$$= \left[\frac{(0.45)(1.00)}{(5.6)(0.48)}\right] (730 \text{ kips})$$
$$= 122 \text{ kips}$$

Based on Eq. 6.23 [UBC-97 Sec. 1630.2.1 Formula 30-6], the minimum total design base shear is

$$V_{min} = 0.11 C_a I W$$
$$= (0.11)(0.35)(1.00)(730 \text{ kips})$$
$$= 28.1 \text{ kips}$$

In addition, for seismic zone 4, the UBC-97 [Sec. 1630.2.1] requires that the above minimum value of total design base shear be further adjusted according to Eq. 6.25 [UBC-97 Formula 30-7].

$$V_{min,\text{zone 4}} = \left(\frac{0.8 Z N_v I}{R}\right) W$$
$$= \left[\frac{(0.8)(0.4)(1.4)(1.0)}{5.6}\right] (730 \text{ kips})$$
$$= 58.4 \text{ kips}$$

$V_{min,\text{zone 4}} = 58.4$ kips is greater than $V_{min} = 28.1$ kips and becomes the controlling total design base shear minimum value. Thus, the calculated design base shear ($V = 122$ kips) can be used in designing this structure since it exceeds the controlling minimum total design base shear value ($V_{min,\text{zone 4}} = 58.4$ kips).

The UBC-97 [Sec. 1630.2.1] also requires that the total design base shear not exceed the maximum total design base shear calculated from Eq. 6.23 [UBC-97 Formula 30-5]. The maximum total design base shear is

$$V_{max} = \left(\frac{2.5 C_a I}{R}\right) W$$
$$= \left[\frac{(2.5)(0.35)(1.00)}{5.6}\right] (730 \text{ kips})$$
$$= 114 \text{ kips}$$

The calculated design base shear $V = 122$ kips is greater than $V_{max} = 114$ kips. Consequently, $V = 114$ kips is the appropriate total design base shear for this structure.

(b) From Eq. 6.28, the top force is zero since $T = 0.48$ sec < 0.7 sec.

Forming a table is the easiest way to determine the floor forces.

level x	h_x (ft)	w_x (kips)	$h_x w_x$ (ft-kips)
7	70	130	9100
6	60	100	6000
5	50	100	5000
4	40	100	4000
3	30	100	3000
2	20	100	2000
1	10	100	1000
		total	30,100

For the third floor (level 2),

$$\frac{h_x w_x}{\sum h_x w_x} = \frac{(20 \text{ ft})(100 \text{ kips})}{30,100 \text{ ft-kips}} = 0.0665$$

From Eq. 6.29,

$$F_2 = \frac{(V - F_t)w_2 h_2}{\sum w_i h_i}$$
$$= (114 \text{ kips} - 0)(0.0665) = 7.58 \text{ kips}$$

SI Solution

(a) The natural period (T) can be determined from Method A of the UBC-97 [Sec. 1630.2.2, Item 1]. From Eq. 6.8, the natural building period is

$$T = C_t(h_n)^{\frac{3}{4}}$$

For an ordinary steel-braced frame system, $C_t = 0.0488$. $h_n = 21.34$ m. Thus,

$$T = 0.0488(21.34 \text{ m})^{\frac{3}{4}}$$
$$= 0.48 \text{ s}$$

Assuming the penthouse is not part of the structural system and is more an "element" than a building, this building is really a seven-story office building. Therefore, the building weight is

$$W = (7 \text{ stories})\left(445 \frac{\text{kN}}{\text{story}}\right) + 133 \text{ kN}$$
$$= 3248 \text{ kN}$$

From Table 6.3, $Z = 0.40$. From Table 6.5, $I = 1.00$. From Table 6.10 for an ordinary steel-braced frame, $R = 5.6$. From Table 6.9, for soil-profile type S_A and for seismic zone factor $Z = 0.40$, the applicable velocity and acceleration-controlled seismic response coefficients are

$$C_v = 0.32N_v$$
$$C_a = 0.32N_a$$

From Table 6.8, for a site being positioned 7.5 km from a potential seismic source type A, the applicable near-source factors N_v and N_a are 1.4 and 1.1, respectively, based on the linear interpolation of given values for distances other than those indicated in the table. Therefore,

$$C_v = (0.32)(1.4)$$
$$= 0.45$$
$$C_a = (0.32)(1.1)$$
$$= 0.35$$

Thus, from Eq. 6.22, the total design base shear is

$$V = \left(\frac{C_v I}{RT}\right)W$$
$$= \left[\frac{(0.45)(1.00)}{(5.6)(0.48)}\right](3248 \text{ kN})$$
$$= 544 \text{ kN}$$

Based on Eq. 6.23 [UBC-97 Sec. 1630.2.1 Formula 30-6], the minimum total design base shear is

$$V_{\min} = 0.11C_a IW$$
$$= (0.11)(0.35)(1.00)(3248 \text{ kN})$$
$$= 125 \text{ kN}$$

In addition, for seismic zone 4, the UBC-97 [Sec. 1630.2.1] requires that the above minimum value of total design base shear be further adjusted according to Eq. 6.25 [UBC-97 Formula 30-7].

$$V_{\min,\text{zone }4} = \left(\frac{0.8ZN_v I}{R}\right)W$$
$$= \left[\frac{(0.8)(0.4)(1.4)(1.0)}{5.6}\right](3248 \text{ kN})$$
$$= 260 \text{ kN}$$

$V_{\min,\text{zone }4} = 260$ kN is greater than $V_{\min} = 125$ kN and becomes the controlling total design base shear minimum value. Thus, the calculated design base shear ($V = 544$ kN) can be used in designing this structure since it exceeds the controlling minimum total design base shear value ($V_{\min,\text{zone }4} = 260$ kN).

The UBC-97 [Sec. 1630.2.1] also requires that the total design base shear not exceed the maximum total design base shear calculated from Eq. 6.23 [UBC-97 Formula 30-5]. The maximum total design base shear is

$$V_{\max} = \left(\frac{2.5C_a I}{R}\right)W$$
$$= \left[\frac{(2.5)(0.35)(1.00)}{5.6}\right](3248 \text{ kN})$$
$$= 508 \text{ kN}$$

The calculated design base shear $V = 544$ kN is greater than $V_{\max} = 508$ kN. Consequently, $V = 508$ kN is the appropriate total design base shear for this structure.

(b) From Eq. 6.28, the top force is zero since $T = 0.48$ s < 0.7 s.

Forming a table is the easiest way to determine the floor forces.

level x	h_x (m)	w_x (kN)	$h_x w_x$ (kN·m)
7	21.34	578	12335
6	18.29	445	8139
5	15.24	445	6782
4	12.19	445	5425
3	9.14	445	4067
2	6.10	445	2715
1	3.05	445	1357
		total	40820

For the third floor (level 2),

$$\frac{h_x w_x}{\sum h_x w_x} = \frac{(6.10)(445)}{40\,820} = 0.0665$$

From Eq. 6.29,

$$F_2 = \frac{(V - F_t)w_2 h_2}{\sum w_i h_i}$$
$$= (507.5 \text{ kN} - 0)(0.0665) = 33.7 \text{ kN}$$

6. The owner of a building is considering adding a north wing to increase the size of the building. The existing floor plan and planned wing are shown in the figure below, along with structural walls and relative rigidities. All walls have the same thickness and height. The roof mass is to be disregarded. The owner is concerned about the lack of symmetry that the remodeled building will have. It has been determined that the base shear on the existing building is 90,000 lbf (400.3 kN), parallel to the long (150-ft (45.72-m)) dimension. An additional 15,000 lbf (66.7 kN) of base shear will be added by the new wing.

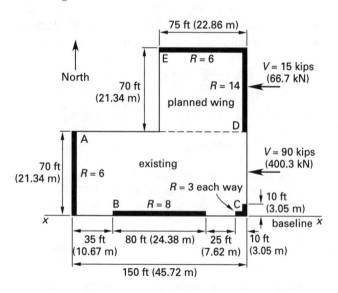

(a) Where will the center of mass be located when the wing is installed?

(b) Where will the center of rigidity be located when the wing is installed?

(c) What will be the torsional moment?

(d) Which active walls experience a negative torsional shear?

Customary U.S. Solution

(a) The wall masses are proportional to their lengths since their thicknesses and heights are all the same. Using the 150-ft east-west wall as the baseline, for earthquakes in the east-west direction, the center of mass is located at a distance of

$$y_{\text{c.m.}} = \frac{\sum m_i y_i}{\sum m_i}$$

$$= \frac{\begin{array}{c}(70 \text{ ft})(35 \text{ ft}) + (80 \text{ ft})(0) + (10 \text{ ft})(0) \\ + (10 \text{ ft})(5 \text{ ft}) + (70 \text{ ft})(70 \text{ ft} + 35 \text{ ft}) \\ + (75 \text{ ft})(70 \text{ ft} + 70 \text{ ft})\end{array}}{70 \text{ ft} + 80 \text{ ft} + 10 \text{ ft} + 10 \text{ ft} + 70 \text{ ft} + 75 \text{ ft}}$$

$$= \frac{20{,}350 \text{ ft}^2}{315 \text{ ft}} = 64.6 \text{ ft (up from the baseline)}$$

(b) Walls running north-south do not contribute to rigidity for east-west earthquakes. Only walls B, C (part), and E are active. The center of rigidity is located at a distance of

$$y_{\text{c.r.}} = \frac{\sum R_i y_i}{\sum R_i}$$

$$= \frac{(8)(0) + (3)(0) + (6)(70 \text{ ft} + 70 \text{ ft})}{8 + 3 + 6}$$

$$= \frac{840 \text{ ft}}{17} = 49.4 \text{ ft (up from the baseline)}$$

(c) The actual eccentricity is

$$e_{\text{actual}} = y_{\text{c.m.}} - y_{\text{c.r.}} = 64.6 \text{ ft} - 49.4 \text{ ft} = 15.2 \text{ ft}$$

The accidental eccentricity required by the UBC-97 is 5% of the transverse building direction.

$$e_{\text{accidental}} = (0.05)(70 \text{ ft} + 70 \text{ ft}) = 7 \text{ ft}$$

The total eccentricity is

$$e = e_{\text{actual}} \pm e_{\text{accidental}} = 15.2 \text{ ft} \pm 7 \text{ ft}$$
$$e_{\text{max}} = 22.2 \text{ ft}$$
$$e_{\text{min}} = 8.2 \text{ ft}$$

The torsional moment is

$$M_{\text{max}} = Ve = (90 \text{ kips} + 15 \text{ kips})(22.2 \text{ ft})$$
$$= 2331 \text{ ft-kips}$$
$$M_{\text{min}} = Ve = (90 \text{ kips} + 15 \text{ kips})(8.2 \text{ ft}) = 861 \text{ ft-kips}$$

(d) Walls B, C, and E resist the direct shear and are active. The base shear acts through the center of mass, tending to cause counterclockwise rotation about the center of rigidity. This is resisted by all walls (A, B,

C, D, and E with clockwise direction forces). For walls B and C the clockwise direction opposes forces due to direct shear. Walls B and C have negative torsional shear components.

SI Solution

(a) The wall masses are proportional to their lengths since their thicknesses and heights are all the same. Using the 45.72-m east-west wall as the baseline, for earthquakes in the east-west direction, the center of mass is located at a distance of

$$y_{\text{c.m.}} = \frac{\Sigma m_i y_i}{\Sigma m_i}$$

$$= \frac{\begin{array}{l}(21.34 \text{ m})(10.67 \text{ m}) + (24.38 \text{ m})(0) \\ + (3.05 \text{ m})(0) + (3.05 \text{ m})(1.52 \text{ m}) \\ + (21.34 \text{ m})(21.34 \text{ m} + 10.67 \text{ m}) \\ + (22.86 \text{ m})(21.34 \text{ m} + 21.34 \text{ m})\end{array}}{\begin{array}{l}21.34 \text{ m} + 24.38 \text{ m} + 3.05 \text{ m} + 3.05 \text{ m} \\ + 21.34 \text{ m} + 22.86 \text{ m}\end{array}}$$

$$= \frac{1891.09 \text{ m}^2}{96.02 \text{ m}}$$

$$= 19.69 \text{ m (up from the baseline)}$$

(b) Walls running north-south do not contribute to rigidity for east-west earthquakes. Only walls B, C (part), and E are active. The center of rigidity is located at a distance of

$$y_{\text{c.r.}} = \frac{\Sigma R_i y_i}{\Sigma R_i}$$

$$= \frac{(8)(0) + (3)(0) + (6)(21.34 \text{ m} + 21.34 \text{ m})}{8 + 3 + 6}$$

$$= \frac{256.08 \text{ m}}{17}$$

$$= 15.06 \text{ m (up from the baseline)}$$

(c) The actual eccentricity is

$$e_{\text{actual}} = y_{\text{c.m.}} - y_{\text{c.r.}} = 19.69 \text{ m} - 15.06 \text{ m} = 4.63 \text{ m}$$

The accidental eccentricity required by the UBC-97 is 5% of the transverse building direction.

$$e_{\text{accidental}} = (0.05)(21.34 \text{ m} + 21.34 \text{ m}) = 2.13 \text{ m}$$

The total eccentricity is

$$e = e_{\text{actual}} \pm e_{\text{accidental}} = 4.63 \text{ m} \pm 2.13 \text{ m}$$

$$e_{\text{max}} = 6.76 \text{ m}$$

$$e_{\text{min}} = 2.15 \text{ m}$$

The torsional moment is

$$M_{\text{max}} = Ve = (400.3 \text{ kN} + 66.7 \text{ kN})(6.76 \text{ m})$$

$$= 3157 \text{ kN·m}$$

$$M_{\text{min}} = Ve = (400.3 \text{ kN} + 66.7 \text{ kN})(2.15 \text{ m})$$

$$= 1004 \text{ kN·m}$$

(d) Walls B, C, and E resist the direct shear and are active. The base shear acts through the center of mass, tending to cause counterclockwise rotation about the center of rigidity. This is resisted by all walls (A, B, C, D, and E with "clockwise" direction forces). For walls B and C, the clockwise direction opposes forces due to direct shear. Walls B and C have negative shear due to torsion.

7. A one-story, 55 ft by 70 ft (16.76 m by 21.34 m) masonry-walled building with a rigid diaphragm roof is constructed in the shape of a rectangle. One side of the building has two doors. There are no other openings. The rigidities of each wall are shown. Each wall is 10 in (254 mm) thick. Disregard the mass of the roof diaphragm.

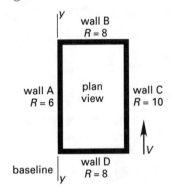

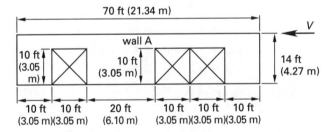

(a) What is the location of the center of rigidity?

(b) What is the torsional moment due to a total lateral force of 50,000 lbf (222.4 kN)?

(c) If 19,000 lbf (84.5 kN) of the lateral load are distributed to wall A, what loads are carried by each of the piers in wall A?

Customary U.S. Solution

(a) For earthquakes in the direction shown, only walls A and C contribute to rigidity. The center of rigidity is located along a line parallel to baseline y-y and located a distance away of

$$x_{\text{c.r.}} = \frac{\sum R_i x_i}{\sum R_i} = \frac{(6)(0) + (10)(55 \text{ ft})}{6 + 10} = \frac{550 \text{ ft}}{16}$$

$$= 34.4 \text{ ft}$$

(b) Although the openings could be disregarded, enough information is given to determine the location of the center of mass considering the openings. The mass of each wall is proportional to its area.

wall	area (mass)
A	$(70 \text{ ft})(14 \text{ ft}) - (3)(10 \text{ ft})(10 \text{ ft}) = 680 \text{ ft}^2$
B	$(55 \text{ ft})(14 \text{ ft}) = 770 \text{ ft}^2$
C	$(70 \text{ ft})(14 \text{ ft}) = 980 \text{ ft}^2$
D	$(55 \text{ ft})(14 \text{ ft}) = 770 \text{ ft}^2$

The center of mass is located at

$$x_{\text{c.m.}} = \frac{\sum m_i x_i}{\sum m_i}$$

$$= \frac{(680 \text{ ft}^2)(0) + (770 \text{ ft}^2)\left(\dfrac{55 \text{ ft}}{2}\right) + (980 \text{ ft}^2)(55 \text{ ft}) + (770 \text{ ft}^2)\left(\dfrac{55 \text{ ft}}{2}\right)}{680 \text{ ft}^2 + 770 \text{ ft}^2 + 980 \text{ ft}^2 + 770 \text{ ft}^2}$$

$$= \frac{96{,}250 \text{ ft}^3}{3200 \text{ ft}^2} = 30.1 \text{ ft}$$

The actual eccentricity is

$$e_{\text{actual}} = x_{\text{c.r.}} - x_{\text{c.m.}} = 34.4 \text{ ft} - 30.1 \text{ ft} = 4.3 \text{ ft}$$

The accidental eccentricity is

$$e_{\text{accidental}} = (0.05)(55 \text{ ft}) = 2.8 \text{ ft}$$

The total eccentricity is

$$e = e_{\text{actual}} \pm e_{\text{accidental}}$$

$$e_{\text{max}} = 4.3 \text{ ft} + 2.8 \text{ ft} = 7.1 \text{ ft}$$

$$e_{\text{min}} = 4.3 \text{ ft} - 2.8 \text{ ft} = 1.5 \text{ ft}$$

The torsional moment is

$$M = Ve$$

$$M_{\text{max}} = (50{,}000 \text{ lbf})(7.1 \text{ ft}) = 355{,}000 \text{ ft-lbf}$$

$$M_{\text{min}} = (50{,}000 \text{ lbf})(1.5 \text{ ft}) = 75{,}000 \text{ ft-lbf}$$

(c) There are three piers in wall A. Due to the effect of the beam running along the top, assume the piers are fixed. The rigidities are found from App. D.

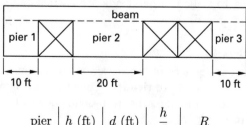

pier	h (ft)	d (ft)	$\dfrac{h}{d}$	R
1	10	10	1	2.5
2	10	20	0.5	6.154
3	10	10	1	2.5

The fraction of the total wall shear taken by piers 1 and 3 each is

$$\frac{2.5}{2.5 + 6.154 + 2.5} = 0.22$$

The fraction taken by pier 2 is

$$1.00 - (2)(0.22) = 0.56$$

The pier shears are

$$V_1 = V_3 = (0.22)(19{,}000 \text{ lbf}) = 4180 \text{ lbf}$$

$$V_2 = (0.56)(19{,}000 \text{ lbf}) = 10{,}640 \text{ lbf}$$

SI Solution

(a) For earthquakes in the direction shown, only walls A and C contribute to rigidity. The center of rigidity is located along a line parallel to baseline y-y and located a distance away of

$$x_{\text{c.r.}} = \frac{\sum R_i x_i}{\sum R_i} = \frac{(6)(0) + (10)(16.76 \text{ m})}{6 + 10} = \frac{167.60 \text{ m}}{16}$$

$$= 10.48 \text{ m}$$

(b) Although the openings could be disregarded, enough information is given to determine the location of the center of mass considering the openings. The mass of each wall is proportional to its area.

wall	area (mass)
A	$(21.34 \text{ m})(4.27 \text{ m}) - (3)(3.05 \text{ m})(3.05 \text{ m})$ $= 63.21 \text{ m}^2$
B	$(16.76 \text{ m})(4.27 \text{ m}) = 71.57 \text{ m}^2$
C	$(21.34 \text{ m})(4.27 \text{ m}) = 91.12 \text{ m}^2$
D	$(16.76 \text{ m})(4.27 \text{ m}) = 71.57 \text{ m}^2$

The center of mass is located at

$$x_{\text{c.m.}} = \frac{\sum m_i x_i}{\sum m_i}$$

$$= \frac{(63.21 \text{ m}^2)(0) + (71.57 \text{ m}^2)\left(\dfrac{16.76 \text{ m}}{2}\right) + (91.12 \text{ m}^2)(16.76 \text{ m}) + (71.57 \text{ m}^2)\left(\dfrac{16.76 \text{ m}}{2}\right)}{63.21 \text{ m}^2 + 71.57 \text{ m}^2 + 91.12 \text{ m}^2 + 71.57 \text{ m}^2}$$

$$= \frac{2726.68 \text{ m}^3}{297.47 \text{ m}^2} = 9.17 \text{ m}$$

The actual eccentricity is

$$e_{\text{actual}} = x_{\text{c.r.}} - x_{\text{c.m.}} = 10.48 \text{ m} - 9.17 \text{ m} = 1.31 \text{ m}$$

The accidental eccentricity is

$$e_{\text{accidental}} = (0.05)(16.76 \text{ m}) = 0.84 \text{ m}$$

The total eccentricity is

$$e = e_{\text{actual}} \pm e_{\text{accidental}}$$
$$e_{\text{max}} = 1.31 \,\text{m} + 0.84 \,\text{m} = 2.15 \,\text{m}$$
$$e_{\text{min}} = 1.31 - 0.84 \,\text{m} = 0.47 \,\text{m}$$

The torsional moment is

$$M = Ve$$
$$M_{\text{max}} = (222.4 \,\text{kN})(2.15 \,\text{m}) = 478.2 \,\text{kN·m}$$
$$M_{\text{min}} = (222.4 \,\text{kN})(0.47 \,\text{m}) = 104.5 \,\text{kN·m}$$

(c) There are three piers in wall A. Due to the effect of the beam running along the top, assume the piers are fixed. The rigidities are found from App. D.

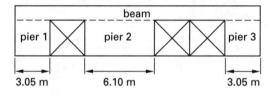

pier	h (m)	d (m)	$\dfrac{h}{d}$	R
1	3.05	3.05	1	2.5
2	3.05	6.10	0.5	6.154
3	3.05	3.05	1	2.5

The fraction of the total wall shear taken by piers 1 and 3 each is

$$\frac{2.5}{2.5 + 6.154 + 2.5} = 0.22$$

The fraction taken by pier 2 is

$$1.00 - (2)(0.22) = 0.56$$

The pier shears are

$$V_1 = V_3 = (0.22)(84.5 \,\text{kN}) = 18.6 \,\text{kN}$$
$$V_2 = (0.56)(84.5 \,\text{k}) = 47.3 \,\text{kN}$$

8. A one-story building with masonry walls is constructed in a box shape. All walls are 24 ft (7.32 m) high. The walls have a weight of 50 lbf/ft² (2.4 kN/m²). The wood structural panel roof and roof skylights have a weight of 25 lbf/ft² (1.2 kN/m²). The roof carries a 20-lbf/ft² (0.96-kN/m²) live load of permanent air-conditioning equipment. The seismic load in the north-south direction is 15% of the participating building weight. Do not consider east-west earthquake performance.

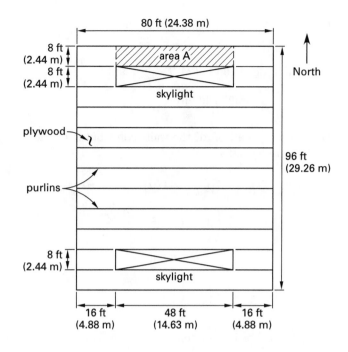

(a) What is the diaphragm force?

(b) What is the maximum wall shear force?

(c) What is the maximum chord force in the diaphragm for a north-sourh earthquake?

Customary U.S. Solution

(a) For a north-south earthquake, the diaphragm force results from the acceleration of the roof mass and half of the short (80-ft) wall masses. The roof weight (including the air-conditioning equipment weight) is

$$W_{\text{roof}} = \text{area} \times \text{loading}$$
$$= (80 \,\text{ft})(96 \,\text{ft}) \left(20 \,\frac{\text{lbf}}{\text{ft}^2} + 25 \,\frac{\text{lbf}}{\text{ft}^2} \right)$$
$$= 345{,}600 \,\text{lbf}$$

The weight of the perpendicular walls is

$$W_{\perp\text{walls}} = (80 \,\text{ft})(24 \,\text{ft})(2 \,\text{walls}) \left(50 \,\frac{\text{lbf}}{\text{ft}^2} \right)$$
$$= 192{,}000 \,\text{lbf}$$

Only the top half of the perpendicular walls contributes to the diaphragm force. Since the seismic load is given as 15% of the building weight,

$$F_{\text{diaphragm}} = (0.15)(W_{\text{roof}} + \tfrac{1}{2} W_{\perp\text{walls}})$$
$$= (0.15) \left(345{,}600 \,\text{lbf} + \frac{192{,}000 \,\text{lbf}}{2} \right)$$
$$= 66{,}240 \,\text{lbf}$$

(b) The weight of the parallel walls is

$$W_{\|\text{walls}} = (96 \text{ ft})(24 \text{ ft})(2 \text{ walls})\left(50 \frac{\text{lbf}}{\text{ft}^2}\right)$$

$$= 230{,}400 \text{ lbf}$$

Only the top half of all the walls contributes to shear in the parallel walls.

$$F_{\text{total}} = (0.15)\left(W_{\text{roof}} + \tfrac{1}{2}W_{\perp\text{walls}} + \tfrac{1}{2}W_{\|\text{walls}}\right)$$

$$= (0.15)\left(345{,}600 \text{ lbf} + \frac{192{,}000 \text{ lbf}}{2}\right.$$

$$\left. + \frac{230{,}400 \text{ lbf}}{2}\right)$$

$$= 83{,}520 \text{ lbf}$$

Since there are two parallel walls, the maximum wall shear force is

$$V_{\|\text{wall}} = \frac{83{,}520 \text{ lbf}}{2 \text{ walls}} = 41{,}760 \text{ lbf/wall}$$

(c) From Eq. 7.9, the chord force in the perpendicular walls is

$$C = \frac{F_{\text{diaphragm}}L}{8b} = \frac{(66{,}240 \text{ lbf})(80 \text{ ft})}{(8)(96 \text{ ft})}$$

$$= 6900 \text{ lbf}$$

SI Solution

(a) For a north-south earthquake, the diaphragm force results from the acceleration of the roof mass and half of the short (24.38-m) wall masses. The roof weight (including the air-conditioning equipment weight) is

$$W_{\text{roof}} = \text{area} \times \text{loading}$$

$$= (24.38 \text{ m})(29.26 \text{ m})\left(0.96 \frac{\text{kN}}{\text{m}^2} + 1.2 \frac{\text{kN}}{\text{m}^2}\right)$$

$$= 1540.9 \text{ kN}$$

The weight of the perpendicular walls is

$$W_{\perp\text{walls}} = (24.38 \text{ m})(7.32 \text{ m})(2 \text{ walls})\left(2.4 \frac{\text{kN}}{\text{m}^2}\right)$$

$$= 856.6 \text{ kN}$$

Only the top half of the perpendicular walls contributes to the diaphragm force. Since the seismic load is given as 15% of the building weight,

$$F_{\text{diaphragm}} = (0.15)(W_{\text{roof}} + \tfrac{1}{2}W_{\perp\text{walls}})$$

$$= (0.15)\left(1540.9 \text{ kN} + \frac{856.6 \text{ kN}}{2}\right)$$

$$= 295.4 \text{ kN}$$

(b) The weight of the parallel walls is

$$W_{\|\text{walls}} = (29.26 \text{ m})(7.32 \text{ m})(2 \text{ walls})\left(2.4 \frac{\text{kN}}{\text{m}^2}\right)$$

$$= 1028.1 \text{ kN}$$

Only the top half of all the walls contributes to shear in the parallel walls.

$$F_{\text{total}} = (0.15)\left(W_{\text{roof}} + \tfrac{1}{2}W_{\perp\text{walls}} + \tfrac{1}{2}W_{\|\text{walls}}\right)$$

$$= (0.15)\left(1540.9 \text{ kN} + \frac{856.6 \text{ kN}}{2}\right.$$

$$\left. + \frac{1028.1 \text{ kN}}{2}\right)$$

$$= 372.5 \text{ kN}$$

Since there are two parallel walls, the maximum wall shear force is

$$V_{\|\text{wall}} = \frac{372.5 \text{ kN}}{2 \text{ walls}} = 186.3 \text{ kN/wall}$$

(c) From Eq. 7.9, the chord force in the perpendicular walls is

$$C = \frac{F_{\text{diaphragm}}L}{8b} = \frac{(295.4 \text{ kN})(24.38 \text{ m})}{(8)(29.26 \text{ m})}$$

$$= 30.8 \text{ kN}$$

9. The plan view of a one-story, 12-ft (3.66-m) high masonry-walled retail shop is shown. Windows cover most of the front and half of one side. The building has a flexible wood structural panel roof diaphragm with continuous roof struts (shown as lighter lines) criss-crossing and dividing the roof into small diaphragms. Each strut is designed to serve as a chord, if necessary. The masonry shear walls have a dead weight of 55 lbf/ft^2 (2.6 kN/m^2). The wood structural panel roof has a dead weight of 15 lbf/ft^2 (0.7 kN/m^2). For the purpose of the UBC-97 base shear equation, $V = 0.183W$.

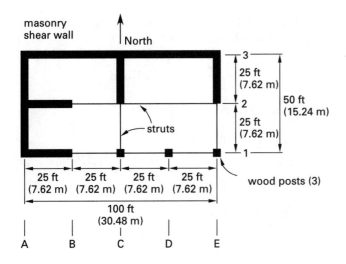

masonry
shear wall

North

struts

wood posts (3)

25 ft
(7.62 m)

25 ft
(7.62 m)

50 ft
(15.24 m)

25 ft
(7.62 m)

25 ft
(7.62 m)

25 ft
(7.62 m)

25 ft
(7.62 m)

100 ft
(30.48 m)

A   B   C   D   E

(a) What is the maximum diaphragm shear along lines A and E?

(b) What is the maximum chord or drag force at point C-1?

(c) What is the maximum drag force at point B-1 due to an east-west earthquake?

Customary U.S. Solution

Notice that the earthquake direction is not given for parts (a) and (b) in this problem and must be considered variable.

(a) For north-south earthquakes, the roof is effectively divided into two subdiaphragms. One diaphragm is bounded by points (moving clockwise) A-1, A-3, C-3, and C-1. The other is bounded by points C-1, C-3, E-3, and E-1. Although each diaphragm is the same size, the accelerating masses are different.

For line A,

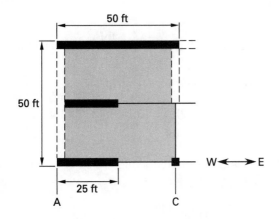

50 ft

50 ft

25 ft

W ← → E

A   C

The weight of the roof is

$$W_{\text{roof}} = (50 \text{ ft})(50 \text{ ft})\left(15 \frac{\text{lbf}}{\text{ft}^2}\right) = 37{,}500 \text{ lbf}$$

The weight of the 50-ft wall (full-height) is

$$W_{50 \text{ ft}} = (12 \text{ ft})(50 \text{ ft})\left(55 \frac{\text{lbf}}{\text{ft}^2}\right) = 33{,}000 \text{ lbf}$$

The weight of the two 25-ft walls totals 33,000 lbf also.

$$W_{25 \text{ ft}} = 33{,}000 \text{ lbf}$$

The seismic effect from the roof and the 50-ft wall is shared equally between walls A and C. Since the two remaining walls extend only halfway between A and C, a rational method of allocating their seismic effects to walls A and C must be used. Consider a simply supported beam loaded uniformly along the first half of its length. The reaction closest to the uniform load will carry $\frac{3}{4}$ of the total load. Therefore, assume that wall A carries $\frac{3}{4}$ of the seismic effect from the two short walls. (Other assumptions may be valid, depending on construction details.) Finally, assume that only the top half of the walls contribute to diaphragm shear.

The diaphragm reaction at line A is

$$
\begin{aligned}
F_A &= 0.183W \\
&= (0.183) \\
&\quad \times \left[\frac{37{,}500 \text{ lbf}}{2} + \frac{\left(\frac{1}{2}\right)(33{,}000 \text{ lbf})}{2} + \frac{\left(\frac{3}{4}\right)(33{,}000 \text{ lbf})}{2}\right] \\
&= 7210 \text{ lbf}
\end{aligned}
$$

For line E,

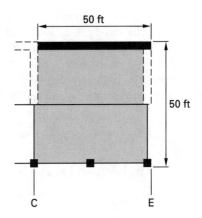

50 ft

50 ft

C   E

$$W_{\text{roof}} = 37{,}500 \text{ lbf}$$

$$W_{\text{E-W wall}} = (12 \text{ ft})(50 \text{ ft})\left(55 \frac{\text{lbf}}{\text{ft}^2}\right) = 33{,}000 \text{ lbf}$$

The east-west wall weight tributary to the roof level is

$$\frac{33,000 \text{ lbf}}{2} = 16,500 \text{ lbf}$$

$$F_E = 0.183W$$
$$= (0.183)\left(\frac{37,500 \text{ lbf}}{2} + \frac{16,500 \text{ lbf}}{2}\right)$$
$$= 4940 \text{ lbf}$$

(b) The earthquake direction is not given.

North-south earthquake

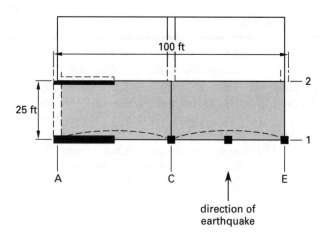

Drag force at point C-1 (line C)

All diaphragm shear would frame into point C-2 since the wooden post at point C-1 can be assumed to have no lateral stiffness. Therefore, the maximum drag force at point C-1 due to a north-south earthquake is zero.

Chord force at point C-1 (line 1)

Half of the roof mass (corresponding to the tributary area bounded by points A-1, A-2, E-2, and E-1) and the upper half of the east-west walls in the tributary area contribute to the chord forces along lines 1 and 2. (Other interpretations might also be justified. For example, the entire roof might be considered.)

$$W_{\text{roof}} = (25 \text{ ft})(100 \text{ ft})\left(15 \frac{\text{lbf}}{\text{ft}^2}\right) = 37,500 \text{ lbf}$$

$$W_{\text{E-W walls}} = (12 \text{ ft})\left[25 \text{ ft} + \left(\tfrac{1}{2}\right)(25 \text{ ft})\right]\left(55 \frac{\text{lbf}}{\text{ft}^2}\right)$$
$$= 24,750 \text{ lbf}$$

Notice that only half of the east-west wall along line 2 is used. This is because the other half is tributary to the adjacent area bounded by points A-2, A-3, C-3, and C-2.

Only the upper half of the east-west walls is effective in loading the diaphragm.

$$\frac{24,750 \text{ lbf}}{2} = 12,375 \text{ lbf}$$

The diaphragm force is

$$F_{\text{diaphragm}} = 0.183W$$
$$= (0.183)(37,500 \text{ lbf} + 12,375 \text{ lbf})$$
$$= 9130 \text{ lbf}$$

From Eq. 7.9, the chord force is

$$C = \frac{F_{\text{diaphragm}}L}{8b} = \frac{(9130 \text{ lbf})(100 \text{ ft})}{(8)(25 \text{ ft})}$$
$$= 4565 \text{ lbf}$$

East-west earthquake

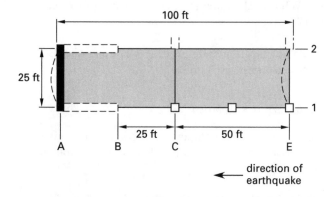

Drag force at point C-1 (line 1)

The collector along line 1 transfers half the diaphragm loading from the area bounded by points C-1, C-2, E-2, and E-1 into point B-1. (The other half is transferred to point B-2.) Since there are no north-south walls in this area, the only contribution to collector force is the tributary roof weight. Between lines C and E, the roof weight is

$$W_{\text{roof}} = (25 \text{ ft})(50 \text{ ft})\left(15 \frac{\text{lbf}}{\text{ft}}\right) = 18,750 \text{ lbf}$$

The diaphragm force is

$$F = 0.183W = (0.183)(18,750 \text{ lbf})$$
$$= 3430 \text{ lbf}$$

Half of this is transferred along the strut on line 1. At point C-1, the drag force is

$$D = \frac{3430 \text{ lbf}}{2} = 1715 \text{ lbf}$$

Chord force at point C-1 (line C)

While there is a chord force along line C, the chord force at C-1 is zero because point C-1 is at the end of the chord.

(c) This is similar to part (b), except that the tributary roof area is 75 ft long instead of 50 ft long. Scaling up from the answer derived in part (b),

$$D = \left(\frac{75 \text{ ft}}{50 \text{ ft}}\right)(1715 \text{ lbf}) = 2570 \text{ lbf}$$

SI Solution

Notice that the earthquake direction is not given for parts (a) and (b) in this problem and must be considered a variable.

(a) For north-south earthquakes, the roof is effectively divided into two diaphragms. One diaphragm is bounded by points (moving clockwise) A-1, A-3, C-3, and C-1. The other is bounded by points C-1, C-3, E-3, and E-1. Although each diaphragm is the same size, the accelerating masses are different.

For line A,

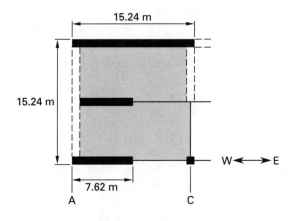

The weight of the roof is

$$W_{\text{roof}} = (15.24 \text{ m})(15.24 \text{ m})\left(0.7 \frac{\text{kN}}{\text{m}^2}\right) = 162.6 \text{ kN}$$

The weight of the 15.24-m wall (full height) is

$$W_{15.24 \text{ m}} = (3.66 \text{ m})(15.24 \text{ m})\left(2.6 \frac{\text{kN}}{\text{m}^2}\right) = 145.0 \text{ kN}$$

The weight of the two 7.62-m walls totals 145.0 kN also.

$$W_{7.62 \text{ m}} = 145.0 \text{ kN}$$

The seismic effect from the roof and the 15.24-m wall is shared equally between walls A and C. Since the two remaining walls extend only halfway between A and C, a rational method of allocating their seismic effects to walls A and C must be used. Consider a simply supported beam loaded uniformly along the first half of its length. The reaction closest to the uniform load will carry $\frac{3}{4}$ of the total load. Therefore, assume that wall A carries $\frac{3}{4}$ of the seismic effect from the two short walls. (Other assumptions may be valid, depending on construction details.) Finally, assume that only the top half of the walls contribute to diaphragm shear.

The diaphragm reaction at line A is

$$
\begin{aligned}
F_A &= 0.183W \\
&= (0.183) \\
&\quad \times \left[\frac{162.6 \text{ kN}}{2} + \frac{\left(\frac{1}{2}\right)(145.0 \text{ kN})}{2} + \frac{\left(\frac{3}{4}\right)(145.0 \text{ kN})}{2}\right] \\
&= 31.5 \text{ kN}
\end{aligned}
$$

For line E,

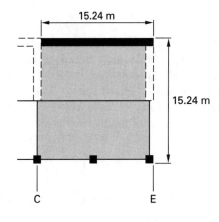

$$W_{\text{roof}} = 162.6 \text{ kN}$$

$$W_{\text{E-W wall}} = (3.66 \text{ m})(15.24 \text{ m})\left(2.6 \frac{\text{kN}}{\text{m}^2}\right) = 145.0 \text{ kN}$$

The east-west wall weight tributary to the roof level is

$$\frac{145.0 \text{ kN}}{2} = 72.5 \text{ kN}$$

$$
\begin{aligned}
F_E &= 0.183W \\
&= (0.183)\left(\frac{162.6 \text{ kN}}{2} + \frac{72.5 \text{ kN}}{2}\right) \\
&= 21.5 \text{ kN}
\end{aligned}
$$

(b) The earthquake direction is not given.

North-south earthquake

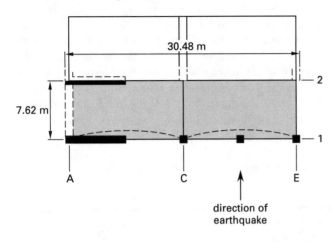

Drag force at point C-1 *(line* C*)*

All diaphragm shear would frame into point C-2 since the wooden post at point C-1 can be assumed to have no lateral stiffness. Therefore, the maximum drag force at point C-1 due to a north-south earthquake is zero.

Chord force at point C-1 *(line* 1*)*

Half of the roof mass (corresponding to the tributary area bounded by points A-1, A-2, E-2, and E-1) and the upper half of the east-west walls in the tributary area contribute to the chord forces along lines 1 and 2. (Other interpretations might also be justified. For example, the entire roof might be considered.)

$$W_{\text{roof}} = (7.62 \text{ m})(30.48 \text{ m})\left(0.7 \frac{\text{kN}}{\text{m}^2}\right) = 162.6 \text{ kN}$$

$$W_{\text{E-W walls}} = (3.66 \text{ m})\left(7.62 \text{ m} + \left(\tfrac{1}{2}\right)(7.62 \text{ m})\right)\left(2.6 \frac{\text{kN}}{\text{m}^2}\right)$$
$$= 108.8 \text{ kN}$$

Notice that only half of the east-west wall along line 2 is used. This is because the other half is tributary to the adjacent area bounded by points A-2, A-3, C-3, and C-2.

Only the upper half of the east-west walls is effective in loading the diaphragm.

$$\frac{108.8 \text{ kN}}{2} = 54.4 \text{ kN}$$

The diaphragm force is

$$F_{\text{diaphragm}} = 0.183W$$
$$= (0.183)(162.6 \text{ kN} + 54.4 \text{ kN})$$
$$= 39.7 \text{ kN}$$

From Eq. 7.9, the chord force is

$$C = \frac{F_{\text{diaphragm}}L}{8b} = \frac{(39.7 \text{ kN})(30.48 \text{ m})}{(8)(7.62 \text{ m})}$$
$$= 19.9 \text{ kN}$$

East-west earthquake

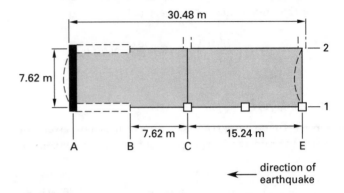

Drag force at point C-1 *(line* 1*)*

The collector along line 1 transfers half the diaphragm loading from the area bounded by points C-1, C-2, E-2, and E-1 into point B-1. (The other half is transferred to point B-2.) Since there are no north-south walls in this area, the only contribution to collector force is the tributary roof weight. Between lines C and E, the roof weight is

$$W_{\text{roof}} = (7.62 \text{ m})(15.24 \text{ m})\left(0.7 \frac{\text{N}}{\text{m}^2}\right) = 81.3 \text{ kN}$$

The diaphragm force is

$$F = 0.183W = (0.183)(81.3 \text{ kN})$$
$$= 14.9 \text{ kN}$$

Half of this is transferred along the strut on line 1. At point C-1, the drag force is

$$D = \frac{14.9 \text{ kN}}{2} = 7.45 \text{ kN}$$

Chord force at point C-1 *(line* C*)*

While there is a chord force along line C, the chord force at C-1 is zero because point C-1 is at the end of the chord.

(c) This is similar to part (b), except that the tributary roof area is 22.86 m long instead of 15.24 m long. Scaling up from the answer derived in part (b),

$$D = \left(\frac{22.86 \text{ m}}{15.24 \text{ m}}\right)(7.45 \text{ kN}) = 11.2 \text{ kN}$$

10. A 120-ft (36.58-m) wide by 240-ft (73.15-m) long warehouse in seismic zone 4 is oriented with its long dimension in the north-south direction. The natural period is 0.38 sec. This structure is underlaid with soil-profile type S_A and is 9.3 mi (15 km) from a potential seismic source type A. The walls are solid cast-in-place concrete, 10 ft (3.05 m) high and 12 in (305 mm) thick, with no significant openings. The warehouse floor has a live load of 200 lbf/ft^2 (9.6 kN/m^2). The roof consists of a 19/32-in (15-mm), C-D plywood blocked diaphragm with an average weight of 15 lbf/ft^2 (0.7 kN/m^2) nailed to a 3-in (76-mm) (nominal) ledger (Douglas fir-larch) with 10d nails. Do not use the simplified lateral-force procedure, and consider only north-south earthquake motions.

(a) What diaphragm edge nail spacing is required along the 240-ft (73.15-m) wall?

(b) Sketch the method of interconnecting the diaphragm, ledger, and wall.

(c) Size the ledger bolts if they are spaced every 3 ft (91.4 cm).

(d) What size grade 60 steel rebar should be used as the diaphragm chord?

(e) Where should the chord be located?

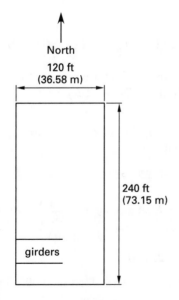

Customary U.S. Solution

From Table 6.3, $Z = 0.40$. From Table 6.5, $I = 1.00$. From Table 6.10 for a concrete shear wall (bearing wall) system, the response modification factor $R = 4.5$. From Table 6.9, for soil-profile type S_A and for seismic zone

factor $Z = 0.40$, the applicable acceleration and velocity-controlled seismic response coefficients are

$$C_a = 0.32N_a$$
$$C_v = 0.32N_v$$

From Table 6.8, for this site 9.3 mi from a potential seismic source type A, the applicable near-source factors, N_a and N_v, are both equal to 1.00. Therefore,

$$C_a = C_v = (0.32)(1.0)$$
$$= 0.32$$

Thus, from Eq. 6.22, the total design base shear is

$$V = \left(\frac{C_v I}{RT}\right) W$$
$$= \left[\frac{(0.32)(1.00)}{(4.5)(0.38)}\right] W$$
$$= 0.187W$$

V_{design} should be greater than the minimum total design base shear obtained from Eq. 6.24 [UBC-97 Sec. 1630.2.1, Formula 30-6].

$$V_{\text{min}} = 0.11 C_a I W$$
$$= (0.11)(0.32)(1.00)W$$
$$= 0.035W$$

For seismic zone 4, the above V_{min} should be further limited according to Eq. 6.25 [UBC-97 Sec. 1630.2.1, Formula 30-7].

$$V_{\text{min,zone 4}} = \left(\frac{0.8ZN_v I}{R}\right) W$$
$$= \left[\frac{(0.8)(0.4)(1.0)(1.0)}{4.5}\right] W$$
$$= 0.071W$$

The calculated $V_{\text{design}} = 0.187W$ can be used since it exceeds the controlling value of $V_{\text{min,zone 4}} = 0.071W$. Based on the UBC-97 [Sec. 1630.2.1], however, this design base shear value should not exceed the maximum total design base shear calculated from Eq. 6.23 [UBC-97 Formula 30-5]. V_{max} is

$$V_{\text{max}} = \left(\frac{2.5C_a I}{R}\right) W$$
$$= \left[\frac{(2.5)(0.32)(1.00)}{4.5}\right] W$$
$$= 0.178W$$

The calculated $V_{\text{design}} = 0.187W$ appears to be greater than $V_{\text{max}} = 0.178W$. Thus, the maximum allowable design base value for this warehouse is

$$V_{\text{design}} = 0.178W$$

(a) The roof weight is

$$W_{\text{roof}} = \frac{(120 \text{ ft})(240 \text{ ft})\left(15 \frac{\text{lbf}}{\text{ft}^2}\right)}{1000 \frac{\text{lbf}}{\text{kip}}} = 432 \text{ kips}$$

The east-west walls contribute to diaphragm loading. Concrete has a weight density of approximately 150 lbf/ft^3.

$$W_{\text{E-W walls}} = \frac{(10 \text{ ft})(2 \text{ walls})(120 \text{ ft})(12 \text{ in})\left(150 \frac{\text{lbf}}{\text{ft}^3}\right)}{\left(12 \frac{\text{in}}{\text{ft}}\right)\left(1000 \frac{\text{lbf}}{\text{kip}}\right)}$$

$$= 360 \text{ kips}$$

Only the top half of the east-west walls are effective in loading the diaphragm.

$$\frac{360 \text{ kips}}{2} = 180 \text{ kips}$$

Since this is a warehouse, a minimum of 25% of the live load must be added to the building weight. (See Sec. 6-29. It could also be argued that the storage sits on grade and does not add to the inertial mass.) Consider instead the case where racks are braced at the top and a portion of the live load is tributary to the roof.

$$W_{\text{live}} = (0.25)(120 \text{ ft})(240 \text{ ft})\left(200 \frac{\text{lbf}}{\text{ft}^2}\right)$$

$$= 1.44 \times 10^6 \text{ lbf} \ (1440 \text{ kips})$$

The base shear is

$$V = 0.178W$$
$$= (0.178)(432 \text{ kips} + 180 \text{ kips} + 1440 \text{ kips})$$
$$= 365 \text{ kips}$$

The shear force is resisted along two sides of the diaphragm, each 240 ft in length. The shear force per unit length along each of the north-south walls is

$$V = \frac{(365 \text{ kips})\left(1000 \frac{\text{lbf}}{\text{kip}}\right)}{(2 \text{ walls})\left(240 \frac{\text{ft}}{\text{wall}}\right)} = 760 \text{ lbf/ft}$$

From Table 12.1, 2-in nail spacing provides 820 lbf/ft of shear resistance.

(b)

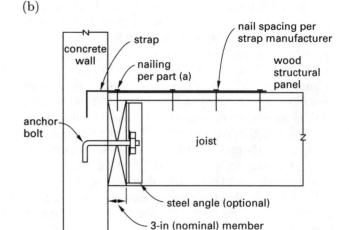

(c) The ledger is nominally 3 in thick. Doubling this gives 6 in. Since the actual thickness is less than the nominal thickness, use the $5\frac{1}{2}$-in row. The shear loading is parallel to the grain. Since the bolts are spaced every 3 ft, the shear load is

$$V_{\text{bolt}} = \left(3 \frac{\text{ft}}{\text{bolt}}\right)\left(760 \frac{\text{lbf}}{\text{ft}}\right) = 2280 \text{ lbf/bolt}$$

Doubling this in order to use the table in a concrete-wood connection, and multiplying by $\frac{3}{4}$ (the reciprocal of $\frac{4}{3}$) to reduce the seismic load to a normal duration load,

$$V = (2)\left(2280 \frac{\text{lbf}}{\text{bolt}}\right)\left(\tfrac{3}{4}\right) = 3420 \text{ lbf/bolt}$$

Use a 1-in diameter bolt that has a table strength of 4090 lbf/bolt. (See Table 12.7.)

(d) From Eq. 7.9, the maximum chord force along the short walls will be

$$C = \frac{F_{\text{diaphragm}}L}{8b}$$

$$= \frac{(365 \text{ kips})\left(1000 \frac{\text{lbf}}{\text{kip}}\right)(120 \text{ ft})}{(8)(240 \text{ ft})}$$

$$= 22{,}800 \text{ lbf}$$

Since the allowable stress for grade 60 rebar is 24,000 psi (see Sec. 7-18), the required bar size is given by Eq. 7.12.

$$A = \frac{C}{\frac{4}{3} \times \text{allowable tensile stress}}$$

$$= \frac{22{,}800 \text{ lbf}}{\left(\frac{4}{3}\right)\left(24{,}000 \frac{\text{lbf}}{\text{in}^2}\right)} = 0.713 \text{ in}^2$$

Use a no. 8 bar (diameter, 1 in; area, 0.78 in^2).

(e) The chord should be located in the plane of the diaphragm. See Fig. 12.2.

SI Solution

From Table 6.3, $Z = 0.40$. From Table 6.5, $I = 1.00$. From Table 6.10 for a concrete shear wall (bearing wall) system, the response modification factor $R = 4.5$. From Table 6.9, for soil-profile type S_A and for seismic zone factor $Z = 0.40$, the applicable acceleration and velocity-controlled seismic response coefficients are

$$C_a = 0.32N_a$$
$$C_v = 0.32N_v$$

From Table 6.8, for this site 15 km from a potential seismic source type A, the applicable near-source factors, N_a and N_v, are both equal to 1.00. Therefore,

$$C_a = C_v = (0.32)(1.0)$$
$$= 0.32$$

Thus, from Eq. 6.22, the total design base shear is

$$V = \left(\frac{C_v I}{RT}\right) W$$
$$= \left[\frac{(0.32)(1.00)}{(4.5)(0.38)}\right] W$$
$$= 0.187W$$

V_{design} should be greater than the minimum total design base shear obtained from Eq. 6.24 [UBC-97 Sec. 1630.2.1, Formula 30-6].

$$V_{\text{min}} = 0.11C_a IW$$
$$= (0.11)(0.32)(1.00)W$$
$$= 0.035W$$

For seismic zone 4, the above V_{min} should be further limited according to Eq. 6.25 [UBC-97 Sec. 1630.2.1, Formula 30-7].

$$V_{\text{min,zone 4}} = \left(\frac{0.8ZN_v I}{R}\right) W$$
$$= \left[\frac{(0.8)(0.4)(1.0)(1.0)}{4.5}\right] W$$
$$= 0.071W$$

The calculated $V_{\text{design}} = 0.187W$ can be used since it exceeds the controlling value of $V_{\text{min,zone 4}} = 0.071W$. Based on the UBC-97 [Sec. 1630.2.1], however, this design base shear value should not exceed the maximum total design base shear calculated from Eq. 6.23 [UBC-97 Formula 30-5]. V_{max} is

$$V_{\text{max}} = \left(\frac{2.5C_a I}{R}\right) W$$
$$= \left[\frac{(2.5)(0.32)(1.00)}{4.5}\right] W$$
$$= 0.178W$$

The calculated $V_{\text{design}} = 0.187W$ appears to be greater than $V_{\text{max}} = 0.178W$. Thus, the allowable design base value for this warehouse is

$$V_{\text{design}} = 0.178W$$

(a) The roof weight is

$$W_{\text{roof}} = (36.58 \text{ m})(73.15 \text{ m})\left(0.7 \frac{\text{kN}}{\text{m}^2}\right) = 1873 \text{ kN}$$

The east-west walls contribute to diaphragm loading. Concrete has a density of approximately 2400 kg/m^3.

$$W_{\text{E-W walls}} = \frac{\begin{array}{c}(3.05 \text{ m})(2 \text{ walls})(36.58 \text{ m}) \\ \times (305 \text{ mm})\left(2400 \frac{\text{kg}}{\text{m}^3}\right)\left(9.81 \frac{\text{m}}{\text{s}^2}\right)\end{array}}{\left(1000 \frac{\text{mm}}{\text{m}}\right)\left(1000 \frac{\text{N}}{\text{kN}}\right)}$$
$$= 1602 \text{ kN}$$

Only the top half of the east-west walls are effective in loading the diaphragm.

$$\frac{1602 \text{ kN}}{2} = 801 \text{ kN}$$

Since this is a warehouse, a minimum of 25% of the live load must be added to the building weight. (See Sec. 6-29. It could also be argued that the storage sits on grade and does not add to the inertial mass.) Consider instead the case where racks are braced at the top and a portion of the live load is tributary to the roof.

$$W_{\text{live}} = (0.25)(36.58 \text{ m})(73.15 \text{ m})\left(9.6 \frac{\text{kN}}{\text{m}^2}\right)$$
$$= 6422 \text{ kN}$$

The base shear is

$$V = 0.178W$$
$$= (0.178)(1873 \text{ kN} + 801 \text{ kN} + 6422 \text{ kN})$$
$$= 1619 \text{ kN}$$

The shear force is resisted along two sides of the diaphragm, each 73.15 m in length. The shear force per unit length along each of the north-south walls is

$$V = \frac{(1619 \text{ kN}) \left(1000 \frac{\text{N}}{\text{kN}}\right)}{(2 \text{ walls}) \left(73.15 \frac{\text{m}}{\text{wall}}\right)}$$
$$= 11\,100 \text{ N/m} \quad (11.1 \text{ N/mm})$$

From Table 12.1, 50.8 mm nail spacing provides 11.97 N/mm of shear resistance.

(b)

(c) The ledger is nominally 76 mm thick. Doubling this gives 152 mm. Since the actual thickness is less than the nominal thickness, use the 140-mm row. The shear loading is parallel to the grain. Since the bolts are spaced every 91.4 cm, the shear load is

$$V = \frac{\left(91.4 \frac{\text{cm}}{\text{bolt}}\right) \left(11\,100 \frac{\text{N}}{\text{m}}\right)}{\left(100 \frac{\text{cm}}{\text{m}}\right)} \approx 10\,100 \text{ N/bolt}$$

Doubling this in order to use the table in a concrete-wood connection, and multiplying by $\frac{3}{4}$ (the reciprocal of $\frac{4}{3}$) to reduce the seismic load to a normal duration load,

$$V = (2) \left(10\,100 \frac{\text{N}}{\text{bolt}}\right) \left(\tfrac{3}{4}\right) = 15\,200 \text{ N/bolt}$$

Use a 25-mm diameter bolt that has a table strength of 18 200 N/bolt. (See Table 12.7.)

(d) From Eq. 7.9, the maximum chord force along the short walls will be

$$C = \frac{F_{\text{diaphragm}} L}{8b}$$
$$= \frac{(1619 \text{ kN}) \left(1000 \frac{\text{N}}{\text{kN}}\right) (36.58 \text{ m})}{(8)(73.15 \text{ m})}$$
$$= 101\,000 \text{ N}$$

Since the allowable stress for grade 60 rebar is 165.5 MPa (see Sec. 7-18), the required bar size is given by Eq. 7.12.

$$A = \frac{C}{\frac{4}{3} \times \text{allowable tensile stress}}$$
$$= \frac{101\,000 \text{ N}}{\left(\frac{4}{3}\right) (165.5 \text{ MPa}) \left(1\,000\,000 \frac{\text{Pa}}{\text{MPa}}\right)}$$
$$= 4.6 \times 10^{-4} \text{ m}^2 = 460 \text{ mm}^2$$

Use a no. 8 bar (diameter, 25 mm; area, 503 mm^2).

(e) The chord should be located in the plane of the diaphragm. See Fig. 12.2.

15-9 QUALITATIVE PROBLEMS

1. List the numbers corresponding to the types of individuals who have the legal authority in California to perform the functions listed.

1) any licensed civil engineer

2) civil engineer licensed before January 1, 1982

3) civil engineer licensed after January 1, 1982

4) licensed civil engineer specializing in structures

5) licensed (California) soils engineer

6) licensed structural engineer

7) engineer licensed in any field

8) unlicensed civil engineer specializing in structures, under the responsible charge of a licensed civil engineer

9) any unlicensed civil engineer

10) licensed architect

11) licensed land surveyor

12) licensed photogrammetric surveyor

13) licensed contractor

14) licensed building designer

15) any member of the general public

16) no one

(a) use the title "consulting engineer" publicly

(b) use the title "civil engineer" publicly

(c) use the title "structural engineer" publicly

(d) use the title "soils engineer" publicly

(e) use the title "land surveyor" publicly

(f) personally perform civil engineering work

(g) solicit civil engineering work for others

(h) perform architectural work

(i) sign, stamp, and seal civil engineering design plans developed by an unlicensed engineering subordinate under the individual's direct engineering control

(j) sign, stamp, and seal civil engineering design plans developed by an unlicensed engineer in the individual's company whose paycheck the individual signs

(k) sign, stamp, and seal civil engineering design plans developed by a qualified, unlicensed, moonlighting engineer who pays the individual

(l) sign, stamp, and seal civil engineering design plans developed by a qualified, unlicensed, engineer who does not pay the individual

(m) allow a qualified, unlicensed person to use a registered civil engineer's stamp, seal, or registration number

(n) design a single-story, wood-framed residence

(o) design a two-story, wood-framed residence

(p) design a five-story, concrete-framed building

(q) design a five-story, steel-framed building

(r) design an above-ground water tower structure

(s) design a hospital building

(t) design a new public school building

(u) design a new private school building

(v) design a steel bridge

(w) inspect an earthquake-damaged building within 30 days of the event, without payment, when requested by the local building official

(x) supervise the construction of designed structures

(y) perform land surveying work for hire

(z) solicit land surveying work for others

(aa) perform a survey of public lands

(bb) perform a survey of private lands to be subdivided

(cc) lay out a construction site using surveying knowledge, methods, and equipment

(dd) gather in the field information to be placed on a deed or record-of-survey map

(ee) file a record-of-survey map with the county

(ff) administer oaths, certify oaths, and take testimony under oath to identify lost corners

Answers

Unless noted otherwise, references in parentheses are to the California Business and Professions Code, Chapters 7 (Professional Engineers) and 15 (Professional Land Surveyors).

(a) 1, 5, 6, 7, 12 (6704, 6732)

(b) 1, 5, 6 (6704, 6732, 6734)

(c) 6 (6703, 6704, 6732, 6736)

(d) 5 (6704, 6732, 6736.1, 6763)

(e) 11 (6731, 8708, 8725, 8731)

(f) 1, 5, 6 (6731.2, 8726.1)

(g) 15

(h) 10 (6737)

(i) 1, 5, 6 (6730.2, 6735, 6740)

(j) 16 (6703, 6735)

(k) 16 (6703, 6735)

(l) 16 (6703, 6735)

(m) 16 (6732, 6735)

(n) 15 (6737, 6737.1)

(o) 15 (6737, 6737.1)

(p) 1, 5, 6, 8, 10 (6737)

(q) 1, 5, 6, 8, 10 (6737)

(r) 1, 5, 6, 8, 10 (6737)

(s) 6 (Refer to Health and Safety Code, Div. 12.5, Chap. 1 Hospitals, Sec. 15048), 10

(t) 6, 10

(u) 6, 10 (6737)

(v) 1, 5, 6, 8

(w) 1, 5, 6, 7 (6706)

(x) 1, 5, 6, 10, 13 (6731, 6731.3, 6735.1)

(y) 2, 11, 12 (6731.2, 8726.1, 8731, 8775)

(z) 15

(aa) 2, 11, 12 (8708, 8731, 8775)

(bb) 2, 11, 12 (8708, 8726, 8731, 8775)

(cc) 12, 15

(dd) 2, 11, 12 (8726, 8731, 8775)

(ee) 2, 11, 12 (8731, 8762, 8775)

(ff) 1, 2, 5, 6, 11, 12 (8760, 8775)

2. How does the UBC-97 cover the design of bridges?

Answer

The UBC-97 covers only the design of buildings and some other building-like structures. Design of bridges is not covered in the UBC-97. This subject is covered in CALTRANS and American Association of State and Highway Transportation Officials (AASHTO) publications.

3. What provisions does the UBC-97 make for buildings subject to landslides, liquefaction, subsidence, gross differential settlement, or for those built close to a major ground-breaking fault?

Answer

None. The UBC-97 provides rules for the design of buildings that will resist typical ground shaking. The UBC-97 assumes the engineer will use good judgment in avoiding inherently dangerous locations.

4. What type of information will generally be supplied by the geotechnical engineer working on a building design team?

Answer

The geotechnical engineer will determine the (a) type of soil (i.e., sand, clay, or rock); (b) depth of water table; (c) depth to bedrock; (d) proximity to a fault; (e) maximum credible earthquake; and (f) likelihood of liquefaction, slides, subsidence, and differential settlement.

5. Which structural system resists lateral loads by flexure in members and joints?

Answer

Only moment-resisting frames resist lateral loads in this manner.

6. What is the meaning of the term "secondary stress" as it relates to a moment-resisting frame?

Answer

Primary stresses are the compressive and tensile forces that act uniformly on the cross section of the member. Secondary stresses are bending stresses that result from distortion of the frame when resisting lateral loads by flexure.

7. (a) What structural elements transfer lateral loads to vertical elements? (b) What structural elements transfer lateral loads to lower levels and the foundation?

Answer

(a) Horizontal elements such as diaphragms, horizontal bracing, and beams in moment-resisting frames transfer horizontal loads to vertical elements. (b) Vertical elements such as shear walls, braced frames, and columns in moment-resisting frames transfer lateral loads to lower levels.

8. A building site with soil-profile S_B is located 2.17 mi (3.5 km) from a potential seismic source type A. In calculating the seismic source of the building,

(a) What is the maximum moment magnitude, M, for seismic source type A?

(b) What acceleration-controlled near-source factor should be used?

(c) What velocity-controlled near-source factor should be used?

Answer

(a) For seismic source type A, the maximum value of the moment magnitude is equal to or greater than 7.0. (See Table 6.7.)

(b) The values of the near-source factors that are acceleration-controlled (N_a) vary based on the seismic source type and the closest distance to known seismic source. From Table 6.8 [UBC-97 Table 16-S], N_a is equal to 1.35 based on the linear interpolation of values (1.5 and 1.2) for distances other than those indicated in the table.

(c) N_v represents velocity-controlled near-source factor. N_v values also vary depending on the seismic source type and the closest distance to a known fault. From Table 6.8 [UBC-97 Table 16-T], N_v is equal to 1.8 based on the linear interpolation of values of 2.0 and 1.6 for the seismic source type A because the indicated 2.2 mi (3.5 km) distance is not shown in the table.

9. A site contains soil that is vulnerable to potential failure or collapse under seismic loading.

(a) What types of soils are vulnerable to potential failure or collapse under seismic loading?

(b) What is the soil-profile type for this site?

(c) Should a site-specific evaluation be conducted for this site?

(d) Under what circumstance can soil-profile type S_D be assigned to this site?

Answer

(a) Under seismic loading, liquefiable soil, quick and highly sensitive clays, collapsible weakly cemented soils, peats and highly organic clays of 10 ft (305 cm) or more in thickness, very high plasticity clays of 25 ft (762 cm) or more in thickness and having plasticity index greater than 75, and very thick soft/medium stiff clays of 120 ft (36.6 m) or more in thickness are types of soils vulnerable to potential failure or collapse [UBC-97 Sec. 1636.2].

(b) According to the UBC-97 [Sec. 1636.2], this site corresponds to the definition of soil-profile type S_F.

(c) Based on the criteria given, soil-profile type S_F requires site-specific evaluation [UBC-97 Sec. 1629.3.1].

(d) Each site should be assigned a soil-profile type according to appropriately documented geotechnical data. The UBC-97 site categorization procedure of Sec. 1636, Div. V, and Table 16-J should be used for that purpose. In determining the soil-profile type when the soil properties are not identified in adequate detail, the soil-profile type S_D can be presumed. The building official or others may "determine" that the site contains soil-profile type S_E or S_F.

10. What is the meaning of the term "soft (weak) story"? Give an example.

Answer

A soft story does not have as much lateral force resistance as the stories above. An example is a moment-resisting frame supported by long columns over an open plaza below.

11. What restriction does the UBC-97 place on situations where the type of structural system is different for different levels of a multistory building? What is the exception?

Answer

The value of R used in the design of one level must be less than or equal to the value of R used to design the level above. An exception is where the story above constitutes less than 10% of the total structure weight (i.e., is very light) [UBC-97 Sec. 1630.4.2].

12. In determining the total design lateral seismic force (F_p) on elements of structures, nonstructural components, and equipment supported by structures, the UBC-97 has introduced the a_p coefficient.

(a) What does this coefficient represent?

(b) How are the values of this coefficient obtained?

(c) When determining the anchorage force of a concrete or masonry wall to a flexible diaphragm, what value for the a_p coefficient should be applied?

Answer

(a) a_p is a numerical coefficient representing the in-structure component amplification factor that varies from 1.0 to 2.5. The minimum value of this factor is equal to 1.0. (See Sec. 6-44.)

(b) Table 6.14 [UBC-97 Table 16-O] provides values for this coefficient. Dynamic properties or empirical data of the component and the structure that supports it can determine this factor as well.

(c) Based on the UBC-97 [Sec. 1633.2.8.1], in seismic zones 3 and 4 where flexible diaphragms provide lateral support for the walls, elements of the wall anchorage system should be designed using an a_p factor equal to 1.5.

13. What is the meaning of the term "irregular building"? Give two examples.

Answer

For the purpose of the UBC-97, an irregular building meets one or more of the characteristics in UBC-97 Table 16-L. Examples are (1) a three-story, L-shaped building and (2) a five-story, square building with an open plaza comprising 60% of the floor area on level 3. (See Sec. 6-23.)

14. What is the meaning of the word "pounding"?

Answer

Pounding refers to adjacent buildings coming into contact with each other. One building can sway into another and pound it. The danger is greater when floor slabs of one building pound the columns of another; the danger is less when the floor slabs are at the same elevation. Up to 20% of the building failures in the 1985 Mexico City earthquake are thought to have been caused by pounding. Some of the buildings damaged in the 1989 Loma Prieta earthquake in the Watsonville-Santa Cruz area were only 6 in (152 mm) apart and were damaged because they pounded each other. (See Sec. 6-41.)

15. Consider determining the seismic force on a building using the UBC-97's simplified static lateral-force procedure.

(a) Which structures qualify for use of this design method?

(b) When using this design procedure for structures in seismic zones 3 and 4, what soil-profile type should be used assuming the soil properties are not known in sufficient detail?

(c) When using this design method in seismic zone 4, what are the limitations for the values of near-source factor N_a?

Answer

(a) The simplified static lateral-force procedure is given in the UBC-97 [Sec. 1630.2.3]. This design method can be used for light-frame structures of any occupancy with the maximum height of three stories excluding basements. Single-family dwellings can be included when they conform to this criterion. Other structures that are designated as standard occupancy (category 4) and miscellaneous occupancy (category 5) can also be designed by this procedure when they are no more than two stories in height excluding basements. (See Sec. 6-34.)

(b) In determining the seismic force on a structure in seismic zones 3 and 4 using the UBC-97's simplified static lateral-force procedure, soil-profile type S_D should be used when the soil properties are not known in sufficient detail to classify the soil-profile type. In seismic zones 1, 2A, and 2B, however, soil-profile type S_E should be used when the above condition exists.

(c) With this procedure in seismic zone 4, the values of near-source factor N_a need not be greater than 1.3 unless types 1, 4, or 5 of UBC-97 Table 16-L (vertical structural irregularities), or types 1 or 4 of UBC-97 Table 16-M (plan structural irregularities) are present. (See Table 6.11.)

16. When can the UBC-97's dynamic analysis method be used to determine the seismic force on a building? When can it not?

Answer

The dynamic method described by the UBC-97 [Sec. 1631] can always be used. It is the static method that is limited and that must satisfy certain conditions [UBC-97 Sec. 1629.8]. (See Sec. 6-33.)

17. What is the maximum span-to-width ratio for a wood structural panel roof diaphragm?

Answer

The maximum span-to-width ratio for a roof diaphragm is 4:1 [UBC-97 Table 23-II-G].

18. When using ASD, what factor should be applied to the dead load when designing for overturning effects caused by earthquake forces?

Answer

Every structure should be designed to resist the overturning effects caused by earthquake forces. Based on the UBC-97 [Secs. 1630.8 and 1612.4], when designing for overturning effects, a factor of 0.9 should be applied to the dead load when using ASD.

19. It is generally stated and understood that flexible diaphragms cannot transmit torsional shear stress to vertical resisting elements. Is this true for a flexible diaphragm that is cantilevered off of a vertical wall?

Answer

This is a tricky question. Any eccentric mass can cause torsion. A cantilevered flexible diaphragm, when acted upon by a seismic force perpendicular to its cantilevered dimension, will cause the wall to twist. However, this is different than transmitting torsion caused by one component to another. A cantilevered flexible diaphragm can cause torsion; it cannot transmit torsion.

20. It is generally stated and understood that the lateral loads resisted by vertical elements attached to rigid diaphragms are proportional to the element rigidities, and the lateral loads resisted by vertical elements attached to flexible diaphragms are proportional to tributary areas. How are lateral loads resisted by closely placed vertical elements that are arranged in-line, are parallel to an earthquake's motion, and are attached to a single flexible diaphragm?

Answer

Since all of the elements have the same tributary area, they will resist the lateral load in proportion to their relative rigidities.

21. What is the maximum allowable height-to-width ratio for a vertical wood structural shear wall panel?

Answer

According to UBC-97 Table 23-II-G, the maximum allowable height-to-width ratio for a wood structural shear wall panel is 2:1.

22. What are the minimum and maximum limits on force F_{px} that floors and diaphragms should be designed for?

Answer

Based on the UBC-97 [Sec. 1633.2.9], floors and diaphragms should be designed to resist forces (F_{px}) calculated from Eq. 7.5 [UBC-97 Formula 33-1]. F_{px} should not be less than $0.5C_aIw_{px}$, but F_{px} need not exceed $1.0C_aIw_{px}$.

23. Which is more life-threatening: shear cracking in a seismically-detailed concrete column or flexural cracking of a seismically-detailed concrete shear wall?

Answer

Cracking in a shear wall is probably more serious than cracking in a column. A properly-detailed column should not lose its loadbearing capacity merely because of cracking. The strict seismic detailing is intended to keep concrete in a column intact and confined even if it cracks. However, such confinement is not as complete in shear walls.

24. A tank on the roof of a building contains hazardous chemicals.

(a) What importance value, I, should be used in calculating the seismic force on the building?

(b) What importance value, I_p, should be used in calculating the seismic force on the tank anchorage?

(c) What importance value, I, should be used in calculating the seismic force on the roof diaphragm-to-wall connectors?

Answer

(a) From Table 6.5 [UBC-97 Table 16-K], $I = 1.25$.

(b) From Table 6.5 [UBC-97 Table 16-K], $I_p = 1.50$.

(c) From Table 6.5 [UBC-97 Table 16-K], $I = 1.25$. The 1.50 value only applies to the tank and its connections.

25. What is the basic distinction between ordinary and special moment-resisting frames?

Answer

A special moment-resisting frame has been carefully detailed to remain ductile. An ordinary moment-resisting frame does not have this detailing. (See Sec. 6-21.)

26. Two buildings have the same mass, but one building has a shorter natural period than the other building.

All other factors being equal, which building will experience the larger seismic force?

Answer

Most response spectra show that the lower the natural period, the higher the acceleration experienced by the building. Therefore, the building with the shorter period will probably experience the larger seismic force.

27. In UBC-97 seismic zone 4, which material is most likely to be less expensive when building a 30-story moment-resisting frame: steel or concrete?

Answer

This is a controversial question whose answer may depend on loyalties. However, more high-rise buildings seem to be built out of steel than out of concrete. All things being equal, steel is probably less expensive.

28. What consideration should be given to the design of a building that resists lateral force by a combination of braced frame and shear wall action?

Answer

Braced frames and shear walls have different stiffnesses and may deflect different amounts. This will cause a separation where the two resisting systems meet. The resisting elements must be proportioned so that the deflections are equal for both resisting systems.

29. In an extreme earthquake, what type of fascia would sustain the most damage: glass or concrete?

Answer

This is a fairly vague question since only the type of fascia material (and not the mounting method) is indicated. Glass has no ductility, so glass fascia probably would not fare well in an extreme earthquake. Concrete fascia would probably have been cast with continuous bar or mesh reinforcing. This reinforcing would help the concrete fascia remain intact when flexed.

30. For small buildings with only one or two floors, which of the different structural systems are more cost-effective? (Limit your discussion to wood structural panel shear wall construction, masonry wall systems, steel braced frames, stiff-redundant steel systems, concrete moment-resisting frames, steel moment-resisting frames, and dual systems.)

Answer

Small buildings with only one or two floors can be built using any of the structural systems listed, although dual, redundant, and moment-resisting frame systems probably would not be used. The systems in order of increasing cost are

1. wood structural panel shear wall construction

2. masonry wall systems

3. ordinary steel braced frames

4. dual systems

5. stiff redundant steel systems

6. concrete moment-resisting frames

7. steel moment-resisting frames

31. For tall buildings with more than ten floors, which of the different structural systems are more cost-effective? (Limit your discussion to wood structural panel shear wall construction, masonry shear wall (box) systems, ordinary steel braced frames, stiff-redundant steel systems, concrete moment-resisting frames, steel moment-resisting frames, and dual systems.)

Answer

Wood structural panel, masonry, and dual systems would not be used for a building with ten floors. Exceptionally tall buildings must be built either exceptionally stiff (e.g., the Empire State Building) or must use moment-resisting frames. Most modern tall buildings in California are constructed of steel. The logical conclusion is that these are less expensive than concrete buildings. Stiffness achieved through multiple redundancy is the most expensive. Though necessary in the early history of tall building construction, designing stiffness through redundancy is no longer practiced.

32. The floors in the top half of a tall multistory building are much smaller (in plan view) than the floors in the bottom half of the building.

(a) What are the problems associated with this design?

(b) How would you counteract the problems?

Answer

(a) This question is essentially about setbacks. The main problem is that the upper half would have a different period and different mode shape than the lower half. The upper floors could oscillate out-of-phase with the upper floors. This is referred to as "whipping action." Large stresses would be generated when the two sections were 180° out-of-phase. The stress would be most severe at the setback points.

(b) The upper half of the building must be designed so that, though smaller, it is as stiff or flexible as the lower half. There are many ways of increasing stiffness,

including adding bracing, changing the spacing or number of interior members, and increasing member sizes. (Since the mass of the upper stories is reduced, just keeping the column and beam sizes the same as in the lower stories would help.) In some cases, a different construction material could be used. It is not generally practical to add stiffness by starting new columns at an upper floor.

33. An air conditioning unit is placed on a wood structural panel roof diaphragm. What effect does the new unit have on the damping ratio of the roof?

Answer

None. The damping ratio of the roof is a function of the roof material, design, and quality of construction.

34. Draw simple diagrams that show how a three-story frame would fail in (a) beam-hinge mode, (b) column-hinge mode, and (c) soft-story (also known as "weak-story") mode. Show all plastic hinge points.

Answer

Plastic hinges are shown as solid bullets.

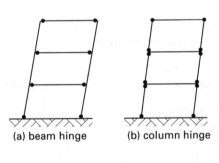

(a) beam hinge   (b) column hinge

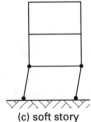

(c) soft story

35. Draw simple diagrams that show how a four-story frame would fail in (a) weak-column mode (i.e., when the beams were stronger than the columns) and (b) weak-beam mode (i.e., when the columns were stronger than the beams). Show all plastic hinge points.

Answer

Plastic hinges are shown as solid bullets.

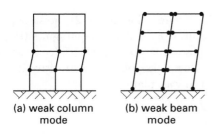

(a) weak column mode   (b) weak beam mode

36. Draw the connections necessary to anchor the floor diaphragms shown to the side of a CMU (concrete masonry unit) wall. Show and label all connectors and other elements. No calculations are necessary and no specific spacings need to be specified. Assume positive attachment to the wall is spaced approximately every 4 ft (102 mm).

(a) wood structural panel floor on 2× joists attached to 4 × 10 ledger

(b) wood structural panel supported directly by 4 × 10 ledger

(c) corrugated steel decking supported by steel ledger angle

(d) steel plate supported by steel ledger angle

(e) poured gypsum deck on metal form deck supported on steel ledger angle

Answer

(a)

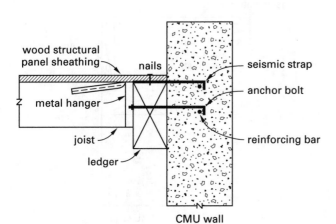

(b)

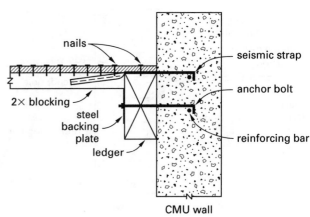

(c)

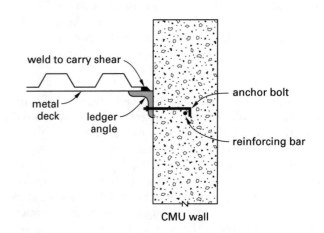

(d)

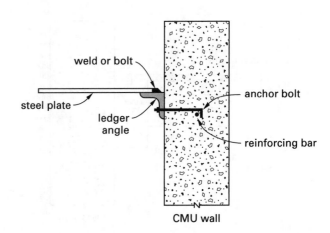

(e)

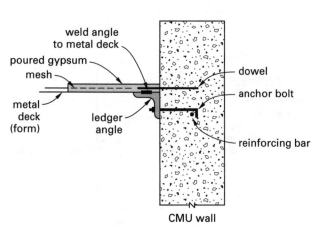

CMU wall

37. A wood structural panel diaphragm is supported by 2× joists. The joists are supported by a wood ledger attached to a CMU (concrete masonry unit) wall. Detail the connection between a joist and the ledger if the joist does not coincide with a wall anchor bolt or seismic strap. Assume the ledger strength is adequate.

Answer

This problem is somewhat contrived because if the application is really critical, the joint should be connected directly to the wall. The intent of this problem is to design a positive connection between the joist and ledger. In doing so, it must be recognized that (a) provisions must be made to avoid tension splits in the joist, and (b) connector pull-out strengths must be considered in attaching the joist to the ledger. Toe-nailing is obviously inadequate. If the joist is attached to the ledger with a commercial hanger having a row of vertical nails, the transmitted force will be limited by edge distance. The detail shown uses (a) nailing or bolting along the joist to avoid tension tear-out and (b) lag bolting to avoid connector pull-out.

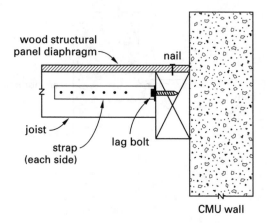

CMU wall

38. Detail a connection for a 2× joist supporting a wood structural panel floor diaphragm sitting directly on a wall plate on top of a CMU (concrete masonry unit) wall. How does your connection avoid cross-grain tension in the wall plate?

Answer

The connection shown avoids cross-grain tension in the wall plate by avoiding any connection to the wall plate. Lateral loads are transmitted in bearing through the anchor bolt. The wall plate remains in vertical compression at all times.

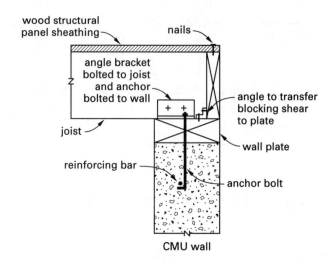

CMU wall

39. Describe how the connection between the joist and masonry unit wall shown may fail in cross-grain tension. How could you retain the basic design and eliminate the cross-grain tension?

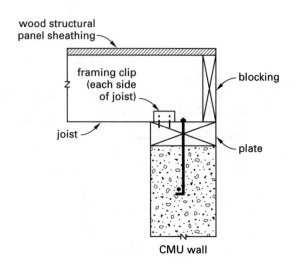

CMU wall

Answer

If the lateral load is from left to right, the plate will be placed in compression by the anchor bolt reaction acting to the left. However, if the lateral load is from right to left, the plate will be placed in cross-grain tension by the anchor bolt reaction acting to the right. This design can be "fixed" by adding a second framing clip to the right of the anchor bolt.

40. Illustrate how wood structural panel shear walls directly above each other on two different levels could be interconnected. (Exterior sheathing cannot be used to interconnect the walls.)

Answer

The most common method is to use a connecting tierod (tie down), as shown.

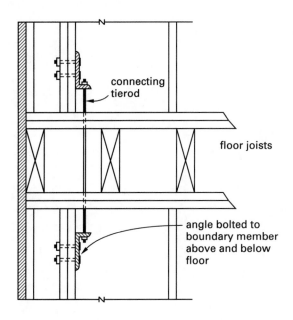

41. Draw and detail three types of column-girder joints for special moment-resisting frames constructed from steel.

Answer

The three types shown are (a) a butt-welded joint, (b) a fillet-welded joint, and (c) a bolted joint. Type (a) is basically the design that performed poorly in the Northridge earthquake.

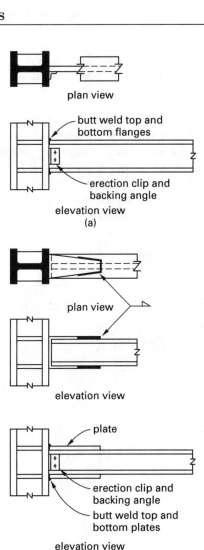

plan view

butt weld top and bottom flanges

erection clip and backing angle

elevation view
(a)

plan view

elevation view

plate

erection clip and backing angle

butt weld top and bottom plates

elevation view
(b)

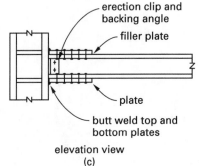

plan view

erection clip and backing angle

filler plate

plate

butt weld top and bottom plates

elevation view
(c)

42. Draw a typical section showing how the end of a
bridge section would be supported by an abutment.

Answer

The most important element in a bridge-to-abutment
connection is a positive connection that prevents the
two pieces from separating. Elements of secondary importance are the bearing, expansion joint, and shock-
absorbing element (i.e., the rubber rings). Secondary
cabling (not shown) may be provided as a back-up to
the primary positive connection.

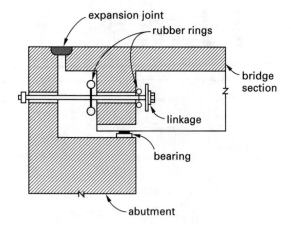

APPENDICES

APPENDIX A
Conversion Factors

Multiply	By	To Obtain	Multiply	By	To Obtain
acre	43,560	ft^2	kg	2.20462	lbm
Btu	778.17	ft-lbf	kg	0.06852	slug
Btu	1.055	kJ	kip	1000	lbf
Btu/h	0.293	W	kJ	0.9478	Btu
Btu/lbm	2.326	kJ/kg	kJ	737.56	ft-lbf
Btu/lbm-°R	4.1868	kJ/kg·K	kJ/kg	0.42992	Btu/lbm
cm	0.3937	in	kJ/kg·K	0.23885	Btu/lbm-°R
cm^3	0.061024	in^3	km	3280.8	ft
erg	7.376×10^{-8}	ft-lbf	km	0.6214	mi
ft	0.3048	m	km/h	0.62137	mi/h
ft^3	7.481	gal	kPa	0.14504	lbf/in^2
ft^3	0.028317	m^3	kW	737.6	ft-lbf/sec
ft-lbf	1.356×10^7	erg	kW	1.341	hp
ft-lbf	1.35582	J	L	0.03531	ft^3
ft/sec^2	0.0316	gravities	L	0.001	m^3
in/sec^2	0.002591	gravities	lbf	4.4482	N
gal	0.13368	ft^3	lbf/ft	14.5938	N/m
gal	3.7854×10^{-3}	m^3	lbf/ft^2	144	lbf/in^2
gal/min	0.002228	ft^3/sec	lbf/in^2	6894.8	Pa
g/cm^3	1000	kg/m^3	lbm	0.4536	kg
g/cm^3	62.428	lbm/ft^3	lbm/ft^3	0.016018	g/cm^3
gravities	32.2	ft/sec^2	lbm/ft^3	16.018	kg/m^3
gravities	386	in/sec^2	m	3.28083	ft
gravities	9.81	m/s^2	m^3	35.3147	ft^3
hp	2545	Btu/h	mm	0.03937	in
hp	33,000	ft-lbf/min	m/s^2	0.1019	gravities
hp	550	ft-lbf/sec	mi	1.609	km
hp	0.7457	kW	mi/h	1.6093	km/h
in	2.54	cm	N	0.22481	lbf
in	25.4	mm	Pa	1.4504×10^{-4}	lbf/in^2
in^3	16.387	cm^3	slug	32.174	lbm
J	0.73756	ft-lbf	W	3.413	Btu/h

APPENDIX B
Definitions

Base: The level at which the earthquake motions are imparted to the structure.

Bearing wall system: A structural system without a complete vertical load-carrying frame. In this system, the lateral forces are resisted by shear walls or braced frames.

Braced frame: A vertical truss system that is provided to resist lateral forces and in which the members are subjected primarily to axial stresses.

Confined concrete: Concrete confined by closely-spaced ties restraining it in a direction perpendicular to the applied stress.

Critical damping: The amount of damping that results in the system recovering from an initial deflection in the minimum amount of time, without an amplitude reversal.

Damping: The characteristic that reduces the vibrational energy, primarily by friction.

Dip: The angle that a stratum or fault makes with the horizontal.

Dip slip: The component of the slip parallel with the dip of the fault.

Ductile moment-resisting frame: A frame with rigid connections between columns and girders that is ductile at potential yielding points. See also *Moment-resisting frame*.

Elastic rebound theory: A seismic theory, based on the tectonic plate concept, that proposes that stresses are created in fault lines by shifting of the tectonic plates, and that faults resist motion until the accumulated stress overcomes the internal friction.

Equivalent static load: A single horizontal load, as defined in the UBC, for which an earthquake-resistant building should be designed.

Essential facility: A facility that must remain functional after a major earthquake.

Fault: A fracture or fracture zone along which the sides can move relative to one another and parallel to the fracture.

Fault creep: Continuous displacement along a fault at a slow and varying rate that is usually not accompanied by noticeable earthquakes.

Fault displacement: Relative movement of the two sides of a fault, measured in any specified direction (usually parallel to the fault).

Fault gouge: Filler material that forms between two plates sliding against each other.

Fault sag: A narrow tectonic (generally earth-filled) depression common in strike-slip fault zones, less than a few hundred feet wide and approximately parallel to the fault zone. See also *Sag pond*.

Fault scarp: A cliff or steep slope formed by displacement of the ground surface.

Fracture: A general term for a break, joint, or fault in the Earth.

Frame: A two-dimensional structural system without bearing walls that is composed of interconnected laterally-supported members and that functions as a self-contained unit.

Gouge: See *Fault gouge*.

Graben (plural, *graben*): A fault block, generally long and narrow, that has dropped down relative to the adjacent blocks.

Hoop: A one-piece closed tie or continuously wound tie that encloses longitudinal reinforcement.

Hypocenter: The actual location of the earthquake beneath the Earth's surface.

Igneous rock: Rocks formed by the solidification of molten magma.

Lateral force-resisting system: The part of the building that resists earthquake and wind forces.

Left-normal slip: Fault displacement consisting of nearly equal components of left and normal slips.

Left slip: Strike-slip displacement in which the block across the fault moves to the left.

Liquefaction: The loss of load-carrying ability in loose, usually saturated, soil or sand.

Moment-resisting frame: A vertical load-carrying frame in which the members and joints are capable of resisting forces primarily by flexure.

Normal fault: Any fault (including those with vertical slip) in which the block above an inclined surface moves downward relative to the block below the fault surface.

Normal slip: Vertical displacement of a fault.

Oblique slip: A combination of strike slip and reverse slip.

APPENDIX B (continued)
Definitions

Parapet: A low wall or railing (including decorative panels) extending, usually vertically, above the roof line.

Plastic hinge: A region where the yield moment strength of a flexural member is exceeded and that experiences significant rotation.

Reserve energy: Energy that a ductile system is capable of absorbing in the inelastic region.

Resonance: A condition existing when the frequency of excitation is the same as the natural frequency of the building or soil.

Reverse fault: A fault in which the block above an inclined fault surface moves upward relative to the block below the fault surface.

Right-normal slip: Fault displacement consisting of nearly equal components of right and normal slips.

Right slip: Strike-slip displacement in which the block across the fault from an observer moves to the right.

Rigid frame: A vertical load-carrying frame in which the members and joints resist forces by rotation and flexure. See also *Moment-resisting space frame*.

Sag pond: A fault sag that has filled with water.

Shear wall: A wall designed to resist lateral forces parallel to the plane of the wall.

Slip: The relative displacement, measured on the surface, of two points on opposite sides of a fault.

Space frame: A three-dimensional structural system without bearing walls that is composed of interconnected laterally supported members and that functions as a self-contained unit.

Special ductile frame: A structural frame designed to remain vertically functional after the formation of plastic hinges from reversed lateral displacements.

Special shear wall: A reinforced concrete shear wall designed and detailed in accordance with the special UBC provisions.

Stirrup tie: A closed stirrup completely encircling the longitudinal members of a beam or column and conforming to the definition of a hoop.

Strike: The horizontal direction or bearing of the fault on the surface.

Strike-slip: The horizontal component of slip, parallel to the strike of the fault.

Strike-slip fault: A fault in which the slip is approximately in the direction of the strike.

Supplementary crosstie: A tie with a standard 180-degree hook at each end.

Tectonic: Pertaining to or designating the internal and external rock structures and features caused from crustal and subcrustal activity deep in the Earth.

Tectonic creep: Fault creep of tectonic origin.

Transcurrent fault: See *Strike-slip fault*.

Wrench fault: See *Strike-slip fault*.

APPENDIX C
Chronology of Important California Earthquakes

Date	Fault	Richter Magnitude	Surface Effects and Significance
1836	Hayward	7.0 (est.)	Ground breakage
1838	San Andreas	7.0 (est.)	Ground breakage
1852	Big Pine		Possible ground breakage
1857	San Andreas	8.0 (est.)	Right-lateral slip, possibly as much as 30 ft (914 cm)
1861	Calaveras		Ground breakage
1868	San Andreas		Long fissure in earth at Dos Palmas
1868	Hayward	7.0 (est.)	Strike-slip
1872	Owens Valley zone	8.3 (est.)	Right-lateral slip of 16–20 ft (488–610 cm). Left-lateral slip may also have occurred. Vertical slip, down to east, of 23 ft (701 cm).
1890	San Andreas		Fissures in fault zone; railroad tracks and bridge displaced
1899	San Jacinto	6.6 (est.)	Possible surface evidence
1906	San Andreas	8.3	Known as the "San Francisco earthquake." Right-lateral slip up to 21 ft (640 cm). Resulted in the formation of the California State Earthquake Investigation Commission.
1922	San Andreas	6.5	Ground breakage
1925	Mesa/Santa Ynez	6.3	Known as the "Santa Barbara earthquake of 1925." U.S. Coast and Geodetic Society was directed to study the field of seismology.
1927	Santa Ynez	7.5	Occurred offshore in a submarine trench and was felt on land
1933	Newport-Inglewood	6.3	Known as the "Long Beach earthquake of 1933." Extensive property damage and loss of life. Many school buildings were destroyed. Resulted in the passage of the Field Act. The Division of Architecture of the State Department of Public Works was assigned responsibility to approve new buildings used for schools. The Riley Act was also passed, which set minimum requirements for lateral force design.
1934	San Andreas	6.0	Ground breakage
1934	San Jacinto	7.1	Ground breakage
1940	Imperial	7.1	Known as the "El Centro earthquake." 40 mi (64.4 km) of surface faulting. 80% of Imperial buildings were damaged. However, no Field Act school buildings were damaged. This was the first major earthquake to yield accelerograph data on building periods. A maximum acceleration (ground) of 0.33 g was experienced.
1947	Manix	6.4	Left-lateral slip of 3 in (76 mm)
1950	(unnamed)	5.6	Vertical slip, down to west, of 5–8 in (127–203 mm) along the west edge of Fort Sage Mountains
1951	Superstition Hills	5.6	Slight right-lateral slip
1952	White Wolf	7.7	Known as the "Kern County earthquake," and the "Arvin-Tehachapi earthquake." Extensive building damage to old buildings. Little or none to properly designed and Field Act buildings. Confirmed the requirement for proper design.
1956	San Miguel	6.8	Right-lateral slip, 3 ft (91 cm); vertical slip, down to southwest, 3 ft (91 cm)
1966	Imperial	3.6	Right-lateral slip, 1/2 in (13 mm)
1966	San Andreas	5.5	Known as the "Parkfield earthquake." Right-lateral slip of several inches. Maximum ground acceleration of 0.50 g—highest recorded to date.
1968	Coyote Creek	6.4	Right-lateral slip up to 15 in (381 mm)
1971	San Fernando	6.4–6.6	Known as the "San Fernando earthquake," or "Sylmar earthquake." Left-lateral slip up to 5 ft (152 cm); north-side thrusting up 3 ft (91 cm). Massive instrumentation due to 1965 Los Angeles building code resulted in more than 300 accelerograph plots. 1.24 g experienced at Pacoima dam.
1979	Imperial	6.6	Known as the "Imperial Valley earthquake." Right-lateral slip up to 21.6 in (55 cm) with more than 18.6 mi (30 km) of surface rupture. Extensive accelerograph data collected. Resulted in the first accelerograph from an extensively damaged building (Imperial County Services Building).
1987	Whittier	6.1	Known as the "Whittier earthquake." Epicenter 10 mi (16 km) east of downtown Los Angeles. 0.45 g maximum lateral acceleration; 0.20 g vertical acceleration typical. Strong shaking duration of 4 sec. Six fatalities; unreinforced masonry structures damaged significantly.
1989	Loma Prieta	7.1	Primarily noted as causing the collapse of the Oakland Interstate 880 Cypress structure and homes in the San Francisco Marina district, both due to soil amplification effects despite the large distance from the epicenter. Ground breakage at epicenter located in Santa Cruz Mountains; 62 fatalities.

APPENDIX C (continued)
Chronology of Important California Earthquakes

Date	Fault	Richter Magnitude	Surface Effects and Significance
1990	Upland	5.5	Acceleration of 0.23 g horizontally and 0.13 g vertically. Dozens of after-shocks. Occurred on the San Antonio Canyon fault, east of downtown Los Angeles. Most damage in Pomona; some damage to reinforced masonry structures.
1992	Yucca Valley	7.4	One fatality, hundreds of injuries. Buckled and displaced roads, damaged 1400 structures. 43-mi (69.2 km) ground rupture. Followed by magnitude 6.5 quake at Big Bear resort area.
1992	Cape Mendocino	7.0	Offshore quake notable for largest-yet recorded ground acceleration of 1.85 g.
1994	Northridge	6.6	Previously unknown thrust fault. 40 s of shaking. Extensive damage to sections of the Santa Monica freeway and freeway overpasses not yet retro-fitted. Unreinforced masonry buildings damaged, as expected. 60 fatalities, thousands of injuries. Numerous failures in steel moment-resisting connections.
2000	West Napa	5.2	Known as the "Yountville earthquake" and the "Napa earthquake." Source was 3 mi west of the West Napa fault on an unknown northwest-oriented, right-lateral stroke-slip fault. Ground acceleration approaching 0.5 g horizontally recorded in town was higher than expected from this magnitude and is attributed to amplification by young sediments along the Napa river.

APPENDIX D
Rigidity of Fixed Piers

(Calculated with $F = 100{,}000$ lbf ($445\,000$ N), $t = 1.0$ in (25 mm), and $E = 1{,}000{,}000$ psi (6.9×10^6 kPa).)

(h/d)	0.00	0.01	0.02	0.03	0.04	0.05	0.06	0.07	0.80	0.09
0.10	33.223	30.181	27.645	25.497	23.655	22.057	20.657	19.421	18.321	17.335
0.20	16.447	15.643	14.911	14.242	13.267	13.061	12.538	12.053	11.602	11.181
0.30	10.788	10.419	10.073	9.747	9.440	9.150	8.876	8.616	8.369	8.135
0.40	7.911	7.699	7.496	7.302	7.117	6.939	6.769	6.606	6.449	6.299
0.50	6.154	6.015	5.880	5.751	5.626	5.506	5.389	5.277	5.168	5.062
0.60	4.960	4.862	4.766	4.673	4.583	4.495	4.410	4.328	4.247	4.169
0.70	4.093	4.019	3.948	3.877	3.809	3.743	3.678	3.615	3.553	3.493
0.80	3.434	3.377	3.321	3.266	3.213	3.160	3.109	3.060	3.011	2.963
0.90	2.916	2.871	2.826	2.782	2.739	2.697	2.656	2.616	2.577	2.538
1.00	2.500	2.463	2.427	2.391	2.356	2.322	2.288	2.255	2.222	2.191
1.10	2.159	2.129	2.099	2.069	2.040	2.012	1.984	1.956	1.929	1.903
1.20	1.877	1.851	1.826	1.802	1.777	1.753	1.730	1.707	1.684	1.662
1.30	1.640	1.619	1.598	1.577	1.556	1.536	1.516	1.497	1.478	1.459
1.40	1.440	1.422	1.404	1.386	1.369	1.352	1.335	1.318	1.302	1.286
1.50	1.270	1.254	1.239	1.224	1.209	1.194	1.180	1.166	1.152	1.138
1.60	1.124	1.111	1.098	1.085	1.072	1.059	1.047	1.034	1.022	1.010
1.70	0.999	0.987	0.976	0.965	0.954	0.943	0.932	0.921	0.911	0.901
1.80	0.890	0.880	0.870	0.861	0.851	0.842	0.832	0.823	0.814	0.805
1.90	0.796	0.788	0.779	0.771	0.762	0.754	0.746	0.738	0.730	0.722
2.00	0.714	0.707	0.699	0.692	0.685	0.677	0.670	0.663	0.656	0.649
2.10	0.643	0.636	0.629	0.623	0.617	0.610	0.604	0.598	0.592	0.586
2.20	0.580	0.574	0.568	0.562	0.557	0.551	0.546	0.540	0.535	0.530
2.30	0.525	0.519	0.514	0.509	0.504	0.499	0.495	0.490	0.485	0.480
2.40	0.476	0.471	0.467	0.462	0.458	0.453	0.449	0.445	0.441	0.437
2.50	0.432	0.428	0.424	0.420	0.417	0.413	0.409	0.405	0.401	0.398
2.60	0.394	0.391	0.387	0.383	0.380	0.377	0.373	0.370	0.367	0.363
2.70	0.360	0.357	0.354	0.350	0.347	0.344	0.341	0.338	0.335	0.332
2.80	0.330	0.327	0.324	0.321	0.318	0.316	0.313	0.310	0.307	0.305
2.90	0.302	0.300	0.297	0.295	0.292	0.290	0.287	0.285	0.283	0.280
3.00	0.278	0.276	0.273	0.271	0.269	0.267	0.264	0.262	0.260	0.258
3.10	0.256	0.254	0.252	0.250	0.248	0.246	0.244	0.242	0.240	0.238
3.20	0.236	0.234	0.232	0.231	0.229	0.227	0.225	0.223	0.222	0.220
3.30	0.218	0.217	0.215	0.213	0.212	0.210	0.208	0.207	0.205	0.204
3.40	0.202	0.201	0.199	0.198	0.196	0.195	0.193	0.192	0.190	0.189
3.50	0.187	0.186	0.185	0.183	0.182	0.181	0.179	0.178	0.177	0.175
3.60	0.174	0.173	0.172	0.170	0.169	0.168	0.167	0.166	0.164	0.163
3.70	0.162	0.161	0.160	0.159	0.157	0.156	0.155	0.154	0.153	0.152
3.80	0.151	0.150	0.149	0.148	0.147	0.146	0.145	0.144	0.143	0.142
3.90	0.141	0.140	0.139	0.138	0.137	0.136	0.135	0.134	0.133	0.132
4.00	0.132	0.131	0.130	0.129	0.128	0.127	0.126	0.126	0.125	0.124
4.10	0.123	0.122	0.122	0.121	0.120	0.119	0.118	0.118	0.117	0.116
4.20	0.115	0.115	0.114	0.113	0.112	0.112	0.111	0.110	0.110	0.109
4.30	0.108	0.108	0.107	0.106	0.106	0.105	0.104	0.104	0.103	0.102
4.40	0.102	0.101	0.100	0.100	0.099	0.099	0.098	0.097	0.097	0.096
4.50	0.096	0.095	0.094	0.094	0.093	0.093	0.092	0.092	0.091	0.091
4.60	0.090	0.089	0.089	0.088	0.088	0.087	0.087	0.086	0.086	0.085
4.70	0.085	0.084	0.084	0.083	0.083	0.082	0.082	0.081	0.081	0.080
4.80	0.080	0.080	0.079	0.079	0.078	0.078	0.077	0.077	0.076	0.076
4.90	0.076	0.075	0.075	0.074	0.074	0.073	0.073	0.073	0.072	0.072
5.00	0.071	0.071	0.071	0.070	0.070	0.069	0.069	0.069	0.068	0.068
5.10	0.068	0.067	0.067	0.066	0.066	0.066	0.065	0.065	0.065	0.064
5.20	0.064	0.064	0.063	0.063	0.063	0.062	0.062	0.062	0.061	0.061
5.30	0.061	0.060	0.060	0.060	0.059	0.059	0.059	0.058	0.058	0.058
5.40	0.058	0.057	0.057	0.057	0.056	0.056	0.056	0.056	0.055	0.055
5.50	0.055	0.054	0.054	0.054	0.054	0.053	0.053	0.053	0.052	0.052
5.60	0.052	0.052	0.051	0.051	0.051	0.051	0.050	0.050	0.050	0.050
5.70	0.049	0.049	0.049	0.049	0.048	0.048	0.048	0.048	0.048	0.047
5.80	0.047	0.047	0.047	0.046	0.046	0.046	0.046	0.045	0.045	0.045
5.90	0.045	0.045	0.044	0.044	0.044	0.044	0.044	0.043	0.043	0.043
6.00	0.043	0.043	0.042	0.042	0.042	0.042	0.042	0.041	0.041	0.041
6.10	0.041	0.041	0.040	0.040	0.040	0.040	0.040	0.039	0.039	0.039
6.20	0.039	0.039	0.039	0.038	0.038	0.038	0.038	0.038	0.038	0.037
6.30	0.037	0.037	0.037	0.037	0.037	0.036	0.036	0.036	0.036	0.036
6.40	0.036	0.035	0.035	0.035	0.035	0.035	0.035	0.034	0.034	0.034
6.50	0.034	0.034	0.034	0.034	0.033	0.033	0.033	0.033	0.033	0.033
6.60	0.033	0.032	0.032	0.032	0.032	0.032	0.032	0.032	0.031	0.031
6.70	0.031	0.031	0.031	0.031	0.031	0.031	0.030	0.030	0.030	0.030
6.80	0.030	0.030	0.030	0.029	0.029	0.029	0.029	0.029	0.029	0.029
6.90	0.029	0.029	0.028	0.028	0.028	0.028	0.028	0.028	0.028	0.028
7.00	0.027	0.027	0.027	0.027	0.027	0.027	0.027	0.027	0.027	0.026
7.10	0.026	0.026	0.026	0.026	0.026	0.026	0.026	0.026	0.026	0.025
7.20	0.025	0.025	0.025	0.025	0.025	0.025	0.025	0.025	0.025	0.024
7.30	0.024	0.024	0.024	0.024	0.024	0.024	0.024	0.024	0.024	0.023
7.40	0.023	0.023	0.023	0.023	0.023	0.023	0.023	0.023	0.023	0.023
7.50	0.023	0.022	0.022	0.022	0.022	0.022	0.022	0.022	0.022	0.022
7.60	0.022	0.022	0.021	0.021	0.021	0.021	0.021	0.021	0.021	0.021
7.70	0.021	0.021	0.021	0.021	0.021	0.020	0.020	0.020	0.020	0.020
7.80	0.020	0.020	0.020	0.020	0.020	0.020	0.020	0.020	0.019	0.019
7.90	0.019	0.019	0.019	0.019	0.019	0.019	0.019	0.019	0.019	0.019
8.00	0.019	0.019	0.019	0.018	0.018	0.018	0.018	0.018	0.018	0.018
8.10	0.018	0.018	0.018	0.018	0.018	0.018	0.018	0.018	0.017	0.017
8.20	0.017	0.017	0.017	0.017	0.017	0.017	0.017	0.017	0.017	0.017

APPENDIX E
Rigidity of Cantilever Piers

(Calculated with $F = 100,000$ lbf $(445\,000$ N$)$, $t = 1.0$ in $(25$ mm$)$, and $E = 1,000,000$ psi $(6.9 \times 10^6$ kPa$)$.)

(h/d)	0.00	0.01	0.02	0.03	0.04	0.05	0.06	0.07	0.08	0.09
0.10	32.895	29.822	27.255	25.076	23.203	21.575	20.146	18.880	17.752	16.738
0.20	15.823	14.992	14.233	13.538	12.898	12.308	11.761	11.252	10.778	10.335
0.30	9.921	9.531	9.165	8.820	8.495	8.187	7.895	7.618	7.356	7.106
0.40	6.868	6.642	6.425	6.219	6.021	5.833	5.652	5.479	5.313	5.153
0.50	5.000	4.853	4.712	4.576	4.445	4.319	4.197	4.080	3.968	3.859
0.60	3.754	3.652	3.555	3.460	3.369	3.280	3.195	3.112	3.032	2.955
0.70	2.880	2.808	2.738	2.670	2.604	2.540	2.478	2.418	2.360	2.303
0.80	2.248	2.195	2.143	2.093	2.045	1.997	1.952	1.907	1.864	1.822
0.90	1.781	1.741	1.702	1.665	1.628	1.593	1.558	1.524	1.492	1.460
1.00	1.429	1.398	1.369	1.340	1.312	1.285	1.259	1.233	1.208	1.183
1.10	1.160	1.136	1.114	1.092	1.070	1.049	1.028	1.008	0.989	0.970
1.20	0.951	0.933	0.916	0.898	0.881	0.865	0.849	0.833	0.818	0.803
1.30	0.788	0.774	0.760	0.746	0.733	0.720	0.707	0.695	0.683	0.671
1.40	0.659	0.648	0.636	0.626	0.615	0.604	0.594	0.584	0.575	0.565
1.50	0.556	0.546	0.537	0.529	0.520	0.512	0.503	0.495	0.487	0.480
1.60	0.472	0.465	0.457	0.450	0.443	0.436	0.430	0.423	0.417	0.410
1.70	0.404	0.398	0.392	0.386	0.380	0.375	0.369	0.364	0.358	0.353
1.80	0.348	0.343	0.338	0.333	0.329	0.324	0.319	0.315	0.310	0.306
1.90	0.302	0.298	0.294	0.290	0.286	0.282	0.278	0.274	0.270	0.267
2.00	0.263	0.260	0.256	0.253	0.250	0.246	0.243	0.240	0.237	0.234
2.10	0.231	0.228	0.225	0.222	0.219	0.216	0.214	0.211	0.208	0.206
2.20	0.203	0.201	0.198	0.196	0.194	0.191	0.189	0.187	0.184	0.182
2.30	0.180	0.178	0.176	0.174	0.172	0.170	0.168	0.166	0.164	0.162
2.40	0.160	0.158	0.156	0.155	0.153	0.151	0.149	0.148	0.146	0.145
2.50	0.143	0.141	0.140	0.138	0.137	0.135	0.134	0.132	0.131	0.129
2.60	0.128	0.127	0.125	0.124	0.123	0.121	0.120	0.119	0.118	0.116
2.70	0.115	0.114	0.113	0.112	0.111	0.109	0.108	0.107	0.106	0.105
2.80	0.104	0.103	0.102	0.101	0.100	0.099	0.098	0.097	0.096	0.095
2.90	0.094	0.093	0.092	0.091	0.091	0.090	0.089	0.088	0.087	0.086
3.00	0.086	0.085	0.084	0.083	0.082	0.082	0.081	0.080	0.079	0.079
3.10	0.078	0.077	0.076	0.076	0.075	0.074	0.074	0.073	0.072	0.072
3.20	0.071	0.071	0.070	0.069	0.069	0.068	0.067	0.067	0.066	0.066
3.30	0.065	0.065	0.064	0.063	0.063	0.062	0.062	0.061	0.061	0.060
3.40	0.060	0.059	0.059	0.058	0.058	0.057	0.057	0.056	0.056	0.055
3.50	0.055	0.055	0.054	0.054	0.053	0.053	0.052	0.052	0.052	0.051
3.60	0.051	0.050	0.050	0.050	0.049	0.049	0.048	0.048	0.048	0.047
3.70	0.047	0.046	0.046	0.046	0.045	0.045	0.045	0.044	0.044	0.044
3.80	0.043	0.043	0.043	0.042	0.042	0.042	0.041	0.041	0.041	0.040
3.90	0.040	0.040	0.040	0.039	0.039	0.039	0.038	0.038	0.038	0.038
4.00	0.037	0.037	0.037	0.037	0.036	0.036	0.036	0.035	0.035	0.035
4.10	0.035	0.034	0.034	0.034	0.034	0.034	0.033	0.033	0.033	0.033
4.20	0.032	0.032	0.032	0.032	0.031	0.031	0.031	0.031	0.031	0.030
4.30	0.030	0.030	0.030	0.030	0.029	0.029	0.029	0.029	0.029	0.028
4.40	0.028	0.028	0.028	0.028	0.028	0.027	0.027	0.027	0.027	0.027
4.50	0.026	0.026	0.026	0.026	0.026	0.026	0.025	0.025	0.025	0.025
4.60	0.025	0.025	0.024	0.024	0.024	0.024	0.024	0.024	0.024	0.023
4.70	0.023	0.023	0.023	0.023	0.023	0.023	0.022	0.022	0.022	0.022
4.80	0.022	0.022	0.022	0.021	0.021	0.021	0.021	0.021	0.021	0.021
4.90	0.021	0.020	0.020	0.020	0.020	0.020	0.020	0.020	0.020	0.020
5.00	0.019	0.019	0.019	0.019	0.019	0.019	0.019	0.019	0.019	0.018
5.10	0.018	0.018	0.018	0.018	0.018	0.018	0.018	0.018	0.017	0.017
5.20	0.017	0.017	0.017	0.017	0.017	0.017	0.017	0.017	0.017	0.016
5.30	0.016	0.016	0.016	0.016	0.016	0.016	0.016	0.016	0.016	0.016
5.40	0.015	0.015	0.015	0.015	0.015	0.015	0.015	0.015	0.015	0.015
5.50	0.015	0.015	0.015	0.014	0.014	0.014	0.014	0.014	0.014	0.014
5.60	0.014	0.014	0.014	0.014	0.014	0.014	0.013	0.013	0.013	0.013
5.70	0.013	0.013	0.013	0.013	0.013	0.013	0.013	0.013	0.013	0.013
5.80	0.013	0.012	0.012	0.012	0.012	0.012	0.012	0.012	0.012	0.012
5.90	0.012	0.012	0.012	0.012	0.012	0.012	0.012	0.012	0.011	0.011
6.00	0.011	0.011	0.011	0.011	0.011	0.011	0.011	0.011	0.011	0.011
6.10	0.011	0.011	0.011	0.011	0.011	0.011	0.010	0.010	0.010	0.010
6.20	0.010	0.010	0.010	0.010	0.010	0.010	0.010	0.010	0.010	0.010
6.30	0.010	0.010	0.010	0.010	0.010	0.010	0.010	0.009	0.009	0.009
6.40	0.009	0.009	0.009	0.009	0.009	0.009	0.009	0.009	0.009	0.009
6.50	0.009	0.009	0.009	0.009	0.009	0.009	0.009	0.009	0.009	0.009
6.60	0.009	0.009	0.008	0.008	0.008	0.008	0.008	0.008	0.008	0.008
6.70	0.008	0.008	0.008	0.008	0.008	0.008	0.008	0.008	0.008	0.008
6.80	0.008	0.008	0.008	0.008	0.008	0.008	0.008	0.008	0.008	0.008
6.90	0.007	0.007	0.007	0.007	0.007	0.007	0.007	0.007	0.007	0.007
7.00	0.007	0.007	0.007	0.007	0.007	0.007	0.007	0.007	0.007	0.007
7.10	0.007	0.007	0.007	0.007	0.007	0.007	0.007	0.007	0.007	0.007
7.20	0.007	0.007	0.007	0.007	0.006	0.006	0.006	0.006	0.006	0.006
7.30	0.006	0.006	0.006	0.006	0.006	0.006	0.006	0.006	0.006	0.006
7.40	0.006	0.006	0.006	0.006	0.006	0.006	0.006	0.006	0.006	0.006
7.50	0.006	0.006	0.006	0.006	0.006	0.006	0.006	0.006	0.006	0.006
7.60	0.006	0.006	0.006	0.006	0.006	0.006	0.005	0.005	0.005	0.005
7.70	0.005	0.005	0.005	0.005	0.005	0.005	0.005	0.005	0.005	0.005
7.80	0.005	0.005	0.005	0.005	0.005	0.005	0.005	0.005	0.005	0.005
7.90	0.005	0.005	0.005	0.005	0.005	0.005	0.005	0.005	0.005	0.005
8.00	0.005	0.005	0.005	0.005	0.005	0.005	0.005	0.005	0.005	0.005
8.10	0.005	0.005	0.005	0.005	0.005	0.005	0.005	0.005	0.005	0.005
8.20	0.004	0.004	0.004	0.004	0.004	0.004	0.004	0.004	0.004	0.004

APPENDIX F
Accelerogram of 1940 El Centro Earthquake
(North-South Component)

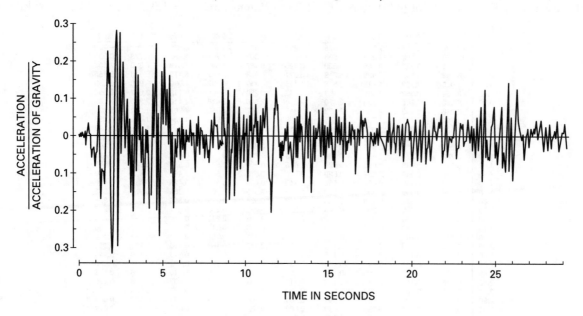

Source: Donald E. Hudson, "Ground Motion Measurements," in *Earthquake Engineering*, Robert L. Wiegel, ed., © 1970, p. 113.

APPENDIX G
Seismic Acronyms

AASHTO	American Association of State Highway Transportation Officials
ACI	American Concrete Institute
ADAS	added damping and stiffness (element)
AFM	Air Force Manual (military)
AISC	American Institute of Steel Construction
ANSI	American National Standards Institute
API	American Petroleum Institute
ARS	Acceleration Response Spectrum
ASCE	American Society of Civil Engineers
ASD	allowable stress design
ASTM	American Society for Testing and Materials
AITC	American Institute of Timber Construction
ATC	Applied Technology Council
AWWA	American Water Works Association
AZG	acceleration zone graph
BOCA	Building Officials and Code Administrators International
CALTRANS	California Department of Transportation
CIP	cast-in-place
CMU	concrete masonry unit
CQC	complete quadratic combination
DAF	dynamic amplification factor
DM	design manual (military)
EBF	eccentrically braced frame
EPA	effective peak ground acceleration
EPV	effective peak ground velocity
EQ	earthquake
EUS	eastern United States
FEA	finite element analysis
FEMA	Federal Emergency Management Agency
FP	full-penetration (welding)
HVAC	heating ventilating and air conditioning
IBC	International Building Code
IC	important classification
ICBO	International Conference of Building Officials
ICC	International Code Council
IDRS	inelastic design response spectra
IEEE	Institute of Electrical and Electronic Engineers
IMRF	intermediate moment-resisting frame
IMRSF	intermediate moment-resisting space frame (obsolete)
LRFD	load resistance factor design
MDOF	multiple degree of freedom
MM	modified Mercalli (intensity)
MMRWF	masonry moment-resisting wall frame
NAVFAC	Naval Facilities Engineering Command
NBC	National Building Code
NCMA	National Concrete Masonry Association
NDS	National Design Specification
NEHRP	National Earthquake Hazards Reduction Program
NRC	Nuclear Regulatory Commission
NSF	National Science Foundation
OMF	ordinary moment frame
OMRF	ordinary moment-resisting frame
OMRSF	ordinary moment-resisting space frame (obsolete)
OTM	overturning moment
PCA	Portland Cement Association
PGA	peak ground acceleration
PGD	peak ground displacement
PGV	peak ground velocity
SBC	Standard Building Code
SBCCI	Southern Building Code Congress International
SCBF	special concentrically braced frames
SD	strength design
SDOF	single degree of freedom
SEAOC	Structural Engineers Association of California
SEAOCC	Structural Engineers Association of Central California
SEAONC	Structural Engineers Association of Northern California
SEAOSC	Structural Engineers Association of Southern California
SH	shear horizontal
SMRF	special moment-resisting frame
SMRSF	special moment-resisting space frame (obsolete)
SPC	seismic performance category
SRSS	square root of the sum of the squares
SSRC	Structural Stability Research Council
STMF	special truss moment frames
SV	shear vertical
T&B	top and bottom (welding)
TM	training manual (military)
TMD	tuned mass damper
TMS	The Masonry Society
UBC	Uniform Building Code
USGS	United States Geological Society
VLLR	vertical and lateral load-resisting elements
WUS	western United States
YDS	yield displacement spectrum

APPENDIX H
Standard Welding Symbols of the AISC/AWS

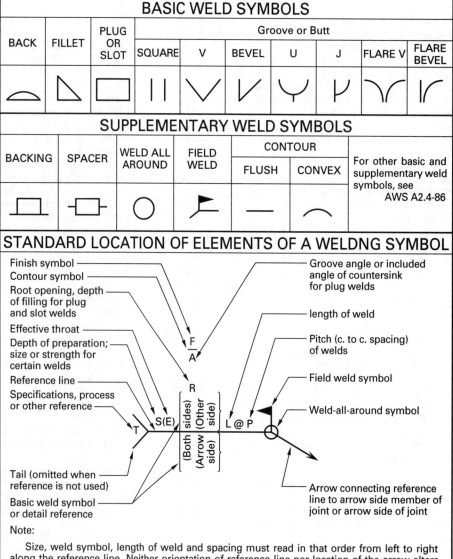

BASIC WELD SYMBOLS

BACK	FILLET	PLUG OR SLOT	Groove or Butt						
			SQUARE	V	BEVEL	U	J	FLARE V	FLARE BEVEL

SUPPLEMENTARY WELD SYMBOLS

BACKING	SPACER	WELD ALL AROUND	FIELD WELD	CONTOUR		For other basic and supplementary weld symbols, see AWS A2.4-86
				FLUSH	CONVEX	

STANDARD LOCATION OF ELEMENTS OF A WELDNG SYMBOL

Finish symbol
Contour symbol
Root opening, depth of filling for plug and slot welds
Effective throat
Depth of preparation; size or strength for certain welds
Reference line
Specifications, process or other reference

Groove angle or included angle of countersink for plug welds
length of weld
Pitch (c. to c. spacing) of welds
Field weld symbol
Weld-all-around symbol

F
A
R
T   S(E)   (Both sides) (Arrow side) (Other side)   L @ P

Tail (omitted when reference is not used)
Basic weld symbol or detail reference

Arrow connecting reference line to arrow side member of joint or arrow side of joint

Note:

Size, weld symbol, length of weld and spacing must read in that order from left to right along the reference line. Neither orientation of reference line nor location of the arrow alters this rule.

The perpendicular leg of weld symbols must be at left.

Arrow and Other Side welds are of the same size unless otherwise shown. Dimensions of fillet welds must be shown on both the Arrow Side and the Other Side Symbol.

Flag of field-weld symbol shall be placed above and at right angle to reference line of junction with the arrow.

Symbols apply between abrupt changes in welding unless governed by the "all around" symbol or otherwise dimensioned.

These symbols do not explicitly provide for the case that frequently occurs in structural work, where duplicate material (such as stiffeners) occurs on the far side of a web or gusset plate. The fabricating industry has adopted this convention: that when the billing of the detail material discloses the existence of a member on the far side as well as on the near side, the welding shown for the near side shall be duplicated on the far side.

APPENDIX I

Commercial Wood Framing Straps and Holdowns

HDA/HD Holdowns

Holdowns are used to transfer tension loads between floors, to tie purlins to masonry or concrete, etc. Use HDAs and HDs for overturn requirements and other applications to transfer tension loads. All HDAs and the HD15 are self-jigging, ensuring code-required minimum 7 bolt diameter spacing from the end of the wood member to the center of the first bolt hole.

HD6A, HD8A, HD10A and HD14A's seat design allows greater installation adjustability. An overall width of 3¼" for the HD6A, HD8A and HD10A, and 3½" for the HD14A provides an easy fit in a standard 4x wall.

HDA SPECIAL FEATURES:
- Single piece non-welded design results in higher capacity.
- Load Transfer Plate eliminates the need for a seat washer.
- Fewer inspection problems.

MATERIAL: See table

FINISH: HD2A, 5A, 6A, 8A, 10A—galvanized. HD8A may be ordered HDG; check with factory. HD14A, HD15, HD20A—Simpson gray paint

INSTALLATION:
- Use all specified fasteners. See General Notes.
- For an improved connection, use a steel nylon locking nut or a thread adhesive on the anchor bolt.
- Bolt holes shall be a minimum of 1/32" to a maximum of 1/16" larger than the bolt diameter (per 1997 NDS, section 8.1.2.1.).
- Standard washers are required between the base plate and anchor nut (HD15 only), and on stud bolt nuts against the wood. The Load Transfer Plate is an integral part of the HDA Holdown and no washer is required. See page 30 for BP/LBP Bearing Plates.
- See SSTB Anchor Bolts, Simpson's Anchoring Systems and Additional Anchorage Designs for anchorage options. The design engineer may specify any alternate anchorage calculated to resist the tension load for a specific job.
- Locate on wood member to maintain a minimum distance of seven bolt diameters, distance is automatically maintained when end of wood member is flush with the bottom of the holdown.
- To tie double 2x members together, the designer must determine the fasteners required to bind members to act as one unit without splitting.
- For holdowns installed on the mudsill, anchor bolt nuts should be finger-tight plus 1/3 to 1/2 turn with a wrench, with consideration given to possible future wood shrinkage. Care should be taken to not over-torque the nut, which may lead to premature anchor bolt failure.
- Stud bolts should be snugly tightened (1997 NDS, section 8.1.2.4).

CODES: BOCA, ICBO, SBCCI NER-393, NER-469; City of L.A. RR 24818, RR 25158 and RR 25293. HD6A and HD14A are not NER listed.

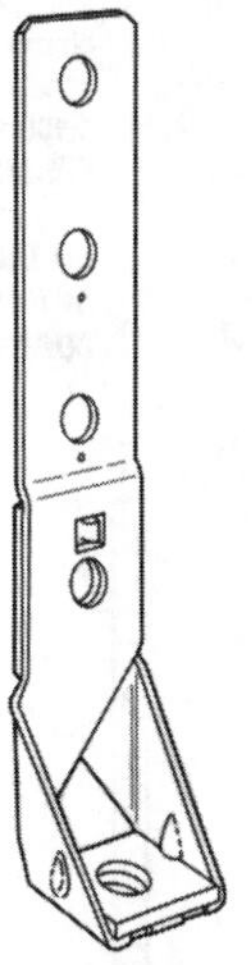

HD10A
HD6A, HD8A
HD14A and
HD20A similar)

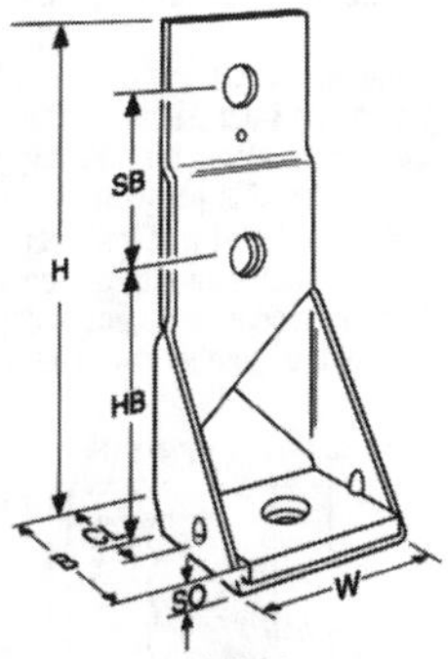

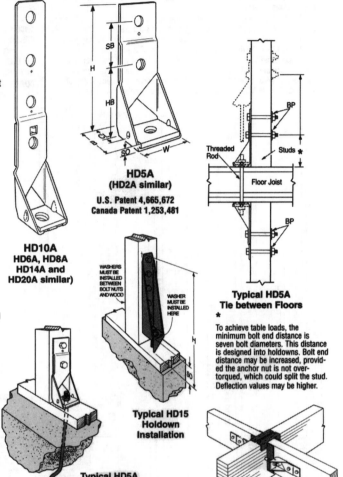

HD5A
(HD2A similar)
U.S. Patent 4,665,672
Canada Patent 1,253,481

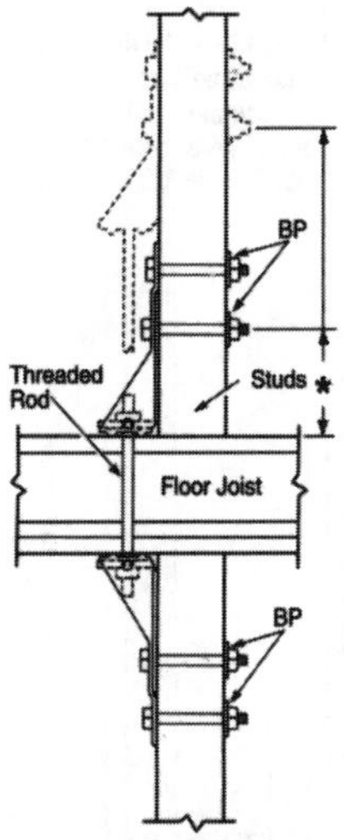

Typical HD5A Tie between Floors

*

To achieve table loads, the minimum bolt end distance is seven bolt diameters. This distance is designed into holdowns. Bolt end distance may be increased, provided the anchor nut is not over-torqued, which could split the stud. Deflection values may be higher.

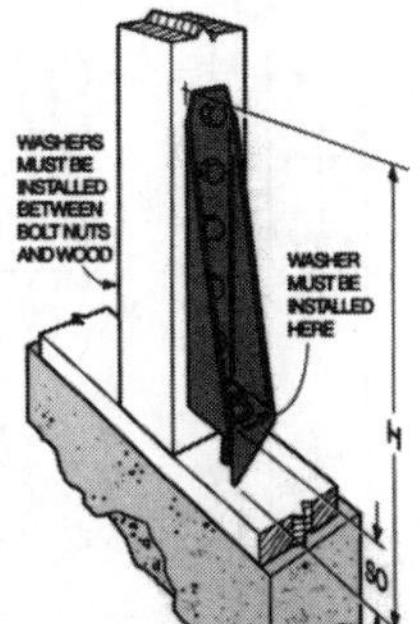

Typical HD15 Holdown Installation

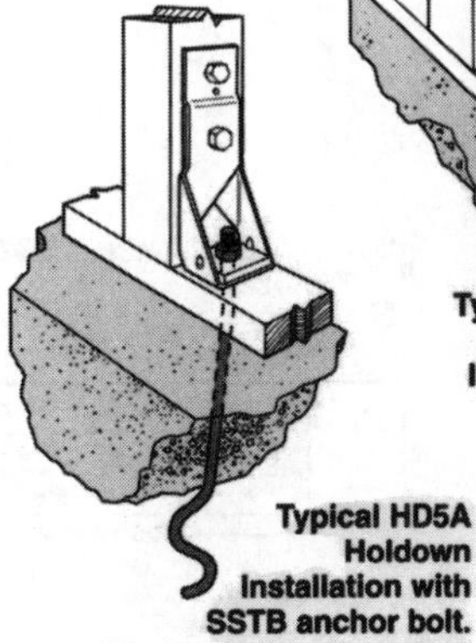

Typical HD5A Holdown Installation with SSTB anchor bolt. Washers are not required at the base.

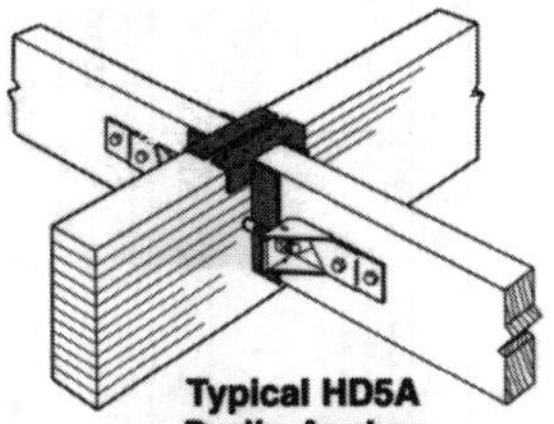

Typical HD5A Purlin Anchor Installation

Model No.	Material: Base Ga	Material: Body Ga	Dimensions: HB [5]	SB	W	H	B	SO	CL	Fasteners: Anchor Dia [6,10]	Stud Bolts: Qty	Stud Bolts: Dia	Avg Ult	Allowable Loads [1,8,9,11] (133) Length of Bolt [2,3,4] in Vertical Wood Member: 1½	2	2½	3	3½	5½	Holdown [12] Deflection at Highest Allowable Design Load
HD2A	7	12	4 9/16	2½	2¾	8	2 9/16	3/8	1 7/16	5/8	2	5/8	12150	1555	2055	2565	2775	2775	2760	0.058
HD5A	3	10	5¼	3	3 1/8	9 7/16	3 9/16	½	2 3/16	5/8 or ¾	2	¾	20767	1870	2485	3095	3705	4010	3980	0.067
HD6A	3/8	7	6 3/16	3½	3¼	11 1/16	3 7/16	9/16	2 1/16	7/8	2	7/8	27333	2275	2980	3685	4405	5105	5510	0.041
HD8A	3/8	7	6 3/16	3½	3¼	14 9/16	3 7/16	9/16	2 1/16	7/8	3	7/8	28667	3220	4350	5415	6465	7460	7910	0.111
HD10A	3/8	7	6 3/16	3½	3¼	18 1/16	3 7/16	9/16	2 1/16	7/8	4	7/8	28667	3945	5540	6935	8310	9540	9900	0.269
HD14A	3/8	3	7	4	3½	20 9/16	3 5/8	5/8	2 3/16	1	4	1	38167	—	—	—	—	11080	13380	0.215
HD20A	3/8	3	7	4	4¼	20 13/16	4 1/8	7/8	2 3/8	1¼	4	1	51333	—	—	—	—	11080	13380	0.250
HD15	3/8	3	7	4	3½	24½	4 7/16	3 5/8	2 1/8	1¼	5	1	55333	—	—	—	—	—	15305	0.082

1. Allowable loads have been increased 33% for earthquake or wind loading with no further increase allowed; reduce where other loads govern.
2. HD15 requires a minimum 6x6 nominal post.
3. Use a minimum 4x6 nominal post for the HD14A and the HD20A.
4. The wood member must be sized for the load-carrying capacity at the critical net section, reducing the gross section area for holes or other removed wood as specified in the code.
5. HB is the required minimum distance from the end of the stud to the center of the first stud bolt hole. End distance may be increased as necessary for installation.
6. The designer must specify anchor bolt type, length and embedment. See SSTB Anchor Bolts and Additional Anchor Designs.
7. See page 30 for anchor bolt retrofit.
8. Lag bolts will not develop the listed loads.
9. Holdowns installed raised off the mudsill may have larger deflection values.
10. Full tension loads apply when HD5A is used with a 5/8" anchor bolt.
11. See pgs 4, 5 for testing and other important information.
12. Deflection at Highest Allowable Design Load: The deflection of a holdown measured between the anchor bolt and the strap portion of the holdown when loaded to the highest allowable load listed in the catalog table. This movement is strictly due to the holdown deformation under a static load test conducted on a steel jig.

APPENDIX I (continued)
ST/FHA/PS Strap Ties

NEW! MSTAM and MSTCM models are designed for wood to masonry applications. See tables on page 101.

The MSTC series, with 2 different strap sizes, has countersunk nail slots for a lower nailing profile. Coined edges ensure safer handling. The RPS meets UBC and City of Los Angeles code requirements for notching plates where plumbing, heating or other pipes are placed in partitions.

Install Strap Ties where plates or soles are cut, at wall intersections, and as ridge ties. LSTA and MSTA straps are engineered for use on 1½" members. The 3" center-to-center nail spacing reduces the possibility of splitting. For the MST, this may be a problem on lumber narrower than

3½"; either fill every nail hole with 10dx1½" nails or fill every other nail hole with 16d commons. Reduce the allowable load based on the size and quantity of fasteners used. The LSTI light strap ties are suitable where gun-nailing is necessary through diaphragm decking and wood chord open web trusses.

FINISH: HST—Simpson gray paint; PS—HDG; all others—galvanized. Some products are available in stainless steel or Z-MAX; see Corrosion-Resistance, page 6.

INSTALLATION: Use all specified fasteners. See General Notes.

OPTIONS: Special sizes can be made to order. See also HCST.

CODES: BOCA, ICBO, SBCCI NER-413, NER-443; ICBO 4935, 5357; Dade County, FL 96-1126.11. City of L.A. RR 25119, RR 25149, RR 25281.

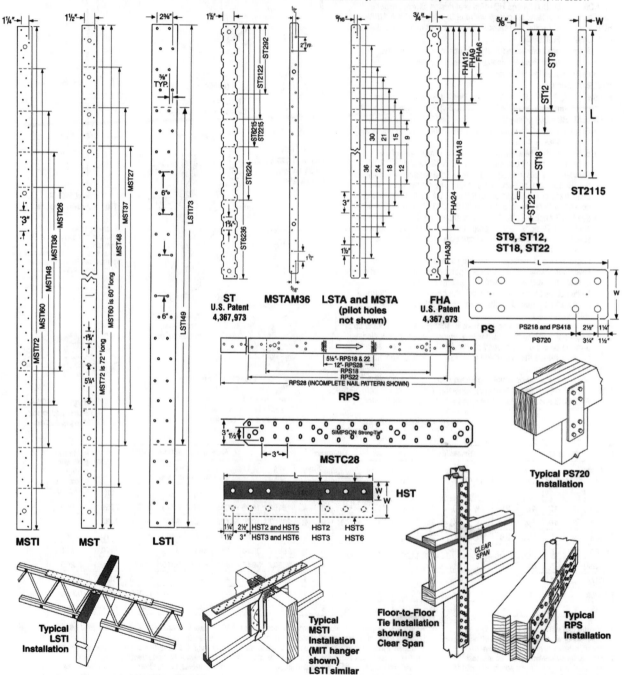

APPENDIX I (continued)
ST/FHA/PS Strap Ties

Model No.	Ga	Dimensions W	Dimensions L	Fasteners (Total) Nails	Allowable Loads Floor (100)	Allowable Loads (133)	Allowable Loads (160)
RPS18		1½	18 5/16	12-16d	810	1080	1295
RPS22	16	1½	22 5/16	16-10d	905	1205	1445
RPS28		1½	28 5/16	12-16d	810	1080	1295
LSTA9		1¼	9	8-10d	450	605	725
LSTA12		1¼	12	10-10d	565	755	905
LSTA15		1¼	15	12-10d	680	905	1085
LSTA18		1¼	18	14-10d	790	1055	1265
LSTA21		1¼	21	16-10d	905	1205	1295
LSTA24	20	1¼	24	18-10d	1015	1295	1295
ST292		2 1/16	9 5/16	12-16d	790	1055	1130
ST2122		2 1/16	12 13/16	16-16d	1070	1425	1505
ST2115		¾	16 5/16	10-16d	450	600	600
ST2215		2 1/16	16 5/16	20-16d	1270	1695	1695
LSTA30		1¼	30	22-10d	1255	1670	1715
LSTA36		1¼	36	26-10d	1480	1715	1715
LSTI49		3¾	49	32-10dx1½	1455	1940	2330
LSTI73		3¾	73	48-10dx1½	2185	2910	3495
MSTA9	18	1¼	9	8-10d	455	610	730
MSTA12		1¼	12	10-10d	570	760	910
MSTA15		1¼	15	12-10d	685	910	1095
MSTA18		1¼	18	14-10d	800	1065	1275
MSTA21		1¼	21	16-10d	910	1215	1460
MSTA24		1¼	24	18-10d	1025	1370	1640
MSTA30		1¼	30	22-10d	1265	1685	2025
MSTA36		1¼	36	26-10d	1495	1995	2135
ST6215		2 1/16	16 5/16	20-16d	1330	1775	2130
ST6224		2 1/16	23 5/16	28-16d	1890	2520	2630
ST9		1¼	9	8-16d	530	705	850
ST12	16	1¼	11⅝	10-16d	665	885	1065
ST18		1¼	17¾	14-16d	900	1200	1200
ST22		1¼	21⅝	18-16d	1025	1370	1370
MSTC28		3	28¼	36-16d sinkers	2070	2760	3310
MSTC40		3	40¼	52-16d sinkers	2990	3985	4740
MSTC52		3	52¼	62-16d sinkers	3555	4740	4740
MSTC66		3	65¾	76-16d sinkers	4390	5855	5855
MSTC78	14	3	77¾	76-16d sinkers	4390	5855	5855
ST6236		2 1/16	33 13/16	40-16d	2575	3430	3430
FHA6		1 7/16	6⅜	8-16d	550	735	885
FHA9		1 7/16	9	8-16d	550	735	885
FHA12		1 7/16	11⅝	8-16d	550	735	885
FHA18		1 7/16	17¾	8-16d	550	735	885
FHA24		1 7/16	23⅞	8-16d	550	735	885
FHA30	12	1 7/16	30	8-16d	550	735	885
MSTI26		2 1/16	26	26-10dx1½	1130	1510	1810
MSTI36		2 1/16	36	36-10dx1½	1565	2090	2505
MSTI48		2 1/16	48	48-10dx1½	2135	2850	3420
MSTI60		2 1/16	60	60-10dx1½	2760	3680	4415
MSTI72		2 1/16	72	72-10dx1½	3310	4415	4725

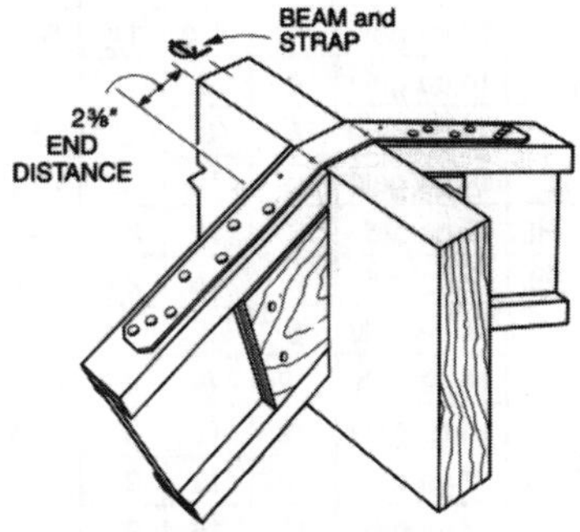

Typical LSTA Installation (hanger not shown)

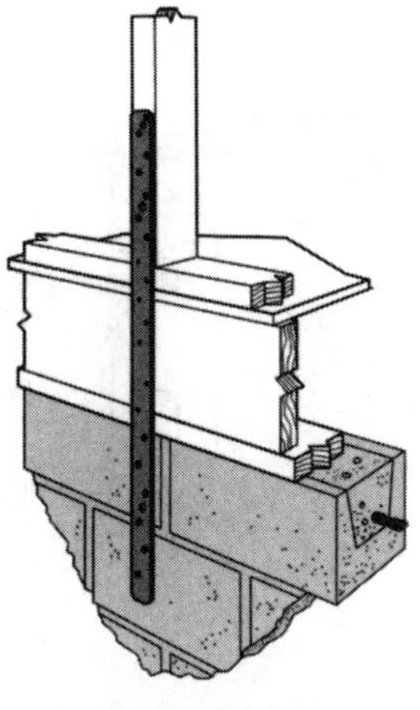

Typical MSTAM36 Installation

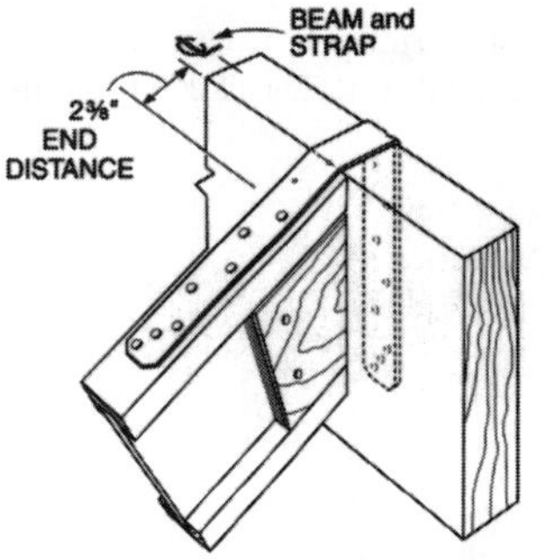

Typical LSTA Installation (hanger not shown)

Floor-to-Floor Clear Span Table

Model No.	Clear Span	Fasteners (Total)	Allowable Load (133)	Allowable Load (160)
MSTC28	18	12-16d sinker	920	1105
	16	16-16d sinker	1225	1470
MSTC40	18	28-16d sinker	2145	2575
	16	36-16d sinker	2455	2945
MSTC52	18	44-16d sinker	3375	4050
	16	48-16d sinker	3680	4415
MSTC66	18	32-16d sinker	5035	5855
	16	34-16d sinker	5350	5855
MSTC78	18	40-16d sinker	5855	5855
	16	40-16d sinker	5855	5855
MST37	18	20-16d	1905	2285
	16	22-16d	2100	2515
MST48	18	32-16d	3135	3765
	16	34-16d	3330	4000
MST60	18	46-16d	4785	5740
	16	48-16d	4990	5800
MST72	18	56-16d	5800	5800
	16	56-16d	5800	5800
MSTI36	18	14-10dx1½	810	975
	16	16-10dx1½	930	1115
MSTI48	18	26-10dx1½	1545	1855
	16	28-10dx1½	1660	1990
MSTI60	18	38-10dx1½	2330	2800
	16	40-10dx1½	2455	2945
MSTI72	18	50-10dx1½	3065	3680
	16	52-10dx1½	3190	3830

Model No.	Ga	Dimensions W	Dimensions L	Bolts Qty	Bolts Dia
PS218[6]		2	18	4	⅝
PS418[6]	7	4	18	4	⅝
PS720[6]		6¾	20	8	½

Masonry Application

Model No.	Ga	Dimensions W	Dimensions L	Fasteners (Total) Nails	Fasteners (Total) Titen Screws	Allowable Loads Floor (100)	Allowable Loads (133)	Allowable Loads (160)
MSTAM24	18	1¼	24	9-10d	5-¼x2¼	1025	1370	1545
MSTAM36	16	1¼	36	13-10d	8-¼x2¼	1435	1915	1915
MSTCM40	16	3	40¼	27-16d sinkers	14-¼x2¼	2990	3985	4395

Model No.	Plate	Notch Width
RPS18	2x4	≤ 5½"
RPS22	2x6	≤ 5½"
RPS28	2x4	≤ 12"

Model No.	Ga	Dimensions W	Dimensions L	Fasteners (Total) Nails	Fasteners (Total) Bolts Qty	Fasteners (Total) Bolts Dia	Allowable Loads Nails Floor (100)	Allowable Loads Nails (133)	Allowable Loads Nails (160)	Allowable Loads Bolts[5] Floor (100)	Allowable Loads Bolts[5] (133)	Allowable Loads Bolts[5] (160)
MST27		2 1/16	27	30-16d	4	½	2070	2760	2790	1295	1725	2070
MST37	12	2 1/16	37½	42-16d	6	½	2860	3815	3815	1825	2435	2920
MST48		2 1/16	48	46-16d	8	½	3345	4460	4460	2225	2970	3560
MST60	10	2 1/16	60	56-16d	10	½	4350	5800	5800	2670	3565	4275
MST72		2 1/16	72	56-16d	10	½	4350	5800	5800	2670	3565	4275
HST2	7	2½	21¼	—	6	⅝	—	—	—	3130	4175	5005
HST5		5	21¼	—	12	⅝	—	—	—	6385	8510	10210
HST3	3	3	25½	—	6	¾	—	—	—	4645	6195	7435
HST6		6	25½	—	12	¾	—	—	—	9350	12465	14955

1. Loads have been increased 33% and 60% for earthquake or wind loading with no further increase allowed. Floor loads may not be increased for other load durations.
2. 10dx1½" nails may be substituted where 16d sinkers are specified at 0.80 of the table loads.
3. 10d commons may be substituted where 16d sinkers are specified at 100% of table loads.
4. 16d sinkers (9 gauge x 3¼") or 10d commons may be substituted where 16d commons are specified at 0.84 of the table loads.
5. Allowable bolt loads are based on parallel-to-grain loading and these minimum member thicknesses: MST–2½"; HST2 and HST5–4"; HST3 and HST6–4½".
6. PS strap design loads must be determined for each installation. Bolts are installed both perpendicular and parallel-to-grain.

Reprinted with permission from *Connectors for Wood Construction*, Product & Information Manual, Catalog C-2000, © 1999, Simpson Strong-Tie Company, Inc.

APPENDIX I (continued)
T and L Strap Ties

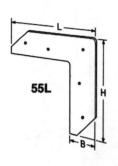

55L

Model No.	Material	Dimensions			Fasteners		
		L	H	B	Nails	Bolts	
						Qty	Dia
55L	16 ga galv	4¾	4¾	1¼	5-10d	—	—
66L	14 ga galv	6	6	1½	10-16d	3	⅜
88L	14 ga galv	8	8	2	12-16d	3	½
1212L	14 ga galv	12	12	2	14-16d	3	½
1212HL	7 ga ptd	12	12	2½	—	4	⅝
1616HL	7 ga ptd	16	16	2½	—	4	⅝
66T	14 ga galv	6	5	1½	8-16d	3	⅜
128T	14 ga galv	12	8	2	12-16d	3	½
1212T	14 ga galv	12	12	2	12-16d	3	½
1212HT	7 ga ptd	12	12	2½	—	6	⅝
1616HT	7 ga ptd	16	16	2½	—	6	⅝

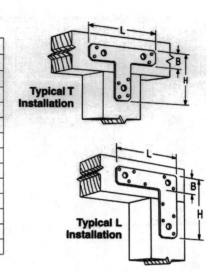

Typical T Installation

Typical L Installation

APPENDIX J
Allowable Shear for Wind or Seismic Forces in Pounds per Foot for
Wood Structural Panel Shear Walls with Framing of Douglas Fir-Larch or Southern Pine[1,2,3]
[UBC Table 23-II-I-1]

PANEL GRADE	MINIMUM NOMINAL PANEL THICKNESS (inches) $\times$ 25.4 for mm	MINIMUM NAIL PENETRATION IN FRAMING (inches)	PANELS APPLIED DIRECTLY TO FRAMING Nail Size (Common or Galvanized Box)[5]	Nail Spacing at Panel Edges (in.) $\times$ 25.4 for mm; 6 $\times$ 0.0146 for N/mm	4	3	2	PANELS APPLIED OVER 1/2-INCH (13 MM) OR 5/8-INCH (16 MM) GYPSUM SHEATHING Nail Size (Common or Galvanized Box)[5]	Nail Spacing at Panel Edges (in.) $\times$ 25.4 for mm; 6 $\times$ 0.0146 for N/mm	4	3	2[3]
Structural I	$5/16$	$1\,1/4$	6d	200	300	390	510	8d	200	300	390	510
	$3/8$		8d	230^4	360^4	460^4	610^4					
	$7/16$	$1\,1/2$	8d	255^4	395^4	505^4	670^4	10d	280	430	550	730
	$15/32$			280	430	550	730					
	$15/32$	$1\,5/8$	10d	340	510	665	870	—	—	—	—	—
C-D, C-C Sheathing, plywood panel siding and other grades covered in UBC Standard 23-2 or 23-3	$5/16$	$1\,1/4$	6d	180	270	350	450	8d	180	270	350	450
	$3/8$		6d	200	300	390	510		200	300	390	510
	$3/8$		8d	220^4	320^4	410^4	530^4					
	$7/16$	$1\,1/2$	8d	240^4	350^4	450^4	585^4	10d	260	380	490	640
	$15/32$			260	380	490	640					
	$15/32$	$1\,5/8$	10d	310	460	600	770	—	—	—	—	—
	$19/32$			340	510	665	870					
			Nail Size (Galvanized Casing)					Nail Size (Galvanized Casing)				
Plywood panel siding in grades covered in UBC Standard 23-2	$5/16$	$1\,1/4$	6d	140	210	275	360	8d	140	210	275	360
	$3/8$	$1\,1/2$	8d	160	240	310	410	10d	160	240	310	410

[1]All panel edges backed with 2-in (51-mm) nominal or wider framing. Panels installed either horizontally or vertically. Space nails at 6 in (152 mm) on center along intermediate framing members for $3/8$-in (9.5-mm) and $7/16$-in (11-mm) panels installed on studs spaced 24 in (610 mm) on center and 12 in (305 mm) on center for other conditions and panel thicknesses. These values are for short-time loads due to wind or earthquake and must be reduced 25 percent for normal loading.

Allowable shear values for nails in framing members of other species set forth in Division III, Park III shall be calculated for all other grades by multiplying the shear capacities for nails in Structural I by the following factors: 0.82 for species with specific gravity greater than or equal to 0.42 but less than 0.49, and 0.65 for species with a specific gravity less than 0.42.

[2]Where panels are applied on both faces of a wall and nail spacing is less than 6 in (152 mm) on center on either side, panel joints shall be offset to fall on different framing members or framing shall be 3-in (76-mm) nominal or thicker and nails on each side shall be staggered.

[3]In seismic zones 3 and 4, where allowable shear values exceed 350 lbf/ft (5.11 N/mm), foundation sill plates and all framing members receiving edge nailing from abutting panels shall not be less than a single 3-in (76-mm) nominal member and foundation sill plates shall not be less than a single 3-in (76-mm) nominal member. In shear walls where total wall design shear does not exceed 600 lbf/ft (8.76 N/mm), a single 2-in (51-mm) nominal sill plate may be used, provided anchor bolts are designed for a load capacity of 50% or less of the allowable capacity and bolts have a minimum of 2-in by 2-in by $3/16$-in (51-mm by 51-mm by 5-mm)-thick plate washers Plywood joint and sill plate nailing shall be staggered in all cases.

[4]The values for $3/8$-in (9.5-mm) and $7/16$-in (11-mm) panels applied direct to framing may be increased to values shown for $15/32$-in (12-mm) panels, provided studs are spaced a maximum of 16 in (406 mm) on center or panels are applied with long dimension across studs.

[5]Galvanized nails shall be hot-dipped or tumbled.

Reproduced from the 1997 edition of the *Uniform Building Code*[TM], copyright © 1997, with the permission of the publisher, the International Conference of Building Officials.

APPENDIX K
Occupancy Category
[UBC Table 16-K]

OCCUPANCY CATEGORY	OCCUPANCY OR FUNCTIONS OF STRUCTURE	SEISMIC IMPORTANCE FACTOR, I	SEISMIC IMPORTANCE[1] FACTOR, I_p	WIND IMPORTANCE FACTOR, I_w
1. Essential facilities[2]	Group I, Division 1 Occupancies having surgery and emergency treatment areas Fire and police stations Garages and shelters for emergency vehicles and emergency aircraft Structures and shelters in emergency-preparedness centers Aviation control towers Structures and equipment in government communication centers and other facilities required for emergency response Standby power-generating equipment for Category 1 facilities Tanks or other structures containing housing or supporting water or other fire-suppression material or equipment required for the protection of Category 1, 2 or 3 structures	1.25	1.50	1.15
2. Hazardous facilities	Group H, Divisions 1, 2, 6 and 7 Occupancies and structures therein housing or supporting toxic or explosive chemicals or substances Nonbuilding structures housing, supporting or containing quantities of toxic or explosive substances which, if contained within a building, would cause that building to be classified as a Group H, Division 1, 2 or 7 Occupancy	1.25	1.50	1.15
3. Special occupancy structures[3]	Group A, Divisions 1, 2 and 2.1 Occupancies Buildings housing Group E, Divisions 1 and 3 Occupancies with a capacity greater than 300 students Buildings housing Group B Occupancies used for college or adult education with a capacity greater than 500 students Group I, Divisions 1 and 2 Occupancies with 50 or more resident incapacitated patients, but not included in Category 1 Group I, Division 3 Occupancies All structures with an occupancy greater than 5000 persons Structures and equipment in power-generating stations; and other public utility facilities not included in Category 1 or Category 2 above, and required for continued operation	1.00	1.00	1.00
4. Standard occupancy structures	All structures housing occupancies or having functions not listed in Category 1, 2 or 3 and Group U Occupancy towers	1.00	1.00	1.00
5. Miscellaneous structures	Group U Occupancies except for towers	1.00	1.00	1.00

[1]The limitation of I_p for panel connections in Section 1631.2.4 shall be 1.0 for the entire connector.
[2]Structural observation requirements are given in Sections 108, 1701 and 1702.
[3]For anchorage of machinery and equipment required for life-safety systems, the value of I_p shall be taken as 1.5.

APPENDIX L
1994 Northridge Earthquake:
Performance of Steel Moment Frame Connections

SUMMARY

The special moment-resisting frame (SMRF) beam-to-column connection depicted in Fig. L-1 (the bolted web/ welded flange design originally developed by seismic pioneer Egor P. Popov in the early 1970s) is now known to be fundamentally flawed. This connection, although 100% consistent with the 1994 UBC (Sec. 2710(g)1B) and previously considered to be seismically invulnerable, does not have sufficient ductility during multiple inelastic cycling. Although this connection has been the "bread-and-butter" design for new construction for many years, it is no longer used in areas of high seismic activity (UBC zones 3 and 4).

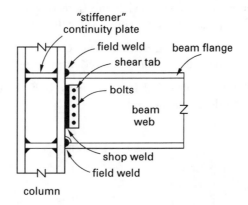

Figure L-1   Connection No Longer Used

NORTHRIDGE EARTHQUAKE

The Northridge earthquake occurred on January 17, 1994. The hypocenter was 11.4 mi (18.4 km) below and 0.6 mi (1 km) to the south of Northridge in Los Angeles. Its magnitude was approximately 6.6. The Northridge thrust fault (previously known as the Pico thrust fault and probably a subsurface extension of the Oak Ridge fault) was involved. This fault was not previously considered to be a seismic danger. Since a thrust fault was involved, the Northridge earthquake had a significant vertical acceleration component.

The duration of the source rupture was 8 sec—longer than the quick jolts of many previous earthquakes. Horizontal ground acceleration was measured as high as 0.93 g at one location. A vertical acceleration of 0.25 g was recorded at one location. Accelerations of 2.3 g horizontally and 1.7 g vertically were recorded at the abutment of the Pacoima Dam. (Horizontally, 2.5 g was measured at the roof of the Olive View Hospital destroyed in the 1971 San Fernando earthquake.)

The Northridge earthquake was the second most expensive U.S. earthquake, with costs and losses exceeding $15 billion. More than 15,000 buildings sustained minor damage, and more than 200 SMRF buildings up to 16 mi (25 km) from the epicenter were damaged. Some of the buildings were brand new, and others were in various stages of completion. (Putting the damage fraction into perspective, there are about 1500 SMRF buildings in Los Angeles.)

There were no building collapses, although a few buildings are said to have been close to collapse. There were 61 fatalities and more than 9000 injuries.

Braced-frame buildings performed well, as did base-isolated structures in distant areas. (There were no base-isolated structures in the vicinity of the actual epicenter.) As expected, reinforced masonry did not perform well. Extensive damage to bridges (most of which were waiting for retrofitting) and infrastructure improvements (e.g., water lines) occurred. Retrofitted bridges fared well, and the viability of carbon-wrapped columns was proven.

APPENDIX L (continued)
1994 Northridge Earthquake:
Performance of Steel Moment Frame Connections

Types of Failures

Although there were some failures in column base plates, this was not a widespread occurrence. Local buckling of beams was minimal or nonexistent. Most of the attention has been focused on the welded beam-column connections that failed in record numbers.

Welded connections failed in a variety of ways. There were some laminar tearouts (delamination) where the weld material pulled out of the column face. There were column-flange divots where the weld was strong enough to rip out full-thickness pieces ("nuggets") of the column.

The most common failure involved cracking of the welded connections to various extents. In some cases, small cracks were observed. In other cases, crack propagation was through the column flange and/or beam web.

Alarmingly, there was no evidence of inelastic beam deformation occurring before the connection failures. Inelastic failure of the beams is at the core of modern-day seismic protection philosophy.

All of the failures are classified as "brittle failures." It is problematic that the connections failed in a brittle mode, since ductility is the primary goal of modern seismic design.

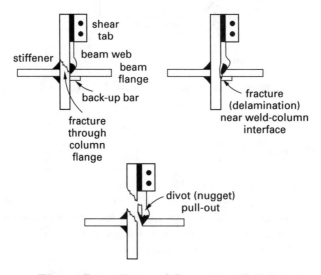

Figure L-2   Types of Connection Failures

Reasons for Failure

The primary reason for the Northridge failures was lack of ductility. The design shown in Fig. L-1 is now known to be "fundamentally flawed" in its ability to perform in a ductile manner. The high degree of ductility that was counted upon to justify the use of an R_w factor of 12 (as specified in the then-current UBC) was not realized. Tests previous and subsequent to the Northridge earthquake indicate that the ductility of this connection design is apparently only 2–4.

Unfortunately, there is no single cause of the connection failures experienced. The poor ductile performance has been blamed on a number of factors: (1) the nature of the earthquake, (2) the materials used, (3) the weld quality, (4) the weld materials, (5) the overall design, and (6) other miscellaneous factors. All of these factors appear to be involved in a complex manner not yet fully understood.

APPENDIX L (continued)
1994 Northridge Earthquake:
Performance of Steel Moment Frame Connections

Cause 1: Nature of the Earthquake

The Northridge earthquake was longer, stronger, and more vertical than had been expected. The horizontal acceleration experienced was, in some cases, twice the design acceleration. The vertical acceleration increased the weight supported by the connections. Vertical acceleration was not considered in the original connection designs.

The large vertical acceleration component of this earthquake placed the connections into a state of triaxial stress. Ductile failure is more difficult to achieve under triaxiality, while brittle failure is relatively unaffected by triaxiality. Therefore, fracture is an easier energy-dissipating mechanism than plastic deformation, leading to brittle failure.

Cause 2: Materials

The second suspected factor centers around the beam and column materials. Anisotropy (i.e., different material properties in different directions) in the ductility and through-thickness strength due to the direction of rolling during beam manufacturing has been accused. Some beams had yield strengths higher than 36-ksi material due to statistical variations in the manufacturing processes. This prevented them from yielding as expected and transferred the earthquake energy into the connection.

The buildup of residual stresses during welding, particularly when the beam and column have different characteristics, is also problematic. Typically, designers use grade-50 columns and grade-36 beams to achieve a strong column-weak beam system where beams yield before the columns yield. (Column yielding and story-mechanism failure is considered to be more life-threatening than beam yielding since the column carries the gravity load of the building.)

The intrinsic weldability of grade-50 steel has been questioned by some, though this may not be a significant factor.

Cause 3: Quality

It is now known that the original welds were notch-sensitive and particularly susceptible to minor imperfections and flaws. Less-than-adequate workmanship and inspection were cited in some cases.

In cases involving laminar tearouts, it is fairly certain that the thick base material did not receive sufficient preheating during welding.

Flaws in root weld passes have also been blamed on the difficulty of doing overhead welds on the lower beam flange.

There appears to have been significant notch effects (notch brittleness) caused by the left-in-place backing bars. Backing bars (placed under the flange when welding from the top) can only be removed by flame cutting, followed by grinding of the exposed area. Since backing bars are expensive to remove, they are usually left in place. Most cracks, however, apparently began in the vicinity of the backing bars.

Welding has always been the whipping boy of connection failures. Even the failure of connections during testing at the University of Texas have been blamed on poor welding. However, although poor quality can contribute to the failure mechanism, it is now known that even perfect assemblies will fail with this type of connection.

Cause 4: Weld Material

Initially, little attention was given to the weld electrode material itself. However, a $1 billion class action lawsuit filed against Lincoln Electric Co. in 1997 questioned the use of certain types of welding electrodes. Lincoln Electric is the manufacturer of E70-T4 electrodes used in a semi-automatic flux-core arc welding process. This process is used to weld the beams to the column after the beams have been bolted to the column's shear tab.

The flux-core welding process may produce weld metals with relatively low notch toughness. (However, high notch toughness was not a requirement of prevailing codes or American Welding Societies specifications at the time.) There may be a possible lack of ductility in the E70-T4 electrodes themselves.

APPENDIX L (continued)
1994 Northridge Earthquake:
Performance of Steel Moment Frame Connections

Cause 5: Overall Design

The connection design has an intrinsic inability to absorb and dissipate energy in an earthquake with sustained strong motion, particularly with a vertical component to acceleration.

The effect of the presence of thick, strong concrete slabs above the beams probably was not considered in most of the designs. However, these slabs substantially increase the strength and stiffness of the beams. The slabs also shift the neutral axis toward the beam top flanges, increasing the stress at the bottom flange. (Most failures occurred at the bottom flanges.)

Cause 6: Other Factors

Early investigations questioned the use of perimeter framing and the lack of redundancy in some of the damaged buildings. In early years of steel design where riveted construction was used, almost every connection was a moment connection. State-of-the-art design prescribes that not all connections need to be moment-resisting, and just a few selected frames are designed as lateral load-resistance systems.

Current design philosophy requires only a few expansive bays supported by gigantic columns and deep beams. Huge moment connections are required. In some buildings, perhaps only two or three columns around the perimeter might have moment connections. Connections of the traditional bolted seated beam variety are used for interior connections. Thus, in an earthquake, the perimeter connections have to do all the work. In some cases, only six welds are available to supply all the moment resistance for a floor. In order to carry all of the moment load, the columns and beams are made even thicker. This increased thickness just compounds the problem, since full penetration welds become all the more difficult.

However, as indicated in the next section, a lack of redundancy was not observed in all cases. Some buildings with SMRF connections across an entire face were damaged.

Similarities in Damage

Most connection failures had the following common elements: (1) The connections were SMRF. (2) Failures initiated or were located in the bottom flange, near the left-in-place backing bar. Columns were typically grade-50 steel. (3) Failures were limited to connections on the building perimeter, at points of plastic hinging.

There has been no statistical correlation between damage and building size, building age, nominal material strength, structural regularity, redundancy, number of bays per frame, frame dimensions, or member sizes. Failures occurred in connections with and without column-flange stiffeners, and with and without return welds on the shear connection plates. Wide flange and box beams were both affected. Interestingly, most buildings were fairly new, with most having been completed in the 1980s.

Most, but not all, buildings were two to four stories in height, and a 22-story building suffered connection failures. The connection failure rate varied from 10 to 100% of the total connections in the building. Also interesting was that most buildings were fairly symmetrical (i.e., rectangular). Building torsion was apparently not an issue.

History

In retrospect, research over more than 20 years has shown that the problem of brittle failure has always existed.

The connection design was originally developed by Egor P. Popov, who released the results of his initial tests in 1972. He had performed tests on both welded-web/welded flange and bolted-web/welded flange connections using a limited number of beam and column sizes. He reported that the all-welded connections showed excellent ductility. The bolted/welded connections performed well, but were less ductile, he reported.

APPENDIX L (continued)
1994 Northridge Earthquake:
Performance of Steel Moment Frame Connections

However, the bolted/welded connection was easier and more economical to make in the field, since the beam could be swung into position and bolted to a shear tab already shop-welded to the column. Therefore, the bolted/welded connection was selected, and Popov's parameters were extrapolated to connections much larger than he had actually tested.

In 1986, Popov reported on subsequent testing he had performed, and the news was bad. However, the announcement did not result in any changes in design methodology. Additional testing in the early 1990s, this time at the University of Texas at Austin by Popov's protégé and UT structural engineering professor, Michael Engelhardt, also indicated that the design was not ductile. However, allegedly sloppy workmanship by Texas welders was cited in explaining the test results. In 1993, Engelhardt repeated the tests with specimens welded in California. He announced more bad news just a month before the Northridge earthquake.

Methods of Repair and Remediation

Repair efforts immediately after the earthquake focused on in-kind repair (cleaning, grinding, rewelding, and ultrasonic testing) without any substantial reinforcement. In some cases, the existing bolted shear plates (bolted beam web plates) were also welded. This approach appears to be justified where only a few of the connections failed.

Although the in-kind repair was permitted by the city of Los Angeles, it is now known to be inadequate. Testing at the University of Texas showed that, although strength is returned to normal, ductility is actually reduced slightly.

Various other repair methods were also tested by the University of Texas, and these tests proved the desirability of bringing the forces and moments out from the area of the welded connection. (These tests have also been criticized due to their low budget, use of slowly-applied static loads, absence of axial loading of the columns, and absence of concrete decking.)

The most common repair method currently being used is to add triangular horizontal cover plates ("flange plates") at the top and bottom flanges. This method, now recommended by the AISC, is the first to be supported by any significant testing. In the University of Texas tests, this method was able to sustain a rotation of 0.0075 radians, which is greater than the 0.005 radians exhibited by the simple-reweld solution, but still far below the requirement of 0.025 radians needed to sustain a magnitude 7 earthquake.

A variation of the flange plate approach is to close the beam sides with vertical plates welded to the column face. After the damage to the original connection has been repaired and the connection has been rewelded, a 1-in thick steel side plate, 5 ft long, is fillet-welded to the column and sides of the beam, creating a 5-ft long "box." (The beam flanges are the top of the box.) This is known as the MNH system, named after Myers, Nelson, Houghton, Inc., which has patented and trademarked a similar "dual strong axes connection" for new construction.

The use of vertical beam-column ribs (gussets), top and bottom, was another connection tested at the University of Texas. However, this method is difficult to apply in situations where access to the top flange of the beam is limited by concrete floor slabs.

A tee-support (haunch support) on the bottom flange only was also tested. Access for welding limits the usefulness of this method for in-field repairs.

Vertical web straps (vertical plates extending from the column face to well into the beam's web) are analogous to flange plates. They also have the ability to move the forces away from the beam-column connection. However, such a repair would be difficult to use where there was a welded shear plate already in place.

Another method, which focuses on the use of bolting instead of welding, has been proposed by the Center for Advanced Technology for Large Structural Systems (ATLSS) at Lehigh University. This method, designed by David Bleiman and Kazuhiko Kasai, replaces broken welds with specifically fabricated brackets using high-strength bolts.

APPENDIX L (continued)
1994 Northridge Earthquake:
Performance of Steel Moment Frame Connections

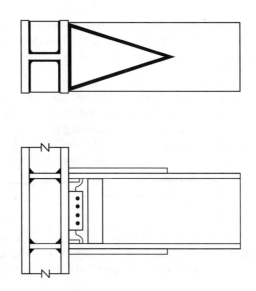

Figure L-3   Types of Connection Repair

Economics of Repair

The costs of "standard" connections, repaired connections, and connections of new designs are quite variable. Costs can be greatly affected by labor rates, overhead and opportunity costs, and other factors. The costs described here have been quoted in the literature following the Northridge earthquake, but should be considered as relative and "flexible" numbers.

Inspection costs can reach $1500 per connection, mostly for removal and replacement of fireproofing. Repair costs are typically $10,000 to $15,000 per connection, though some repairs can reach $20,000, including removal and replacement of finishes. This compares to the approximate $250 cost of making the original moment-resisting connection.

One steel fabricator has estimated the relative costs of repair (using a standard connection as the basis) as 3.7 for the horizontal flange plates, 8.6 for vertical triangular ribs, and 14.2 for vertical web straps.

Interim Design

Current design and construction is still in a state of flux, but flange plates are being used with many new buildings. Two different schools of thought have evolved regarding the design of flange plates. The Englehardt approach or "Texas method" is to use thick and narrow plates, while the Forell-Elsesser approach (nicknamed the "Son of Texas method" by some) uses wider and thinner cover plates.

Myers, Nelson, Houghton, Inc., (MNH) has patented and trademarked a "dual strong axes connection" that uses only fillet welds. Fillet welds are longer, easier, and more redundant than full-penetration welds. The connection incorporates the side-plate approach, as previously described, but for new construction, shop-fabrication is used to achieve greater quality control. Column trees are shop-welded. Column and beam flanges are fillet-welded to steel plates, but not to each other. A "fuse" is created by field-bolting the column tree beam stubs to the beam link between column trees using high-strength energy-absorbing friction bolts in slightly oversized holes. There is no direct connection between the beam and the column.

Another design for new construction is the ATLSS new connection design. The ATLSS design uses specifically fabricated brackets connecting beam and column with high-strength bolts. It has been shown to be as stiff as older-style connections, but with more ductility during overloads. All welds are done in the shop. One very specific advantage cited is that bolted connections are not particularly sensitive to workmanship.

APPENDIX L (continued)
1994 Northridge Earthquake:
Performance of Steel Moment Frame Connections

Beam Weakening Approaches

It is interesting that beam yielding did not occur in any of the connection failures observed. The connections absorbed all of the earthquake energy. Various beam weakening methods have been proposed in order to ensure that beam yielding occurs in the future. (It is popular to refer to a weakened beam as a "structural fuse," since the beam yields before the connection fails.) These beam weakening methods can be used with old and new construction.

In a "dogbone" beam, the beam flange is shaved to reduce the interior cross section of the beam so that it will yield and buckle locally, away from the connection.

Seismic Structural Design Associates (SSDA, Los Angeles) has a different approach for repairing existing connections. The first step is to repair the connection to its original state. The connection is "softened" using horizontal slots in the beam web, top and bottom, and vertical slots in the column web near the beam flanges.

Changes to the Building Code

Until 1988, the UBC did not even specify the type of connection to be used. Rather, the UBC merely specified the force the connection had to support. However, based on the SEAOC's 1985 *Lateral Force Requirements* (the "Blue Book"), the UBC added the design shown in Fig. L-1. For various reasons, including liability to ICBO and SEAOC, the bolted web/welded flange connection detail has been stricken from the UBC. It is likely that future building codes will return to prescribing only the force to be withstood, rather than the method for withstanding the force.

Other Lessons Learned

The Northridge earthquake taught us many important lessons. It may seem obvious, but connections need to be inspected after every major earthquake, regardless of whether there is observed damage. In these inspections, architectural elements and coverings need to be removed. Cracking of the concrete fire-proofing or other telltale signs cannot be counted on to indicate connection failures.

It has also been learned that ultrasonic testing may not be sufficient to detect weld cracks, particularly when the back-up bar has been left in place. Back-up bars are not consistent with the use of flange plates, but there are many thousands of old connections with back-up bars still in place.

CONCLUSION

Although the bolted-web/welded flange connection is no longer used in seismically active zones, ongoing research and development are needed to evaluate the possible replacements. In the short run, economics will have a major impact in determining which method is adopted for the majority of new designs. In the long run, seismic events such as the Northridge earthquake will identify which methods are successful.

INDEX

Entries from footnotes are indicated by page numbers in italics.

INDEX OF FIGURES AND TABLES

INDEX BY UBC SECTION (UBC-97)

Note that the UBC uses the identifier "Formula," not "Equation," in the code.

NECESSARY
IF MAILED
IN THE
UNITED STATES

BUSINESS REPLY MAIL

FIRST CLASS MAIL   PERMIT NO. 33   BELMONT, CA

POSTAGE WILL BE PAID BY ADDRESSEE

PROFESSIONAL PUBLICATIONS INC
1250 FIFTH AVE
BELMONT CA 94002-9979

NO POSTAGE
NECESSARY
IF MAILED
IN THE
UNITED STATES

BUSINESS REPLY MAIL

FIRST CLASS MAIL   PERMIT NO. 33   BELMONT, CA

POSTAGE WILL BE PAID BY ADDRESSEE

PROFESSIONAL PUBLICATIONS INC
1250 FIFTH AVE
BELMONT CA 94002-9979

NO POSTAGE
NECESSARY
IF MAILED
IN THE
UNITED STATES

BUSINESS REPLY MAIL

FIRST CLASS MAIL   PERMIT NO. 33   BELMONT, CA

POSTAGE WILL BE PAID BY ADDRESSEE

PROFESSIONAL PUBLICATIONS INC
1250 FIFTH AVE
BELMONT CA 94002-9979